Lecture Notes in Geoinformation and Cartography

Series editors

William Cartwright, Melbourne, Australia
Georg Gartner, Wien, Austria
Liqiu Meng, München, Germany
Michael P. Peterson, Omaha, USA

The Lecture Notes in Geoinformation and Cartography series provides a contemporary view of current research and development in Geoinformation and Cartography, including GIS and Geographic Information Science. Publications with associated electronic media examine areas of development and current technology. Editors from multiple continents, in association with national and international organizations and societies bring together the most comprehensive forum for Geoinformation and Cartography.

The scope of Lecture Notes in Geoinformation and Cartography spans the range of interdisciplinary topics in a variety of research and application fields. The type of material published traditionally includes:

- proceedings that are peer-reviewed and published in association with a conference;
- post-proceedings consisting of thoroughly revised final papers; and
- research monographs that may be based on individual research projects.

The Lecture Notes in Geoinformation and Cartography series also includes various other publications, including:

- tutorials or collections of lectures for advanced courses;
- contemporary surveys that offer an objective summary of a current topic of interest; and
- emerging areas of research directed at a broad community of practitioners.

More information about this series at http://www.springer.com/series/7418

Paolo Fogliaroni · Andrea Ballatore
Eliseo Clementini
Editors

Proceedings of Workshops and Posters at the 13th International Conference on Spatial Information Theory (COSIT 2017)

 Springer

Editors
Paolo Fogliaroni
Department of Geodesy and Geoinformation
Vienna University of Technology
  (TU-Wien)
Wien
Austria

Eliseo Clementini
Department of Industrial and Information
  Engineering and Economics
University of L'Aquila
L'Aquila
Italy

Andrea Ballatore
Department of Geography
Birkbeck, University of London
London
UK

ISSN 1863-2246   ISSN 1863-2351   (electronic)
Lecture Notes in Geoinformation and Cartography
ISBN 978-3-319-87678-8   ISBN 978-3-319-63946-8   (eBook)
https://doi.org/10.1007/978-3-319-63946-8

Printed on acid-free paper

This Springer imprint is published by Springer Nature
The registered company is Springer International Publishing AG
The registered company address is: Gewerbestrasse 11, 6330 Cham, Switzerland

Preface

The 13th International Conference on Spatial Information Theory (COSIT 2017) is the latest edition of the series of scientific events that have been bringing together leading researchers from the field for more than 20 years. Spatial information theory is concerned with all aspects of space and spatial environments as experienced, represented, and elaborated by humans, other animals, and artificial agents. The scope of the conference includes both the conceptualization of frameworks for specific spatiotemporal domains, as well as the development of general theories of space and time and theories of spatial and temporal information.

The conference was instrumental in promoting interaction between computer science, cognitive psychology, geographic information science, logic, philosophy, and linguistics, around the way we understand, conceptualize, represent, describe, and interact with the geographic space around us. The insights produced at COSIT are now integral part of many of these fields. On its trajectory through time and space, the conference has returned to where it started: Italy. This year, it was held at L'Aquila, the capital city of the Abruzzo region, in Central Italy. The beauty of the city and the region is matched by the diversity and quality of COSIT 2017 scientific program.

While the main conference proceedings have been published in a different volume, the current volume collects 27 short papers, presented at a poster session at the main conference, and 20 papers discussed at four satellite workshops. The short papers section, edited by Christian Kray and May Yuan, gathers a set of contributions covering a wide range of subjects, spanning from classical COSIT topics to emergent research areas.

After the short papers, this volume collects the proceedings of four workshops organized as satellite events to the main conference. These workshops raised an extraordinary interest in the scientific community, significantly broadening the scope of the conference to uncharted territories. Each workshop was a one-day mini-conference with its own program chairs, program committee members, and a peer-reviewed selection process.

The workshop on *Rethinking Wayfinding Support Systems* was chaired by Jakub Krukar, Angela Schwering, Heinrich Löwen, Marcelo De Lima Galvao, and Vanessa Joy Anacta. It focused on the emergent challenges in GPS-powered navigation systems, taking into account the spatial-cognitive dimension beyond traditional usability studies.

The second workshop, entitled *Speaking of Location: Future Directions in Geospatial Natural Language Research*, was chaired by Kristin Stock, Chris Jones, and Maria Vasardani, and tackled facets of the fascinating relation between language and space. These include frames of reference, referencing of landmarks, and spatial mental models.

The workshop on *Spatial Humanities meets Spatial Information Theory: Space, Place, and Time in Humanities Research* was chaired by Ben Adams, Karl Grossner, and Olga Chesnokova. The event opened COSIT to the digital humanities, an interdisciplinary area in which a "spatial turn" has brought researchers to engage with the geographies of their entities. Deep mapping, narratology of space, and modeling of historical data are trends of high relevance to spatial information theory.

Finally, the humanities were also central to the workshop on *Computing Techniques for Spatio-Temporal Data in Archaeology and Cultural Heritage*, chaired by Alberto Belussi, Roland Billen, Pierre Hallot, and Sara Migliorini. It discussed the need to foster more interaction between archaeology and the COSIT community, providing new problem-driven challenges for information representation and discovery.

Organizing an event such as COSIT and making it a success is only possible with the help and commitment of many people. We thank the workshop organizers who proposed the topics, ensured a quality program, and managed the entire editorial process, from the promotion of the events to the peer review. We would also like to thank the local organizers and University of L'Aquila that hosted the conference and its satellite events, offering the location and technical and logistic support. For the first time, this year COSIT was promoted by the IFIP (International Federation for Information Processing) and by the AICA (Italian Association for Informatics and Automatic Calculus), without financial support. Finally, we would like to thank all who attended COSIT 2017 to present their work, to discuss the work showcased at the conference and at the workshops, and to advance the state of the art in the field of spatial information theory.

L'Aquila, Italy   Paolo Fogliaroni
July 2017   Andrea Ballatore
                                                                 Eliseo Clementini

List of Part Editors

Part I COSIT 2017 Short Papers

Christian Kray, University of Münster, e-mail: c.kray@uni-muenster.de
May Yuan, University of Texas at Dallas, e-mail: myuan@utdallas.edu

Part II Rethinking Wayfinding Support Systems

Jakub Krukar, University of Münster, e-mail: krukar@uni-muenster.de
Angela Schwering, University of Münster, e-mail: schwering@uni-muenster.de
Heinrich Löwen, University of Münster, e-mail: loewen.heinrich@uni-muenster.de
Marcelo De Lima Galvao, University of Münster, e-mail: galvao.marcelo@uni-muenster.de
Vanessa Joy Anacta, University of Münster, e-mail: v.anacta@uni-muenster.de

Part III Speaking of Location: Future Directions in Geospatial Natural Language Research

Kristin Stock, Massey University, e-mail: k.stock@massey.ac.nz
Chris Jones, Cardiff University, e-mail: jonescb2@cardiff.ac.uk
Maria Vasardani, University of Melbourne, e-mail: maria.vasardani@unimelb.edu.au

Part IV Spatial Humanities Meets Spatial Information Theory: Space, Place, and Time in Humanities Research

Ben Adams, University of Canterbury, e-mail: benjamin.adams@canterbury.ac.nz
Olga Chesnokova, University of Zurich, e-mail: olga.chesnokova@geo.uzh.ch
Karl Grossner, World Heritage Web, e-mail: karlg@worldheritageweb.org

Part V Computing Techniques for Spatio-Temporal Data in Archaeology and Cultural Heritage

Alberto Belussi, University of Verona, e-mail: alberto.belussi@univr.it
Roland Billen, University of Liège, e-mail: rbillen@ulg.ac.be
Pierre Hallot, University of Liège, e-mail: p.hallot@ulg.ac.be
Sara Migliorini, University of Verona, e-mail: sara.migliorini@univr.it

Contents

Part I
COSIT 2017 Short Papers

COSIT 2017 Short Papers—Introduction

Christian Kray and May Yuan

Established in 1993, COSIT is a biennial international conference series concerned with theoretical aspects of space and spatial information. COSIT was motivated by wanting to establish a counterpoint to several applied GIS conferences at which reports on applications and development in GIS technology were made but often without a contribution to scientific theory and literature. The focus at COSIT from the beginning has been on theories of space and time relevant to the establishment and the advancement of (geo)spatial information science.

COSIT grew out of a series of scientific events, NATO Advanced Study Institutes and NSF specialist meetings, all of which concerned with cognitive and applied aspects of representing large-scale space, particularly geographic space. At these meetings, the need for a well-founded theory on spatial information processing was identified, and COSIT was formed in order to provide the platform for the intensive interdisciplinary scientific exchange on such theories.

This section presents a series of short papers that have been presented at a poster session at the main conference. COSIT 2017 poster session was primarily designed to give the opportunity especially to young researchers to present their work to the community, but, given the exceptional response to the call for papers, it became a chance for all researchers to present high-quality ongoing work.

C. Kray (✉)
Institute for Geoinformatics, University of Münster, Münster, Germany
e-mail: c.kray@uni-muenster.de

M. Yuan
School of Economic Political and Policy Sciences, The University of Texas at Dallas, Richardson, TX, USA
e-mail: myuan@utdallas.edu

© Springer International Publishing AG 2018
P. Fogliaroni et al. (eds.), *Proceedings of Workshops and Posters at the 13th International Conference on Spatial Information Theory (COSIT 2017)*, Lecture Notes in Geoinformation and Cartography, https://doi.org/10.1007/978-3-319-63946-8_1

A quick look at the table of contents is enough to appreciate the interdisciplinary breadth and diversity of the topics covered by these short papers. Several contributions are within the traditional remit of COSIT, such as cognitive maps, spatial semantics, ontology engineering, uncertainty, wayfinding, and spatial language. More recent trends are also covered, including natural language processing, volunteered geographic information, linked data, and social-spatial network analysis. Other contributions focus on emergent research areas such as 3D virtual reconstruction and navigation of unmanned aerial vehicles, engaging with the state of the art of rapidly evolving technical fields.

Many of these new topics expand the theoretical horizon of spatial information and continue pushing the boundaries for new ideas, interpretations, and applications. Most of the lead authors of these short papers are doctoral students or early career scholars. Their studies demonstrate a vibrant research potential for advances in spatial information theories and exciting possibilities to come in the next few years.

Generating Spatial Footprints from Hiking Blogs

Elise Acheson, Flurina M. Wartmann and Ross S. Purves

Abstract Explicitly linking text documents to geographical space is an important processing step for applications such as map visualization, spatial querying, and placename disambiguation. In this work, we present a proof-of-concept processing pipeline to generate spatial footprints for a spatially-rich, manually-annotated corpus of hiking blogs. We present preliminary results obtained by exploiting the spatially-focused nature of our input data and the rich placename resources at our disposal. Future work will fully automate the pipeline and systematically examine the influence of processing decisions on the footprints and downstream tasks.

Keywords Footprints · Toponyms · Document scope · Clustering · Geographic information retrieval

1 Introduction

How can we geographically represent text documents? For documents strongly linked to geographical space, such as hiking blogs, a representative geographical area can be automatically generated by combining natural language processing methods with geographical processing such as spatial filtering or clustering. The resulting document representations, known as 'document scopes' or 'document footprints', are useful in downstream tasks (e.g. spatial queries), upstream tasks (e.g. improving placename disambiguation), and as an end in themselves (e.g. visualizing a text doc-

E. Acheson (✉) · F.M. Wartmann · R.S. Purves
Geography Department, University of Zurich, Winterthurerstrasse 190,
8057 Zurich, Switzerland
e-mail: Elise.Acheson@geo.uzh.ch

F.M. Wartmann
e-mail: Flurina.Wartmann@geo.uzh.ch

R.S. Purves
e-mail: Ross.Purves@geo.uzh.ch

© Springer International Publishing AG 2018
P. Fogliaroni et al. (eds.), *Proceedings of Workshops and Posters at the 13th
International Conference on Spatial Information Theory (COSIT 2017)*, Lecture Notes
in Geoinformation and Cartography, https://doi.org/10.1007/978-3-319-63946-8_2

ument on a mapping interface) (Monteiro et al. 2016; Purves et al. 2007; Quercini et al. 2010).

The context of our work is a project on how people describe landscapes in Switzerland. With the goal of comparing landscape descriptions from different data sources (hiking blogs and Flickr photos), document footprints are generated for a web-crawled corpus of hiking blogs in order to query and select Flickr photos based on location. For this task, we aimed to generate high-precision, geographically focused footprints.

2 Data and Methods

Our corpus consisted of web documents related to ten study sites in the German speaking region of Switzerland in a first-person narrative. Documents were collected by targeted web-crawling, with five texts per site selected by manual triage, for a final corpus of 50 documents.

To generate document footprints, we followed the established three-step processing pipeline (Amitay et al. 2004; Monteiro et al. 2016) consisting of: (1) identifying placenames, (2) grounding placenames, and (3) generating a footprint (geometry) (Fig. 1). Poor placename identification has been identified as a major source of error for document scope propagated downstream (Amitay et al. 2004; Purves et al. 2007). Thus, to obtain precise footprints, we performed step 1 manually, which was feasible due to the small corpus size. Step 2, grounding, involved querying an API to obtain ranked results from the SwissNames3D gazetteer for each placename, after having aggregated placenames repeating within a study site and recording their frequencies. For the final step, footprint generation, we experimented with two approaches: iterative filtering based on the centroid and standard deviation of our candidate points (Smith and Crane 2001), and clustering using DBSCAN to identify one main cluster and discard outliers.

Finally, we experimented with permutations in processing decisions in order to generate optimal footprints which suited our requirements. Decisions included: how many candidates for each placename to retain at the grounding stage; whether to treat with higher priority placenames with exactly one candidate; and whether to use the

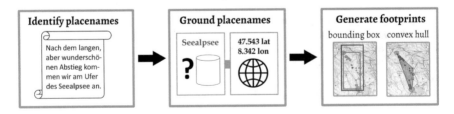

Fig. 1 Processing pipeline for document footprint generation

frequency per site of placenames. We automated the entire processing pipeline, starting from the manually annotated placenames, to output ten convex hulls or bounding boxes on each run.

3 Results and Conclusion

Our preliminary results showed that for placename grounding, simple approaches worked well enough for our purposes: ranked candidate placenames from Swiss-Names3D were sufficiently accurate that we obtained good results by retaining just the top candidate for each placename. For footprint generation, satisfactory results were obtained using both distance-based filtering and DBSCAN clustering, but the DBSCAN results were better suited to more complex geometric arrangements.

Our footprint requirements stemmed from our downstream tasks: performing a spatial query for Flickr photos, and ultimately, comparing datasources about landscapes at our ten study sites. These task-based requirements, along with the availability of quality placename resources for our area of study, influenced our processing decisions at every stage. Future work will fully automate the placename identification stage, and will systematically measure the effects of permutations in processing decisions on the downstream tasks of querying and document comparison.

References

Amitay E, Har'El N, Sivan R, Soffer A (2004) Web a where: geotagging web content. In: Proceedings of the 27th annual international ACM SIGIR conference on research and development in information retrieval, ACM, New York, NY, USA, SIGIR '04, pp 273–280. doi: 10.1145/1008992.1009040

Monteiro BR, Davis CA Jr, Fonseca F (2016) A survey on the geographic scope of textual documents. Comput Geosci. doi:10.1016/j.cageo.2016.07.017. http://www.sciencedirect.com/science/article/pii/S0098300416301972

Purves RS, Clough P, Jones CB, Arampatzis A, Bucher B, Finch D, Fu G, Joho H, Syed AK, Vaid S, Yang B (2007) The design and implementation of SPIRIT: a spatially aware search engine for information retrieval on the Internet. Int J Geogr Inf Sci 21(7):717–745. doi:10.1080/13658810601169840

Quercini G, Samet H, Sankaranarayanan J, Lieberman MD (2010) Determining the spatial reader scopes of news sources using local Lexicons. In: Proceedings of the 18th SIGSPATIAL international conference on advances in geographic information systems, ACM, New York, NY, USA, GIS '10, pp 43–52. doi:10.1145/1869790.1869800

Smith DA, Crane G (2001) Disambiguating geographic names in a historical digital library. In: Constantopoulos P, Slvberg IT (eds) Research and advanced technology for digital libraries, no. 2163 in Lecture notes in computer science. Springer, Berlin, Heidelberg, pp 127–136. http://link.springer.com/chapter/10.1007/3-540-44796-2_12

Validating GEOBIA Based Terrain Segmentation and Classification for Automated Delineation of Cognitively Salient Landforms

Samantha T. Arundel and Gaurav Sinha

Abstract Landform objects extracted from Geographic Object Based Image Analysis (GEOBIA) based terrain segmentation to locations are overlaid and compared to feature types of landforms mapped in the USGS maintained Geographic Names Information System (GNIS) topographic database. GEOBIA terrain objects were found to statistically related to GNIS feature classes. Comparison of GNIS feature classes and GEOBIA landform classes suggests that GEOBIA landform class semantics correspond well with naïve geographic conceptualizations reflected in GNIS feature types.

Keywords Terrain mapping · Semantics · Feature extraction · GEOBIA · GNIS

1 GEOBIA Terrain Classification and Its Validation with GNIS

An important goal of the National Geospatial Program (NGP) of the U.S. Geological Survey (USGS) is to produce semantically derived terrain objects from elevation rasters and point clouds to support automated cartography, natural language querying, and terrain analysis. While various and numerous techniques abound to automate the 'extraction' of landform features, it is unclear as to how, or even if, the results from any of these techniques relate to cognitive representations of the same terrain. Recently, geographic object based image analysis (GEOBIA) has been proposed as an intuitive, automated method for delineating multi-scale landform objects from elevation fields (Drăguţ and Eisank 2012). This paper pre-

S.T. Arundel (✉)
Center of Excellence for Geographic Information Science (CEGIS), USGS, Rolla, MO, USA
e-mail: sarundel@usgs.gov

G. Sinha
Department of Geography, Ohio University, Athens, OH, USA
e-mail: sinhag@ohio.edu

© Springer International Publishing AG 2018   9
P. Fogliaroni et al. (eds.), *Proceedings of Workshops and Posters at the 13th International Conference on Spatial Information Theory (COSIT 2017)*, Lecture Notes in Geoinformation and Cartography, https://doi.org/10.1007/978-3-319-63946-8_3

sents initial results from a project exploring the extent to which GEOBIA terrain segmentation and classification can help delineate cognitively salient landforms (e.g., mountains, hills, ranges, plains, valleys, uplands). The project workflow can be summarized as follows: (i) select GEOBIA model parameters (particularly a multi-resolution scale parameter); (ii) run model to extract landform objects; (iii) test for spatial and conceptual alignment with cognitively salient landforms; and (iv) continue to refine parameters until cognitive validity of extracted objects can-not be improved.

In this paper, results are reported from the comparison of GEOBIA extracted landform objects to locations and type of landform features recorded in the USGS maintained Geographic Names Information System (GNIS) topographic database. The USGS developed GNIS as the official federal repository of geo-graphic feature names and other descriptive information for physical and cultural geographic fea-tures (excluding roads and highways) in the United States. GNIS does not contain delineated extents for features, but supports several descriptive spatial and non-spatial feature attributes. While the generic part of the feature toponym offers complementary information about a feature type, the possible semantics of such terms can vary considerably. Hence, in the 1970s, GNIS developers created a generalized set of 64 (including 41 landform related) non-hierarchical, mutually exclusive feature classes. For most features, the classificatory use of the feature class codes reflects the intuitive, everyday interpretation (dictionaries) of those terms. Despite limitations arising from this simple, due to its comprehensive cov-erage (>2 million features) GNIS is a good resource for making broad interpreta-tions about how people intuitively perceive and categorize geographic features in the United States.

Relating GEOBIA extracted landform objects to GNIS feature types can provide insights for designing automated geomorphometric methods for search and retrieval of landforms from digital elevation models. Since the GEOBIA terrain segmenta-tion workflow used in this study also classifies extracted landform objects, it is possible to calculate co-occurrence frequencies of GNIS landform feature classes for each type of GEOBIA landform class. Therefore, the goals of this study were to (i) find if there is any significant correspondence between GEOBIA landform objects and GNIS feature classes, and (ii) determine whether GEOBIA can reduce computational processing times for landform delineations that are dependent on user specified contextual parameters.

2 Methodology

GEOBIA object segmentation and classification was implemented within the e-Cognition software suite. The methodology for segmentation was based on a previously published GEOBIA terrain segmentation methodology (Drăguţ and Eisank 2012) designed to extract landform objects from global scale digital ele-vation data. The inspiration for this GEOBIA workflow is Iwahashi and Pike's

(2007) pixel-based nested-means unsupervised classification algorithm. The appeal of this GEOBIA approach to terrain segmentation is supposed to be its advanced degree of automation and conceptual simplicity. Only raw elevation values and overall statistics (mean and standard deviation) are analyzed at multiple scales to segment the landscape in-to composite landform objects. Thus, users do not have to worry about data pre-processing, derivation of additional terrain fields (slope, curvature), or parameterizing of different combinations of input variables.

For GEOBIA terrain segmentation, the multi-resolution segmentation (MRS) algorithm is used since it is sensitive to morphological discontinuities in the DEMs, detection of which is critical for deriving landform objects from pixels (Drăguţ and Eisank 2012). The scale parameter for the MRS segmentation process can be determined by an automated optimization strategy. The segmentation process is executed iteratively at three levels of detail to offset problems associated with over-segmentation of large and/or high relief areas and under-segmentation of relatively smaller and/or smoother areas. While it is possible for users to experiment with scale parameters, for this study, the default MRS scale parameters of 100 (Levels 1 and 2) and 10 (Level 3) were used. Users can also specify a minimum mapping unit (MMU) for each level to filter numerous small "island" landform objects considered inconsequential at that level. Thus, this GEOBIA workflow adapts automatically to local morphological complexity.

In this paper, the focus is on the classifications of objects extracted at the three recommended levels. For intuitive appeal, Drăguţ and Eisank (2012) chose the class labels based on a simplified version of the physiographic regions originally proposed by Hammond (1954). At Level 1, mean elevation (elev) of individual objects are compared with the overall elevation mean of the entire image to classify objects as 'High' ($elev_{obj} > mean$) or 'Low' ($elev_{obj} < mean$). At Level 2, standard deviation (SD) of individual objects are compared to the means of Level 1 High and Low objects to further classify 'High' objects as 'Mountains' ($SD_{obj} > mean_{High}$) or 'Tablelands and High Hills' ($SD_{obj} < mean_{High}$), and "Low" objects as 'Low Hills' ($SD_{obj} > mean_{Low}$) or 'Plains' ($SD_{obj} < mean_{Low}$). Finally, at Level 3, eight more classes are defined by comparing objects means or SDs with means of Level 2 objects. Based on comparison of object means to parent Level 2 object means, Mountains are further classified as 'High Mountains' and 'Low Mountains', while Plains are classified further as 'High Plains' and 'Low Plains'. For classification of Tablelands and High Hills and Low Hills, standard deviation of objects is compared to parent Level 2 object means to derive 'High Hills', 'Tablelands', 'Rough Low Hills' and 'Smooth Low Hills'.

For assessing the cognitive validity of GEOBIA object classifications, both Level 2 and Level 3 classifications were subjected to spatial overlay analysis against the GNIS feature classes. Three 1/3rd arc-second resolution elevation datasets were obtained from the USGS 3D Elevation Program for three physiographic provinces (selected from the 25 provinces distributed over 8 regions mapped by Fenneman and Johnson 1946). The provinces were chosen from different physiographic regions to capture sufficient topographic variety, and because they are small, which reduced computational processing times for GEOBIA segmentation. Two provinces (Adirondack, Cascade-Sierra Mountains) have high relief, and the third (Superior Upland), while also an upland area, exhibits relatively lower relief. The DEMs were smoothed once using a mean 3 × 3 mean filter before GEOBIA segmentation and classification.

For validation, classified GEOBIA objects were mapped against GNIS features (see an example for Adirondacks province in Fig. 1), and then cross-tabulations of

Fig. 1 An example from Adirondack province of GNIS landform features overlaid on landform objects extracted from GEOBIA iterative segmentation

GNIS feature counts for Level 2 and Level 3 GEOBIA classes were developed. Finally, Chi-square statistics were calculated to assess if there is any significant correlation between GNIS feature classes and GEOBIA classes.

3   Comparing GEOBIA and GNIS Landform Classes

Results were analyzed for all GNIS feature classes for both Level 2 and 3 objects, but patterns emerged only for some of the most frequent GNIS feature for Level 2 classes. Cascade-Sierra Mountains had a much larger number of GNIS features (~1,500) compared to Adirondack (~2,400) and Superior Upland (~1,000). Across these three provinces, a large majority of GNIS summits (92–97%), *ridges* (81–90%), *ranges* (79–100%) and *gaps* (~87%) co-occur within GEOBIA Hill (Low/High) or Mountain (Low/High) objects. No other feature type exhibits such a high frequency of overlap with Hills and Mountains. This suggests a general agreement between the naïve geographic interpretation of GNIS feature types and GEOBIA classifications (Mountains or Hills) assigned to landforms.

   In contrast, GNIS *valleys* were found to be distributed across many GEOBIA landform object types, but still most commonly found in low mountains and flat plains. This is also to be expected since valleys extend from high to low elevations, and, therefore, likely to overlap with both high and low elevation landform classes. *Swamps* were found to be mostly (72–88%) within the low-lying plains, which also agrees with naïve geographic knowledge about swamps. Similarly, a majority of GNIS *flats* falls within GEOBIA Tablelands and Plains. Chi-square tests further indicated statistically significant relationship between (all) GNIS feature types and both four Level 2 and eight Level 3 GEOBIA landform classes.

   Two broad implications of this analysis of the GEOBIA based landform object extraction workflow is that it (i) offers landform objects that could correspond closely in some cases to topographic eminences (GNIS summits, ranges, ridges) recognized by people and (ii) can likely reduce processing times for on-demand delineations of topographic eminences (GNIS summits, ridges, and ranges). However, this GEOBIA workflow does not offer any computational advantage for delineating depression landforms (GNIS valleys, arroyos, basins, channels etc.)— which is not surprising, considering it has no class specifically representing depression landforms.

   The important next validation step is to compare size and shape correspondence between GEOBIA hill and mountain objects and individual eminence objects delineated from other automated methods or (preferably) human subjects, and then test for statistical Chi-square correlations.

Acknowledgements "Any use of trade, firm, or product names is for descriptive purposes only and does not imply endorsement by the U.S. Government."

References

Drăguţ L, Eisank C (2012) Automated object-based classification of topography from SRTM data. Geomorphology 141–142(4):21–33

Fenneman NM, Johnson DW (1946) Physiographic divisions of the United States. U.S. Geol Surv

Hammond EH (1954) Small-scale continental landform maps. Ann Assoc Am Geogr 44:33–42

Iwahashi J, Pike RJ (2007) Automated classifications of topography from DEMs by an unsupervised nested-means algorithm and a three-part geometric signature. Geomorphology 86 (3–4):409–440

Boundary Based Navigation Is Impaired in Old Age

Rachel Bhushan, Elisabetta Colombari and Sang Ah Lee

Abstract Decline of hippocampal function with age has many effects, including deficits in episodic memory and spatial orientation. Motivated by research in cognitive psychology and behavioral neuroscience showing the importance of environmental boundaries in spatial mapping, we tested adult and aged rats in a simple reorientation task involving both environmental boundary geometry and landmarks. We find preliminary evidence that aged rats are impaired in their ability to reorient, particularly with respect to boundary geometry.

Keywords Hippocamus · Aging · Reorientation · Boundary geometry

1 Introduction

Animals can apply various strategies in finding their way towards a target location. To do so they use information from the external environment to create an internal spatial representation to facilitate their navigation. In a controlled laboratory setup, animals can be tested on their use of two main kinds of cues: the geometric properties of the environment, such as the arena shape, or landmark features such as visual colors and patterns. Several studies have shown that a wide range of animal species, including humans, use environmental geometry to encode and find their way to a goal location (Cheng 1986; Hermer and Spelke 1994; Lee et al. 2012). The neural substrates of spatial mapping and navigation are said to be in the

R. Bhushan (✉) · E. Colombari · S.A. Lee
Center for Mind/Brain Sciences, University of Trento,
Corso Bettini 31, 38068 Rovereto, Italy
e-mail: rachel.bhushan@unitn.it

S.A. Lee
e-mail: sang.ah.lee@unitn.it

S.A. Lee
Department of Bio and Brain Engineering, Korea Advanced Institute of Science
and Technology, Daejeon, Korea

© Springer International Publishing AG 2018
P. Fogliaroni et al. (eds.), *Proceedings of Workshops and Posters at the 13th
International Conference on Spatial Information Theory (COSIT 2017)*, Lecture Notes
in Geoinformation and Cartography, https://doi.org/10.1007/978-3-319-63946-8_4

hippocampal formation. Particular spatially specific neurons such as place cells, grid cells, head direction and boundary cells provide the representations that make up an internal map in the animal's brain (Hartley et al. 2014). Out of these cells, the boundary cells have been shown to play a crucial role in defining a location with respect to the environment (Hartley et al. 2000; Lever et al. 2009).

2 Methods

Given the marked changes in hippocampal function with aging, we tested fourteen Lister hooded rats (*Rattus Norvegicus*) at the age of 6 months (adults) and seventeen at the age of 18 months (old) in a working memory task. We trained animals in each trial to explore the arena and find the rewarded corner that contained pieces of chocolate. The animal was then removed, disoriented, and placed back in the empty arena for a one-minute testing period. Each animal received three trials per day for two consecutive days. We later observed the time they spent in each corner. We applied this task in three different environmental conditions: a black rectangular arena, a rectangular arena with a striped wall and a square arena with a striped wall.

3 Results

Young adult rats (n = 5) in a rectangular arena spend more time in the correct and rotationally correct corners showing successful use of geometry. However, old rats (n = 7) in the same condition show clearly no distinction between correct and incorrect corners (Fig. 1).

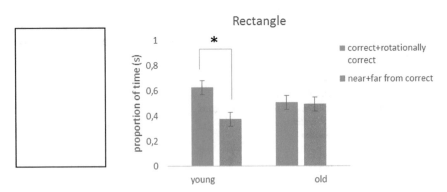

Fig. 1 Rectangular arena (120 × 60 cm) to test the use of boundary geometry. The *bars* indicate the proportion of time they spend in a 2 cm range from each corner during the 60 s testing. *Asterix* indicates $p < 0.05$

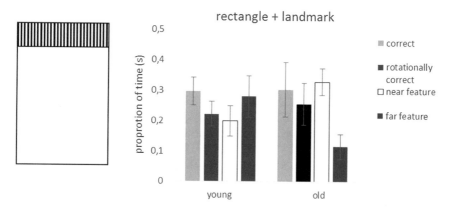

Fig. 2 Rectangular arena (120 × 60 cm) to test the use of boundary geometry and landmark use such as a striped wall (0.5 cm thick polyurethane panels alternating *black* and *white stripes* thick 10 cm). The *bars* indicate the proportion of time they spend in a 2 cm range from each corner during the 60 s testing

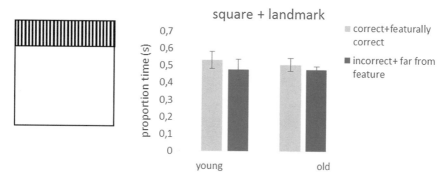

Fig. 3 Square arena (85 × 85 cm²) to test the use of only landmark use such as a striped wall (0.5 cm thick polyurethane panels alternating *black* and *white stripes* (thickness 10 cm). The *bars* indicate the proportion of time they spend in a 2 cm range from the corner during the 60 s testing

The introduction of a striped wall helps searches toward the correct corner in both groups (n = 4 young; n = 6 old) showing that a salient cue like a striped wall coupled with landmark help in reorientation towards the target location (Fig. 2).

In the last condition both young and old rats (n = 5 young; n = 4 old) do not show a statistical preference towards the correct corner showing difficulty in landmark use alone (Fig. 3).

4 Discussion

Our findings suggest that age can alter some of our cognitive abilities such as our ability to reorient in a simple spatial environment. This is consistent with the idea that our brain and specific brain structures such as the hippocampus involved in spatial navigation face physiological changes with age and consequently result in impairments on such tasks. Being able to assess hippocampal, function in spatial cognition using simple noninvasive behavioral tests can benefit patients and better evaluate the cognitive effects of treatments. This is of particular importance in disorders such as Alzheimer's disease, in which early diagnosis is crucial.

References

Cheng K (1986) A purely geometric module in the rat's spatial representation. Cognition 23 (2):149–178

Hartley T, Burgess N, Lever C, Cacucci F, O'Keefe J (2000) Modeling place fields in terms of the cortical inputs to the hippocampus. Hippocampus 10(4):369–379

Hartley T, Lever C, Burgess N, O'Keefe J (2014) Space in the brain: how the hippocampal formation supports spatial cognition. Phillos Trans R Soc B 369(1635):20120510

Hermer L, Spelke ES (1994) A geometric process for spatial reorientation in young children. Nature 370(6484):57

Lee SA, Spelke ES, Vallortigara G (2012) Chicks, like children, spontaneously reorient by three-dimensional environmental geometry, not by image matching. Biol Lett p.rsbl20120067

Lever C, Burton S, Jeewajee A, O'Keefe J, Burgess N (2009) Boundary vector cells in the subiculum of the hippocampal formation. J Neurosci 29(31):9771–9777

Distributing Attention Between Environment and Navigation System to Increase Spatial Knowledge Acquisition During Assisted Wayfinding

Annina Brügger, Kai-Florian Richter and Sara Irina Fabrikant

Abstract Travelers happily follow the route instructions of their devices when navigating in an unknown environment. Navigation systems focus on route instructions to allow the user to efficiently reach a destination, but their increased use also has negative consequences. We argue that the limitation for spatial knowledge acquisition is grounded in the system's design, primarily aimed at increasing navigation efficiency. Therefore, we empirically investigate how navigation systems could guide users' attention to support spatial knowledge acquisition during efficient route following tasks.

Keywords Attention guidance · Decision making · Environmental learning · Mobile eye-tracking · Pedestrian navigation · Spatial cognition

1 Introduction

Navigation systems have become an integral part of our everyday lives. They support us in reaching a destination more quickly, and help to reduce mental effort during wayfinding. Despite their popularity, concerns have been raised in the literature about the negative effects on spatial awareness and spatial knowledge acquisition due to their extensive use. Due to significant technological advancement in positional accuracy or in route calculation (e.g. speed), navigation systems impact human cognitive abilities such as allocation of attention negatively (Gardony et al. 2013). Automated

A. Brügger (✉) · S.I. Fabrikant
Geographic Information Visualization and Analysis (GIVA) Department of Geography, University of Zurich, Winterthurerstr. 190, 8057 Zurich, Switzerland
e-mail: annina.bruegger@geo.uzh.ch

S.I. Fabrikant
e-mail: sara.fabrikant@geo.uzh.ch

K.-F. Richter
Department of Computing Science, Umeå University, 90 187 Umeå, Sweden
e-mail: kai-florian.richter@umu.se

© Springer International Publishing AG 2018
P. Fogliaroni et al. (eds.), *Proceedings of Workshops and Posters at the 13th International Conference on Spatial Information Theory (COSIT 2017)*, Lecture Notes in Geoinformation and Cartography, https://doi.org/10.1007/978-3-319-63946-8_5

navigation guidance splits a navigators' attention between the navigation system and the environment (Gardony et al. 2013; Ishikawa et al. 2008). Attentional split can be induced by a GPS positioning signal, drawing a navigator's continued attention to the display, as it is constantly updated during navigation (Ishikawa et al. 2008). Continued reliance on this kind of positional updates further leads to memory loss of environmental information processing, and respective navigation skill training (Parush et al. 2007). As the engagement of the user with the environment is supposed to strengthen spatial memory (Klippel et al. 2010), navigation systems should (I) draw navigators' attention back to the environment and (II) ensure efficient navigation. They should also pro-actively support a navigator's increased mental effort for this task. This in turn, could mitigate potential limitations to spatial knowledge acquisition (Parush et al. 2007). Navigation systems should not only enable user interaction with the environment (Hirtle and Raubal 2013), but also increase visual attention to relevant attention-grabbing features in the environment, and their respective depiction on the navigation system (e.g. landmarks (Richter 2017)) to increase spatial knowledge acquisition (Kiefer et al. 2013). This is the research problem we aim to tackle with our research program (also see Brügger et al. 2016).

2 Empirical User Study in Outdoor Environments

We empirically investigate how navigation systems could guide users' attention to support spatial knowledge acquisition during efficient outdoor route following tasks. Our user study features two phases (Fig. 1) similar to a previous virtual environment study by Karimpur et al. (2016). First, participants execute a route following task in an unknown urban environment while being assisted by a navigation system. Second, participants are asked to find the same route back from memory, without any navigation assistance. In doing so, we directly assess their spatial knowledge acquired during the first (assisted) phase.

We implemented four different ways with varying levels of user engagement in which the navigation system actively engages the navigator, with both, the traversed environment, and the navigation system during navigation. These four levels of user engagement were deployed in different combinations during the first (learning) phase. Accordingly, participants (N = 64) were divided into four groups (between-subject design) and randomly assigned to four variations of navigation system behaviour. Performance in the second (recall) phase reveals how the acquisition of spatial knowledge varies as a result of the different levels of user engagement with the system or the environment, respectively.

Furthermore, mobile eye-tracking data collected during participants' trials will support understanding how the four assessed ways of user engagement might lead to differences in perceiving both, the environment, and the navigation system (Fig. 2).

Fig. 1 Empirical user study. Task 1: assisted route following (*left*). Task 2: unassisted route reversal (*right*)

Fig. 2 Participant's gaze on a navigation system (*left*) and on the environment (*right*). Data collection and visualization (*yellow–red circle*) with mobile eye-tracking technology

3 Results and Future Work

Our results show that participants who used the navigation system with a higher level of user engagement acquired better spatial knowledge without harming navigation efficiency. Participants who collected their own environmental information simultaneously improved their spatial mental representation during the learning phase. Conversely, those participants whose attention was guided by the navigation system (lower level of user engagement) made significantly more errors in the recall phase.

These findings have significant implications for the understanding of how people should be guided by navigation systems during wayfinding, in terms of balancing efficiency of route following with potential loss of spatial knowledge. Further experiments will have to focus on the distribution of attention between the environment and the navigation system, as to develop a more coherent picture of the trade-offs between efficient navigation performance and effective spatial knowledge acquisition.

References

Brügger A, Richter KF, Fabrikant SI (2016) Walk and learn: an empirical framework for assessing spatial knowledge acquisition during mobile map use. In: International conference on GIScience short paper proceedings, vol 1, no 1

Gardony AL, Brunyé TT, Mahoney CR, Taylor HA (2013) How navigational aids impair spatial memory: evidence for divided attention. Spat Cogn Comput 13(4):319–350

Hirtle SC, Raubal M (2013) Many to many mobile maps. In: Raubal M, Mark DM, Frank AU (eds) Cognitive and linguistic aspects of geographic space: new perspectives on geographic information research. Springer, Berlin, Heidelberg, pp 141–157

Ishikawa T, Fujiwara H, Imai O, Okabe A (2008) Wayfinding with a GPS-based mobile navigation system: a comparison with maps and direct experience. J Environ Psychol 28(1):74–82

Karimpur H, Röser F, Hamburger K (2016) Finding the return path: landmark position effects and the influence of perspective. Front Psychol 7:1–16

Kiefer P, Giannopoulos I, Raubal M (2013) Where Am I? Investigating map matching during self-localization with mobile eye tracking in an urban environment. Trans GIS 18(5):660–686

Klippel A, Hirtle S, Davies C (2010) You-are-here maps: creating spatial awareness through map-like representations. Spat Cogn Comput 10(2–3):83–93

Parush A, Ahuvia S, Erev I (2007) Degradation in spatial knowledge acquisition when using automatic navigation systems. In: Winter S, Duckham M, Kulik L, Kuipers B (eds) Spatial information theory: COSIT 2007. Springer, Berlin, Heidelberg, pp 238–254

Richter KF (2017) Identifying landmark candidates beyond toy examples. KI - Künstliche Intelligenz 31(2):135–139

Lake District Soundscapes: Analysing Aural Experience Through Text

Olga Chesnokova, Joanna E. Taylor and Ross S. Purves

Abstract The importance of perception across all the senses has been recognised in previous studies on landscape preferences. Here, we focus on aural perception, and in a preliminary study explore how references to sounds and their sources can be extracted from descriptions of images in a corpus containing 85,000 documents. We classified references to sounds according to previous work as biophony, geophony and antrophony. As a first step we have extracted descriptions related to sounds associated with verbs. The most common sound emitters in our corpus are wind (geophony), birds (biophony) and traffic (anthrophony) respectively. In future work we will move beyond the sentence level to deal with co-references, and use other parts of speech (e.g. adjectives such as quiet and loud or nouns such as noise, silence, echo, etc.).

Keywords VGI · Crowdsourcing · Semantics · Landscape preference · Soundscapes

1 Introduction

Vision has long dominated humanities studies of historical landscapes; indeed, the picturesque attitude was based upon a way of *seeing* the landscape (Andrews 1989). The importance of the visual in perception is underlined by, for example, the dominance of perceptual verbs related to vision across languages (San Roque et al.

O. Chesnokova (✉) · R.S. Purves
Department of Geography, University of Zurich, Winterthurerstrasse 190,
8057 Zurich, Switzerland
e-mail: olga.chesnokova@geo.uzh.ch

R.S. Purves
e-mail: ross.purves@geo.uzh.ch

J.E. Taylor
Department of History, Lancaster University, Lancaster LA1 4YD, UK
e-mail: j.e.taylor1@lancaster.ac.uk

© Springer International Publishing AG 2018
P. Fogliaroni et al. (eds.), *Proceedings of Workshops and Posters at the 13th International Conference on Spatial Information Theory (COSIT 2017)*, Lecture Notes in Geoinformation and Cartography, https://doi.org/10.1007/978-3-319-63946-8_6

2015). This property makes linking language to visual properties through images and associated descriptions possible, and we have demonstrated such an application in work on *scenicness* (Chesnokova et al. 2017). However, images alone cannot always easily convey other senses, despite the recognised importance of perception across all the senses in landscape preference estimation (Tudor 2014). Sound in particular has emerged in recent years as an important alternative to sight as a means of understanding and negotiating both historical and contemporary landscapes through notions such as the soundscape (Schafer 1993; Krause 2008). In this preliminary study we explore how references to sounds and their sources can be extracted from text. We are especially interested in how the sounds identified in the English Lake District might be classified according to previous work as part of the *biophony* (e.g. gulls calling, bellowing stag), *geophony* (e.g. rustling wind, bubbling waterfall) and *antrophony* (e.g. roaring jets, rumbling traffic) (Krause 2008).

The Lake District's tourist industry became central to the local economy thanks to popular guidebooks such as Thomas West's *A Guide to the Lakes in Cumberland, Westmoreland and Lancashire* (1778) in the late eighteenth century and a distinctive soundscape was recorded from the earliest accounts. We thus begin our analysis by exploring records of some of the regions most important acoustic experiences contained in the Corpus of Lake District Writing, developed at Lancaster University. Applying geographical text analysis methods to this corpus allows us to suggest where in the Lake District offered particularly distinctive acoustic experiences, and to demonstrate how these historical accounts anticipate modern records (Donaldson et al. 2017).

2 Data and Methods

Having established this historical context, we analyse a much larger corpus of textual descriptions associated with some 85,000 georeferenced photographs submitted to the crowdsourced project Geograph.[1] Firstly, we identify all sentences which potentially contain verbs associated with sound. Since many of these verbs are polysemous, we then disambiguate verb usage at the sentence level. On the disambiguated sentences we identify the verb subject, before classifying these subjects into the classes defined above.

Our initial list of verbs was based on those listed by Levin (1993) as verbs of sound emission, verbs of sounds made by animals and verbs of sound existence with the addition of the verbs 'hear' and 'listen'. Our disambiguation was carried out using hypernyms extracted from WordNet[2] which were associated with the verb and its context in a sentence using the Lesk algorithm. We then used a dependency parser to identify the subjects of these sound related verbs.

---

[1]http://www.geograph.org.uk.

[2]https://wordnet.princeton.edu/.

3 Results and Discussion

The precision of word sense disambiguation in the Lake District was 83%. The subjects were identified in 44% of the cases. The most common emitters of sound were wind (geophony), birds (biophony) and traffic (anthrophony) respectively. Initial exploration of the results shows that, unsurprisingly, there is clear connection between the locations of roads and railways and anthrophony. Our results also allow us to map the spatial patterns of both individual sound emitters and their associated classes.

Descriptions relating to sounds associated with verbs are rare; we found 134 in the Lake District. The low figures are to be expected, given that these descriptions are associated with images. However, we expect to increase it by using other parts of speech which are relatively common in our corpus (e.g. adjectives such as quiet, loud, deafening etc.)—for example, the adjective 'quiet' is used 243 times in the Lake District and nouns (e.g. voice, noise, silence, echo). Nevertheless, we could demonstrate that it is possible to extract spatially located perceptual information related to sounds, and that doing so allows us to provide a more detailed understanding of visitors' aural experiences.

Acknowledgements Joanna E. Taylor's contribution was funded by the Leverhulme-Trust project Geospatial Innovation in the Digital Humanities: A Deep Map of the English Lake District (RPG-2015-230).

References

Andrews M (1989) The search for the picturesque. Stanford University Press

Chesnokova O, Nowak M, Purves RS (2017) A crowdsourced model of landscape preference. In: Clementini E, Donnelly M, Yuan M, Kray C, Fogliaroni P, Ballatore A (eds) 13th International conference on spatial information theory (COSIT 2017). Leibniz International Proceedings in Informatics

Donaldson C, Gregory IN, Taylor JE (2017) Locating the beautiful, picturesque, sublime and majestic: spatially analysing the application of aesthetic terminology in descriptions of the English Lake District. J Hist Geogr 56:43–60

Krause B (2008) Anatomy of the soundscape. J Audio Eng Soc 56(1/2)

Levin B (1993) English verb classes and alternations. University of Chicago Press

San Roque L, Kendrick KH, Norcliffe E, Brown P, Defina R, Dingemanse M, Dirksmeyer T, Enfield NJ, Floyd S, Hammond J, Rossi G, Tufvesson S, Van Putten S, Majid A (2015) Vision verbs dominate in conversation across cultures, but the ranking of non-visual verbs varies. Cogn Linguist 26(1):31–60

Schafer RM (1993) The soundscape: our sonic environment and the tuning of the world. Inner Traditions/Bear & Co

Tudor C (2014) An approach to landscape character assessment. Natural England

Do Skyscrapers Facilitate Spatial Learning Under Stress? On the Cognitive Processing of Global Landmarks

Sascha Credé, Sara Irina Fabrikant, Tyler Thrash
and Christoph Hölscher

Abstract Affective states have been found to influence peoples abilities to orient in and to mentally represent large scale spaces. For example, navigators can become stressed when searching for destinations in unfamiliar environments. How then is spatial knowledge acquisition influenced by navigators stress state during assisted wayfinding? We report an ongoing empirical navigation study in which we investigate how acute distress affects spatial knowledge acquisition during navigation, moderated by the use of different landmark types.

Keywords Spatial orientation · Navigation · Spatial knowledge · Acute stress · Global and local landmarks

1 Introduction

Even though map based pedestrian navigation may seem very simple, people often get stressed when searching for destinations in unfamiliar urban environments. According to prior research, working memory formation is frequently impaired as a result of stress in a variety of tasks (Lupien et al. 1999). Especially when learning large-scale spaces, working memory is crucial because navigators are required to mentally integrate visuospatial information from different locations and various

S. Credé (✉) · S.I. Fabrikant
Department of Geography, Geographic Information Visualization and Analysis,
University of Zurich, Zurich, Switzerland
e-mail: sascha.crede@geo.uzh.ch

S.I. Fabrikant
e-mail: sara.fabrikant@geo.uzh.ch

T. Thrash · C. Hölscher
Department of Humanities, Social and Political Science, Chair of Cognitive Science,
ETH, Zurich, Switzerland
e-mail: tyler.thrash@gess.ethz.ch

C. Hölscher
e-mail: christoph.hoelscher@gess.ethz.ch

© Springer International Publishing AG 2018
P. Fogliaroni et al. (eds.), *Proceedings of Workshops and Posters at the 13th International Conference on Spatial Information Theory (COSIT 2017)*, Lecture Notes in Geoinformation and Cartography, https://doi.org/10.1007/978-3-319-63946-8_7

perspectives over the course of time. While a solid basis of prior empirical work provides evidence for the influence of acute stress on spatial knowledge acquisition during navigation, this research procured mixed results. Some studies found evidence that acute stress during navigation improves spatial knowledge acquisition (Duncko et al. 2007; Frei et al. 2016), others show that stress impairs spatial knowledge acquisition (Evans et al. 1984; Mackingtosh et al. 1975). Again other studies did not find a significant relationship between the two variables (Richardson and Tomasulo 2011). In the present study, we argue that diverging results can be partly explained by intermediary cognitive processes through which stress exerts its influence on spatial knowledge acquisition. We report an ongoing empirical navigation study in which we investigate how acute distress affects spatial knowledge acquisition during navigation, moderated by the use of different landmark types.

2 Method

In a virtual reality experiment, two groups of participants (stress|no-stress) solve three map-assisted navigation tasks in an unfamiliar virtual city. During navigation activity, participants learn different sets of landmark configurations: highlighted local landmarks along the route, highlighted local landmarks while non-highlighted global landmarks are present, and highlighted global landmarks visible in the distance (Fig. 1). The landmarks to be learned are highlighted in the virtual environment. By contrast, landmarks are not displayed on the navigation device, which aims to impede landmark learning from the mobile map. Subsequent to each learning phase, participant's spatial knowledge acquisition was assessed in a judgement of relative direction task.

Based on previous work on the construction of survey knowledge (Meilinger et al. 2014), we expect larger pointing errors for local landmarks, which are experienced sequentially during navigation, than for global landmarks, which often can be experienced simultaneously from different locations and thus load less on working memory. We hypothesize that this learning advantage of highlighted global landmarks

(a)   **(b)**   **(c)**

Fig. 1 Participants walk different virtual cities in three conditions. Participants are instructed to memorize the relative positions of **a** *highlighted* local or **b** *highlighted* global landmarks. In condition (**c**), participants are instructed to learn *highlighted* local landmarks, while global landmarks that are not *highlighted* are present

increases when participants cope with limited cognitive resources under distress. However, we expect stressed participants to narrow attention and to be less able to take advantage of non-highlighted global landmarks.

3 Relevance

Memorizing spatial information is an important ability for autonomous spatial navigation. However, detrimental effects of assisted navigation on spatial memory are recognized and agreed upon amongst researchers from different fields (Caquard 2015). Unfortunately, there is little agreement on the underlying mechanisms which lead to the observed memory impairments. We argue that the negative effects of stress on spatial knowledge acquisition have not yet been sufficiently incorporated into the debate. The present study is the first to directly investigate effects of stress states on user's ability to mentally process and represent local and global configurations of landmarks. The expected findings will help to develop guidelines for the design of stress resistant pedestrian navigation assistance. Based on insights we gain from this study, forthcoming empirical studies considering map users' affective states during navigation shall evaluate landmark based map design proposals optimized for spatial knowledge acquisition.

References

Caquard S (2015) Cartography III: a post-representational perspective on cognitive cartography. Prog Hum Geogr 39(2):225–235

Duncko R, Cornwell B, Cui L, Merikangas KR, Grillon C (2007) Acute exposure to stress improves performance in trace eyeblink conditioning and spatial learning tasks in healthy men. Learn Mem 14(5):329–335

Evans GW, Skorpanich MA, Gärling T, Bryant KJ, Bresolin B (1984) The effects of pathway configuration, landmarks and stress on environmental cognition. J Environ Psychol 4(4):323–335

Frei P, Richter KF, Fabrikant SI (2016) Stress supports spatial knowledge acquisition during wayfinding with mobile maps. In: Proceedings of the 9th international conference on geographic information science, Montreal, Canada, pp 100–103

Lupien SJ, Gillin CJ, Hauger RL (1999) Working memory is more sensitive than declarative memory to the acute effects of corticosteroids: a dose-response study in humans. Behav Neurosci Am Psychol Assoc 113(3):420

Mackingtosh E, West S, Saegert S (1975) Two studies of crowding in urban public spaces. Environ Behav 7(2):159–184

Meilinger T, Riecke BE, Bülthoff HH (2014) Local and global reference frames for environmental spaces. Q J Exp Psychol 67(3):542–569. Taylor & Francis

Richardson AE, Tomasulo MMV (2011) Influence of acute stress on spatial tasks in humans. Physiol Behav 103(5):459–466

The Virtual Reconstruction Project of Unavailable Monuments: An Example of the Church of Santa Maria Paganica in L'Aquila

Giovanni De Gasperis, Silvia Mantini and Alessio Cordisco

Abstract This contribution highlights the potential that modern technologies offer for virtual reconstruction and immersive navigation of monuments that are unavailable due to catastrophic events or other causes. A methodology is proposed for the enjoyment of the Cultural Heritage, starting from modeling, three-dimensional and photorealistic, to arrive at visualization in a virtual world enriched with extra content.

Keywords Cultural heritage · Virtual reality · Archeomatics · Public history

1 Introduction

The project of virtual reconstruction of monuments damaged by catastrophic events arises from the need to continue to enjoy the good during the stages of safety-laying, reconstruction and finishing, large-scale interventions that usually take a long time, especially for buildings of the Cultural Heritage. The experience of the earthquake in L'Aquila of 2009 has reflected on the importance of Heritage and collective identity that changes whenever a cultural asset is not used anymore, as in the case of the youngest aquilan population who has no memory of the destroyed heritage after 8 years, except through memories of their parents. In recent years, computer graphics have allowed to create more and more complex three-dimensional models that can be very realistic, built for desktop or mobile devices, but the question is: how to design

G. De Gasperis (✉) · S. Mantini · A. Cordisco
Dipartimento di Ingegneria e Scienze dell'Informazione e Matematica,
Università degli Studi dell'Aquila, Via Vetoio 1, L'Aquila, Italy
e-mail: giovanni.degasperis@univaq.it

S. Mantini
e-mail: silvia.mantini@univaq.it

A. Cordisco
e-mail: alessio.cordisco@univaq.it

© Springer International Publishing AG 2018
P. Fogliaroni et al. (eds.), *Proceedings of Workshops and Posters at the 13th International Conference on Spatial Information Theory (COSIT 2017)*, Lecture Notes in Geoinformation and Cartography, https://doi.org/10.1007/978-3-319-63946-8_8

and implement, with what features, a 3D model for immersive use without losing in definition and detail that characterizes the historical-artistic and archaeological heritage?

2 The Example of the Church of Santa Maria Paganica in L'Aquila

The church of Santa Maria Paganica is one of the monuments of L'Aquila, which has contributed to the history, to the social, economic, political and cultural growth of the city, occupying the role of the church of "Capoquarto". Built in 1308, it is today in its eighteenth-century style, the result of post-earthquake reconstruction of 1703, a single space with lateral chapels and a large transept in the presbytery area, topped by a majestic dome. With the earthquake of April 6, 2009, the structure suffered significant structural damage both to the roof, to the side walls and to the chapels. The 3D reconstruction work and virtualization has seen several steps:

1. the collection of graphic and photographic material;
2. modeling and rendering;
3. virtualization.

For the first step, in addition to the study of existing official and non official documentation, we attempted to record the historical memory from those who knew the building well by working in it, with the typical techniques of the anthropological study; very important was the contribution of the Parish, Priest Don Stefano Rizzo who has made possible not only to reconstruct a significant the photographic archive of more than 200 high-resolution photos of works in the church before the earthquake, but also his visual memory to reproduce small and large elements that had not been photographed and digitized. Before embarking on the modeling process, exterior and interior inspections to the post earthquake ruins were carried out, indispensable for photo shoots, consisting of a set of 380 photos of detail and overall, and measurements of surviving environments with direct metering techniques. At this stage, two still existing elements were the object of a deep photogrammetric acquisition: the side portal and the baptismal font. So in general it was then possible to recover a considerable amount of material, but unfortunately not enough to reconstruct the whole building in detail. The photographic material, published and not, was analyzed and divided into two groups: monochromatic and color, using black and white models as the basis for modeling, while others to reproduce the color range of walls, floors and furnishings. For the modeling phase, several software was used: SketchUp for basic modeling, Cinema 4D for details and Lumion for video rendering. Using a 1: 1 scale plan and a side panel representing the side of the church, the perimeter walls, the openings, the chapels and vaulted roofs were created. The raw model has been iteratively improved to the achievement of a structure as faithful as possible to the former reality. In some cases, using photos as a base, it was possible to reproduce real-life models.

The resulting 3D virtual reality model was presented to the public on 26 September 2016, after 3 months of intensive work, as a dissemination event of the Night of Researcher.

A demonstration video in available in YouTube at: http://youtu.be/G5Tf1Jv88gA.

Socio-spatial Networks, Multilingualism, and Language Use in a Rural African Context

Pierpaolo Di Carlo, Jeff Good, Ling Bian, Yujia Pan
and Penghang Liu

Abstract A GIS spatial perspective can provide important insights into many poorly understood sociolinguistic phenomena such as multilingualism in rural Africa. By relying on ethnographic and individual-based sociolinguistic information as well as on high spatial-temporal resolution data, our interdisciplinary team composed of linguists and geographers aims to (i) make original contributions to the cartographic representation of multilingualism and (ii) develop spatial-analytical models able to capture a complex array of linguistic, cultural, and spatial variables for a compact rural area of Cameroon.

Keywords Language & space · Multilingualism · Ethnographic GIS · Rural Africa

1 Multilingualism, Space, and GIS

The existing literature on the application of GISystems to the study of multilingualism represents the distribution of languages in specific areas—mostly urbanized regions of Western countries—where many languages are spoken by residents (Williams and Van der Merwe 1996; Veselinova and Booza 2009). What has yet to see attention is the spatial analysis of individual patterns of multilingualism, i.e., the ability of a given individual to use multiple languages. Individual multilingualism is a pervasive social feature in many parts of the world, including Sub-Saharan Africa, which is where our area of focus is located. Such an individual-based cognitive phenomenon lacks immediate cartographic representations (Luebbering et al. 2013: 386). In addition, sociolinguistic scholarship on multilingualism has mostly focused

P. Di Carlo (✉) · J. Good
Department of Linguistics, University at Buffalo, 613 Baldy Hall,
Buffalo, NY 14260, USA
e-mail: pierpaol@buffalo.edu

L. Bian · Y. Pan · P. Liu
Department of Geography, University at Buffalo, Buffalo, USA

© Springer International Publishing AG 2018
P. Fogliaroni et al. (eds.), *Proceedings of Workshops and Posters at the 13th International Conference on Spatial Information Theory (COSIT 2017),* Lecture Notes in Geoinformation and Cartography, https://doi.org/10.1007/978-3-319-63946-8_9

on the behaviors of urban migrants, whose multilingual repertoires are characterized by the addition of one or more languages of wider communication—such as, e.g., ex-colonial languages and pidgins—to more localized "heritage" languages. Both limits have made it thus far impossible—in fact, inconceivable—to attempt analyses of how multilingual repertoires pattern in space.

Following a theoretical shift from single languages to communicative practices, some recent language documentation projects have focused on small languages spoken in linguistically highly diverse areas and are now offering novel and more complex views of multilingual behaviors in non-urban regions of the world (see Lüpke 2016 for a review; Woodbury 2011 provides an overview of the practice of language documentation more generally). The multidisciplinary data collected in such projects and the localized nature of the languages documented provide new grounds for the application of GISystems for the study of multilingualism in both geographic and socially-constructed space (Low 2017), and we report on the application of GISystems to a project documenting rural patterns of multilingualism in Sub-Saharan Africa here.[1]

2 The Target Area: Lower Fungom

Our target area, Lower Fungom, lies at the northern edge of the Cameroonian Grassfields, one among the most linguistically dense parts of the world (Stallcup 1980). Many of the region's languages are endangered, and there is increasing consensus that multilingualism in local languages, likely to be an ancient phenomenon, plays a key role in the maintenance of such a diverse linguistic ecology. Within this exceptionally diverse region, Lower Fungom shows the highest degree of language density: in an area of around 200 sq km, one finds eight distinct languages associated with its thirteen villages and roughly 12,000 inhabitants (Good et al. 2011). Moreover, the Cameroonian Grassfields are known to be a "singularity area", i.e., one in which local language ideologies tend to identify a one-to-one relationship between language varieties and traditional political units (i.e., chiefdoms). In other words, locals conceptualize each chiefdom—which in Lower Fungom coincides with a single village—as being the center of a distinct language.

3 The Database

Multidisciplinary field research aimed at developing a holistic documentation of the languages of Lower Fungom has resulted in the collection of linguistic, ethnographic, archaeological, and geographic data. In particular, surveys have been

[1]This paper is based upon data collected during research projects supported by the U.S. NSF under grants BCS#0853981 (2009–2013), BCS#1360763 (2014–2017), and by the Endangered Languages Archive Programme (IPF0180 2012). Interdisciplinary research is funded by the University at Buffalo under IMPACT grant #077.

collected that provide detailed information on the self-reported multilingual repertoires of 206 individuals (ca. 2% of the area population), in addition to information on their social ties and family background. On this basis, Esene Agwara (2013) established that there are essentially no adult monolinguals in Lower Fungom and that the average individual speaks around six languages.

The spatial data at hand include a 1:50,000 topographic map, a high-resolution QuickBird image, aerial photos, DEM, and the locations of streams, roads, and footpaths. Such a wealth of information—linguistic, cultural, historical, and spatial—is highly unusual for rural African contexts.

4 Working Hypotheses

Di Carlo (2016) and Di Carlo et al. (forthc.) have proposed (i) that individuals in Lower Fungom acquire multiple languages primarily in order to gain access to the resources associated with different villages and (ii) that language use is not tied to a deep cultural notion such as ethnicity but, rather, is used to index an individual's participation in different kinds of personal relationships, in particular kinship (cf. Brubaker and Cooper 2000). This is different from what is known from Western societies (see, e.g., Fishman 1967, 1977; Irvine and Gal 2000) where languages are seen to be associated with cultural "essences".

5 Expected Outcomes

Ongoing research in the context of an interdisciplinary collaboration including linguists and geographers has three different, but tightly interrelated, goals: (i) transform qualitative data—in particular ethnographic data—into formats that can be effectively used for spatial analysis; (ii) adapt existing cartographic representation techniques to a new domain in order to represent multilingual repertoires and behaviors in space; and (iii) attempt spatial analyses of both individual-based and aggregate data concerning the size and nature of multilingual repertoires (see Sect. 3).

We have created a fine-grained spatial model that can support the exploration of the relationship between individual-based sociolinguistic and ethnographic information and the locations in which individuals reside and have lived in the past. In parallel to this work, we have also developed models for quantifying qualitative data that can minimize the loss of information via a system of weighted variables. This has allowed us to carry out socio-spatial analyses using a range of methods and to create visualizations of linguistic, sociolinguistic, and cultural information in geographic space, building on work representing epidemics in space (see, e.g., Zhong and Bian 2016) as well as economic patterns (Buys et al. 2006).

Preliminary results of this work have allowed for consideration of socio-spatial patterns of language "on the ground" and provide new insights into how the behavior of individuals patterns with observed linguistic-spatial patterns. These results suggest that geographical proximity plays a key role in shaping an individual's multilingual repertoire, with kinship networks also playing an important role. However, neither factor seems to account for the overwhelming majority of the individuals examined, thus suggesting the need to explore additional factors to understand multilingual patterns (see Sect. 4).

The high spatial-temporal resolution available to us, along with individual-level data, is playing a crucial role in uncovering precolonial, *longue durée* sociolinguistic and spatial patterns still at work in rural Africa that might be significant for the maintenance of local languages and that would be otherwise impossible to retrieve. In addition, this work is able to inform our goals for future fieldwork, directing us, in particular, towards the identification of new kinds of sociocultural and economic information to collect which will support the development of more adequate analytical models.

References

Brubaker R, Cooper F (2000) Beyond "identity". Theory Soc 29(1):1–47

Buys P, Deichmann U, Wheeler D (2006) Road network upgrading and overland trade expansion in Sub-Saharan Africa. Policy Research Working Paper. World Bank. doi:10.1596/1813-9450-4097

Di Carlo P (2016) Multilingualism, affiliation and spiritual insecurity: from phenomena to processes in language documentation. In Seyfeddinipur M (ed) African language documentation: new data, methods and approaches, pp 71–104. Language Documentation and Conservation special publication no. 10. http://hdl.handle.net/10125/24649. Accessed 31 May 2017

Di Carlo P, Good J, Ojong RA (forthcoming) Multilingualism in Rural Africa. In: Aronoff M (ed). Oxford Research Encyclopedia of Linguistics

Esene Agwara AD (2013) Rural multilingualism in the North West Region of Cameroon: the case of Lower Fungom. Buea, Cameroon: University of Buea MA thesis. http://buffalo.edu/~jcgood/EseneAgwara-2013-RuralMultilingualism.pdf. Accessed 31 May 2017

Fishman J (1967) Bilingualism with and without diglossia; diglossia with and without bilingualism. J Soc Issues 23(2):29–38

Fishman J (1977) Language and ethnicity. In: Giles H (ed) Language, ethnicity, and intergroup relations. Academic Press, New York, pp 15–58

Good J, Lovegren J, Mve JP, Tchiemouo CN, Voll R, Di Carlo P (2011) The languages of the Lower Fungom region of Cameroon: Grammatical overview. Africana Linguistica 17:101–164

Irvine JT, Gal S (2000) Language ideology and linguistic differentiation. In: Kroskrity PV (ed) Regimes of language. Ideologies, polities, and identities. SAR Press/ James Currey, Santa Fe/ Oxford, pp 35–84

Low S (2017) Spatializing culture. The ethnography of space and place. Routledge, London and New York

Luebbering CR, Kolivras KN, Prisley SP (2013) The lay of language: surveying the cartographic characteristics of language maps. Cartogr Geogr Inf Sci 40(3):383–400

Lüpke F (2016) Uncovering small-scale multilingualism. Crit Multilingualism Stud 4(2):35–74

Stallcup K (1980) La géographie linguistique des Grassfields. In Hyman L, Voorhoeve J (eds) L'expansion Bantoue: Actes du Colloque International du CNRS, Viviers (France) 4–16 avril 1977. Volume I: Les Classes Nominaux dans le Bantou des Grassfields. SELAF, Paris, pp 43–57

Veselinova L, Booza J (2009) Studying the multilingual city: a GIS-based approach. J Multilingual Multicult Dev 30(2):145–165

Williams CH, Van der Merwe I (1996) Mapping the multilingual city: a research agenda for urban geolinguistics. J Multilingual Multicult Dev 17(1):49–66

Woodbury AC (2011) Language documentation. In: Austin PK, Sallabank J (eds) The Cambridge handbook of endangered languages. Cambridge University Press, Cambridge, pp 159–186

Zhong S, Bian L (2016) A location-centric network approach to analyzing epidemic dynamics. Ann Am Assoc Geogr 106(2):480–488. doi:10.1080/00045608.2015.1113113

Reactive Obstacle Avoidance for Multicopter UAVs via Evaluation of Depth Maps

Luca Di Stefano, Eliseo Clementini and Enrico Stagnini

Abstract Reacting to unforeseen obstacles is a major issue in the field of autonomous navigation. In the context of Unmanned Aerial Vehicles, an "obstacle" is any object that stands between the UAV and its desired position (*waypoint*). Therefore, obstacle detection can be reduced to the problem of assessing the visibility of the waypoint from the point of view of the drone. In this work, data acquired from an onboard depth camera are used to describe the visibility of the target waypoint in a qualitative framework, and to plan a new route when obstacles are detected.

Keywords UAV · Obstacle avoidance · Qualitative spatial reasoning · Depth map · Motion planning

1 Introduction

Unmanned Aerial Vehicles (UAVs) have a wide range of applications. Multicopters, in particular, have found great popularity due to their maneuverability and relative low cost (Cai et al. 2014). However, the state of the art only provides solutions for point-to-point navigation in free space: therefore their application in cluttered environments still depends heavily on remote control by a human operator.

In most cases, UAV missions consist of a sequence of movements interleaved with operations that involve the payload, such as taking photographs or dropping a package. To increase the degree of automation in such missions, systems are needed that can adapt the flight plan to newly-discovered information about the environment.

L. Di Stefano (✉)
Gran Sasso Science Institute, L'Aquila, Italy
e-mail: luca.distefano@gssi.it

E. Clementini
University of L'Aquila, L'Aquila, Italy
e-mail: eliseo.clementini@univaq.it

E. Stagnini
DroniAbruzzo, L'Aquila, Italy
e-mail: info@droniabruzzo.it

© Springer International Publishing AG 2018
P. Fogliaroni et al. (eds.), *Proceedings of Workshops and Posters at the 13th International Conference on Spatial Information Theory (COSIT 2017)*, Lecture Notes in Geoinformation and Cartography, https://doi.org/10.1007/978-3-319-63946-8_10

Many strategies found in the literature (see Goerzen et al. 2010 for a broad classification; Kendoul 2012 for a more recent and comprehensive review) rely on the creation and periodic update of a detailed model of the drone surroundings. For concrete examples of this approach, see Nieuwenhuisen and Behnke (2014), Hrabar (2011).

Qualitative approaches, on the other hand, could mimic the adaptability of human operators without storing such a great amount of information.

To this effect, we propose an algorithm that can guide a multicopter UAV towards a destination waypoint, adapting the route as new obstacles are detected. Both obstacle detection and path replanning are based on the evaluation of GPS information together with a depth map, i.e. raster data acquired from an on- board depth camera. The algorithm is *reactive*: no model of the environment is stored in memory, and decisions only rely on the currently available sensor data.

2 Assessing Visibility Relationships

We can use our knowledge of both the UAV and the target's GPS coordinates to point the depth camera towards the target and then "project" the latter on a pixel p of the depth map.

We can then compute the waypoint-UAV visibility relation according to the qualitative framework presented in Fogliaroni and Clementini (2014). This is achieved by evaluating a neighborhood of p whose size depends on the distance between the UAV and the target.

3 Wayfinding

When an obstacle occludes the target, we need to find an *escape waypoint*. This is an intermediate position that is already visible from the current position of the UAV, and from which the original target should be visible unless more obstacles are detected along the way.

Wayfinding is a two-step process: first we have to rule out points that are visible but too close to obstacles. This can be done by applying a "depth-aware" dilation filter to the depth map.

We then choose the candidate that minimizes the overall distance from the target, and apply a transformation to obtain the GPS coordinates of the escape waypoint.

4 Simulation

Currently the algorithm is implemented as a set of Python 3 scripts that communicate with a simulated quadcopter inside the Coppelia V-REP robotic platform (Rohmer et al. 2013).

The tests show good performance even with lower depth map resolutions, as the effective path taken by the UAV is well below 1.1x the distance between the start position and the target waypoint.

References

Cai G, Dias J, Seneviratne L (2014) A survey of small-scale unmanned aerial vehicles: recent advances and future development trends. Unmanned Syst 02(02):175–199. doi:10.1142/S2301385014300017

Fogliaroni P, Clementini E (2014) Modeling visibility in 3D space: a qualitative frame of reference. In 9th International 3DGeoInfo 2014—Lecture Note in Geoinformation and Cartography, pp 1–19. doi:10.1007/978-3-319-12181-9_15

Goerzen C, Kong Z, Mettler B (2010) A survey of motion planning algorithms from the perspective of autonomous UAV guidance, vol 57. doi:10.1007/s10846-009-9383-1

Hrabar S (2011) Reactive obstacle avoidance for rotorcraft UAVs. In: IEEE International Conference on Intelligent Robots and Systems (August), pp 4967–4974. doi:10.1109/IROS.2011.6048312

Kendoul F (2012) Survey of advances in guidance, navigation, and control of unmanned rotorcraft systems. J Field Robot 29(2):315–378. doi:10.1002/rob.20414,10.1.1.91.5767

Nieuwenhuisen M, Behnke S (2014) Hierarchical planning with 3D local multiresolution obstacle avoidance for micro aerial vehicles. In: Joint 45th international symposium on robotics (ISR) and 8th German conference on robotics (ROBOTIK), University of Bonn, Germany

Rohmer E, Singh SPN, Freese M (2013) V-REP : a versatile and scalable robot simulation framework. In: Proceedings of The International Conference on Intelligent Robots and Systems (IROS)

New and Given Information in Alpine Route Directions

Ekaterina Egorova

Abstract The paper reports on an exploratory study where we examine alpine route directions from the perspective of new and given information. The first results reveal traces of a "mountain scenario"—we encounter geographic entities that are introduced into the discourse as assumingly familiar to the reader, signalling a certain structure associated with alpine space.

Keywords Spatial cognition · Landmarks · Pedestrian navigation · Familiarity · Personalization

1 Motivation

Route directions have long been seen as an ideal laboratory for the investigation of mental representations of space (Tversky 2000). As any communication act, they require the provider to follow certain principles—in particular, that of relevance (Grice 1975; Sperber and Wilson 1986). In line with the latter, route providers adjust their instructions to spatial knowledge of recipients, as well as the purpose of the activity that wayfinding is embedded in (Hirtle et al. 2011; Tenbrink 2012; Tomko 2007).

The specifics of online alpine route directions lies in the asynchronous, non-co-located communication situation (Janelle 2004), where a route provider has to *make assumptions* about the spatial knowledge of a potential recipient. This offers a unique opportunity to investigate if any spatial information is introduced as assumingly

E. Egorova (✉)
Department of Geography, University of Zurich, Winterthurerstrasse 190,
8057 Zurich, Switzerland
e-mail: ekaterina.egorova@geo.uzh.ch

E. Egorova
URPP Language and Space, University of Zurich, Winterthurerstrasse 190,
8057 Zurich, Switzerland

© Springer International Publishing AG 2018
P. Fogliaroni et al. (eds.), *Proceedings of Workshops and Posters at the 13th International Conference on Spatial Information Theory (COSIT 2017)*, Lecture Notes in Geoinformation and Cartography, https://doi.org/10.1007/978-3-319-63946-8_11

familiar to the recipient. Revealing a certain scenario associated with alpine space would contribute to our knowledge of spatial cognition of continuous and seemingly unstructured natural space (Smith and Mark 2003; Egorova et al. 2015).

2 Methodology

We rely on the taxonomy of given-new information that reflects the degrees of assumed familiarity of information to the recipient and includes the following types of entities (Prince 1981). For brand-new entities a hearer has to create new entities in mind, as in (1a); linguistically, they are mostly indefinite, marked by indefinite articles and quantifiers, including numerals. Unused entities are assumed to be known to the hearer, although not activated in his consciousness at the time of the utterance, as in (1b); these are expressed by personal pronouns and proper nouns or marked by definite or demonstrative articles, possessive adjectives or certain quantifiers. Further, textually evoked entities are present in discourse on textual ground, as in (1c), while situationally evoked ones represent discourse participants and salient features of the extratextual context, as in (1d). Finally, inferrable are entities that, assumingly, can be inferred via logical reasoning from discourse entities already introduced. In (1e), the door is introduced in the text for the first time, but is treated as something already known—the speaker assumes that the reader knows that Bastille is a building and that buildings typically have doors.

1. a. In the park yesterday, **a kid** threw up on me. (brand-new)
    b. **Noam Chomsky** went to Penn. (unused)
    c. A guy I work with says **he** knows your sister. (textually evoked)
    d. Pardon, would **you** have change of a quarter? (situationally evoked)
    e. He passed by the Bastille and **the door** was painted purple. (inferrable)

Relying on linguistic markers, we annotated the information status of 353 references to geographic objects that we identified in a corpus of 10 online alpine route directions from an online platform.[1]

3 Current Results and Future Work

Two main observations can be made based on this first exploratory study.

First, we see traces of a mountain scenario in the form of inferrable entities. Route providers indeed assume the knowledge of certain alpine space structure from the recipients—for example, that ridges have crests and slopes, chutes and couloirs have tops and bottoms, that snowfields are to be expected in the mountains, as in (2a-b).

---

[1] http://www.summitpost.org/.

Second, revealing about spatial thinking involved in route direction production are cases such as (2c-d). Though linguistically similar to inferrable entities, they represent geographic objects that *will be* salient or easily identifiable in the environment when the route recipient finds himself in the location described (note "on your left" in (2c)). This calls for the adjustment of the original taxonomy to the specifics of the route directions style of discourse.

2.   a. The Chute itself is about 40° easing off near **the top**.
     b. There are a couple of ridges on sand and gravel which take you onto **the snow fields** without too much difficulty.
     c. Follow the signs toward **the prominent buttress**.
     d. There is **this obvious chute on your left**.

In future work, we aim to look closer into the linguistic and conceptual distinctions between inferrable entities and those of the type (2c-d), followed by the measure of the inter-annotator agreement. Working with a larger corpus we will further expand our understanding of objects associated with alpine space, as part of our broader goal of contributing to the knowledge of spatial cognition in natural, non-urban spaces.

References

Egorova E, Tenbrink T, Purves RS (2015) Where snow is a landmark: route direction elements in alpine contexts. In: Fabrikant S, Raubal M, Bertolotto M, Davies C, Freundschuh S, Bell S (eds) COSIT 2015. Springer, LNCS, pp 175–195

Grice P (1975) Logic and conversation. In: Cole P, Morgan J (eds) Syntax and semantics. 3: Speech acts. Academic Press, pp 41–58

Hirtle S, Timpf S, Tenbrink T (2011) The effect of activity on relevance and granularity for navigation. In: Egenhofer M, Giudice N, Moratz R, Worboys M (eds) COSIT 2011, vol 6899. LNCS, Springer, pp 73–89

Janelle D (2004) Impact of information technologies. Geogr Urb Transp 86–112

Prince E (1981) Toward a taxonomy of given-new information. Radical Pragmat 223–255

Smith M, Mark D (2003) Do mountains exist? toward an ontology of landforms. Plan Des Env Plan B 411–428

Sperber D, Wilson D (1986) Relevance: communication and cognition. Blackwell

Tenbrink T (2012) Relevance in spatial navigation and communication. In: Stachniss C, Schill K, Uttal D (eds) Spatial cognition VIII. Springer, LNCS, pp 1358–377

Tomko M (2007) Destination descriptions in urban environments. University of Melbourne

Tversky B (2000) Levels and structure of cognitive mapping. In: Kitchin R, Freundschuh S (eds) Cognitive mapping: past, present and future. Routledge, pp 24–43

Defining Spatial Boundaries:
A Developmental Study

Eugenia Gianni and Sang Ah Lee

Abstract Although the capacity to navigate by environmental boundaries has been widely documented, the perceptual and physical factors that define a boundary have yet to be defined. In this study, we tested children's navigation in spatial arrays consisting of 20 freestanding objects with varied inter-object spacing and length. Children begin to successfully compute locations using aligned (but discontinuous) object arrays around the seventh year of age. Our results suggest a late-emerging capacity of extrapolating geometric information from discontinuous structures.

Keywords Boundaries · Navigation · Freestanding objects · Geometric information

1 Introduction

Humans and animals possess impressive capacities of navigating and orienting in familiar and unfamiliar environments. It is generally believed that they do this by means of storing global spatial representations of the surrounding environment, so-called cognitive maps (O'keefe and Nadel 1978). A wide range of studies, from behavioral to electrophysiological, to computational models (Hartley et al. 2014) have shown that one of the primary inputs to place coding are environmental boundaries. When rats found food in one particular corner of a rectangular arena, upon disorientation they tended to limit their search to the target corner and the

E. Gianni (✉)
Center for Mind/Brain Sciences, University of Trento, Corso Bettini 31,
38068 Rovereto, TN, Italy
e-mail: eugenia.gianni@unitn.it

S.A. Lee
Department of Bio and Brain Engineering, Korea Advanced Institute of Science and Technology, 291 Daehak-ro, Yuseong-gu, Daejeon 34141, South Korea
e-mail: sangah.lee@kaist.ac.kr

© Springer International Publishing AG 2018
P. Fogliaroni et al. (eds.), *Proceedings of Workshops and Posters at the 13th International Conference on Spatial Information Theory (COSIT 2017)*, Lecture Notes in Geoinformation and Cartography, https://doi.org/10.1007/978-3-319-63946-8_12

geometrically identical (rotationally symmetric) diagonal corner with the same frequency, even when the target corner could be disambiguated by means of a featural cue such as color or visual pattern (Cheng 1986). Interestingly, almost all animals have shown to behave similarly, including human children (Hermer and Spelke 1994; Lee and Spelke 2010) Even in human adults, boundaries and object/landmark features are processed separately during navigation and respond to different neural and behavioral mechanisms (Doeller and Burgess 2008; Doeller et al. 2008). While boundary-based navigation is well-documented even across species, the factors that define surfaces as boundaries have yet to be determined. Studies with children have started to answer this question by manipulating the physical and visual properties of spatial boundaries. They have shown that pre-school children can use boundaries to navigate as long as they are 3D and extended on the ground-plane, while they cannot use 2D forms and geometric arrays made up of three or four free-standing objects (Gouteux et al. 2001; Lee and Spelke 2008; Lee and Spelke 2011).

What is the fundamental difference between an array of objects and an array of walls? Previous work has shown that even two-year-old children succeed in using surfaces even if they are segmented into four distinct 100 or 80-cm-long walls in a rectangular array (Lee et al. 2012) as long as they are opaque (Gianni and Lee (in preparation)). However, it is still not clear how their use in navigation relates to their capacity of preventing movement (Kosslyn et al. 1974) or how their length and solidity/continuity factor plays a role into their conceptualization as boundaries (Lee et al. 2012), and finally if their conceptualization is submitted to fundamental changes over the course of development (Fig. 1).

In our study we addressed these questions by testing children from 4 to 9 years old in four different arrays consisting of twenty free-standing objects (see Fig. 2). In Experiment 1, the objects were arranged in a rectangular fashion with an inter-object spacing of 16 cm. In Experiment 2, the objects were arranged more densely together to form four segments with an inter-object space of 8 cm. In these two conditions the objects were freestanding and discontinuous, but sufficiently dense to underline the geometric figure and to prevent children's movement. In Experiment 3, objects were aligned to form a rectangular array of 4 continuous walls of 50 cm, and in Experiment 4 they were rearranged into two longer con-tinuous walls (100 cm long).

2   Methods

2.1   Experimental Setting

Experiments took place within a round room formed out of black curtains (2.10 m diameter) hanging from the ceiling on a circular track. The entrance to the room (created by the curtain's latch) was masked in order to avoid spatial cues.

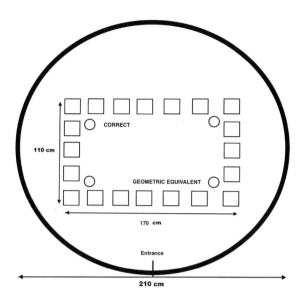

Fig. 1 Experimental setting, schematic view from *above*. Inside a *black round curtain* 20 free-standing objects (here schematically arranged as in Exp. 1, Fig. 2a) were placed as to form a *rectangular arena* (170 × 110) cm, outside perimeter). If during the game the sticker was placed in the *upper-left corner* (Correct), the opposite *diagonal corner* was indistinguishable for a disoriented subject and therefore labeled as 'Geometric Equivalent.' A majority of choices for both correct and geometric equivalent corners, after disorientation suggested the subject had correctly encoded and used the geometric properties of the array to reorient

4 symmetrically LEDs on the ceiling provided uniform lighting. The experimental arrays were placed at the center of the room, on a grey-colored non-slip floor. 4 inverted cups were placed at the four corners of the arena and served as hiding places for stickers (Fig. 1).

2.2 Experimental Apparatus

20 while plastic parallelepipeds, 30 cm high, 10 × 10 cm wide, were arranged as to create a rectangular arena (170 × 110 cm, outer perimeter). In Experiment 1, they were uniformly aligned along the perimeter with an inter-object space of 16 cm (Fig. 2a). In Experiment 2, the objects were arranged along the same 170 × 110 cm perimeter as to create four distinct 80 cm segments, with an inter-object space of 8 cm (Fig. 2b). In Experiment 3, objects were arranged to create four distinct compact walls of equal length (50 cm) (Fig. 2c). Finally, in Experiment 4, the objects were compactly arranged to form two 100 cm walls (Fig. 2d).

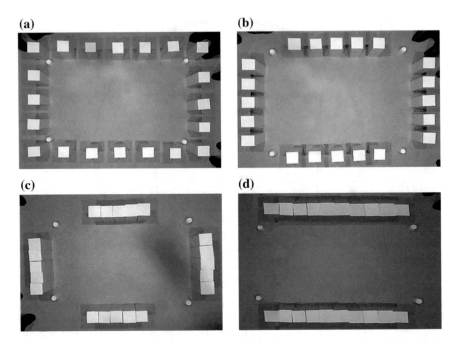

Fig. 2 Arrays made up of twenty freestanding objects: **a** Uniformly arranged, 16 cm apart. **b** Arranged more densely, 8 cm apart. **c** Continuous, four walls 50 cm long. **d** Continuous, two walls 100 cm long

2.3 Participants and Experimental Procedures

89 healthy children, 48 males and 41 females, ranging from 48 to 119 months (4–9 years old) were tested in this study. They were recruited from daycares and recreational centers in the area of Rovereto and the surrounding area of the Province of Trento. They voluntarily came to the laboratory accompanied by their parents. Before the test they were let playing in the toy-area of the laboratory for about 10 min. During the test they were accompanied by the experimenter to the experimental room and let step into the center of the arena. The experimenter then hid a sticker in one of the four cups placed at each arena's corner, taking care the child was paying attention. The child was instructed to be about to play a hiding and finding game where he had to exactly locate the sticker after disorientation in order to win it. The experimenter then blindfolded the child and let him rotate on its place for 10 s in either directions for disorienting. He then removed the blindfold and let the child search for the sticker. First choices were recorded. The procedure was repeated for four trials.

3   Methods

3.1   *Experiments 1 and 2. Discontinuous Objects*

Analyses were conducted by computing the average proportion of geometric searches (at the correct corner + geometric equivalent corner; see Fig. 1). Since no significant effect of condition across the first and the second experiment was found, data from experiment 1 and experiment 2 were collapsed. A univariate ANOVA was used to compare the performance of 4–6 year-olds (48 subjects) with performances of 7–9 year-olds (34 subjects). A significant main effect of age group was found (F(2) = 4.234; p = 0.018). T-tests against the level of chance (0.5) showed that 4–6 years old children did not perform above the level of chance (T (47) = 1.091; p = 0.281), while children ranging from 6 to 9 did (T(33) = 5.583; p < 0.001). There were no significant effects of sex or condition (F < 1, n.s.) (Fig. 3).

3.2   *Experiment 2 and 3. Continuous Surfaces*

Experiment 3 and Experiment 4 were designed to test continuous surfaces of two different lengths, while keeping total number of objects (surface area) equal across all four conditions. Subjects were tested in a within-subjects design and the order was counterbalanced across subjects (each subject was alternatively submitted to either experiment 3 or 4 first). So far only 7 children between the age of 4 and 5 years old have been tested. A repeated measures ANOVA looking at geometric search in the two conditions (within-subjects factor) and two between-subjects factors—sex and experimental order (either experiment 3 or 4 first)—was used to explore these preliminary data. A significant main effect of condition was found (F(1) = 15,364; p = 0.017). Exploratory tests against the level of chance revealed that children did not perform significantly above chance in Experiment 3 (T(6) = 1.922; p = 0.103) while they did perform above the level of chance in Experiment 4. (T(6) = 4.804; p = 0.003) (Fig. 4).

Fig. 3 Mean performance for 4, 5, 6 years old and 7, 8, 9 years old groups. *Error bars* represent standard errors. ** p ≤ 0.01

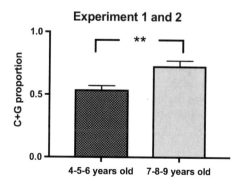

Fig. 4 Mean performance in
Experiments 3 and 4. *Error
bars* represent standard errors.
**p ≤ 0.01

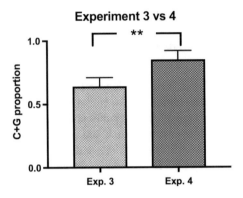

4 Discussion

Considering experiments 1 and 2, our results suggest that children develop the capacity of using discontinuous walls around the 7th year of age, considerably later then continuous walls (Hermer and Spelke 1994). Indeed previous studies have shown that children developed the capacity to reorient using continuous boundaries at the age of two years-old, even if the boundaries were segmented (Lee et al. 2012) and of low visibility (Lee and Spelke 2011). Previous studies have shown children fail in orienting towards structures made up of three (Gouteux et al. 2001) or four free-standing objects (Lee and Spelke 2008; Lee and Spelke 2011) but they left room for hypothesizing whether children failed because either the boundary were not sufficiently visible in their geometric shape or because they did not sufficiently prevent movement. In our experiment we show for the first time that, as long as the structure is sufficiently dense to prevent movement, and sufficiently visually robust to underline the geometric shape, children still fail until the age of seven. The failure to navigate by the geometry of the discontinuous object arrays might be explained by a later-emerging capacity of integrating information from qualitatively different sources, such as boundaries and landmarks, which are separately processed in navigation (Doeller et al. 2008). Such a conclusion might be consistent with the finding that the capacity of children to integrate a feature such as color and boundaries' geometric structure to correctly orient is acquired only between the 5th and the 7th year of age (Hermer-Vazquez et al. 2001).

Moreover preliminary data in experiments 3 and 4 show that younger children are not able to use 50 cm walls but perform well with 100 cm walls, putting into context previous work showing success with 100 or 80-cm-long walls (Gianni and Lee (in prep.); Lee and Spelke 2011). The relevance of wall length should be further explored by taking into account factors such as the child's physical size and interaction with the objects/boundaries, as they may reveal the threshold at which children begin to perceive the wall as qualitatively different from an object.

References

Cheng K (1986) A purely geometric module in the rat's spatial representation. Cognition 23 (2):149–178

Doeller CF, Burgess N (2008) Distinct error-correcting and incidental learning of location relative to landmarks and boundaries. Proc Natl Acad Sci 105(15):5909–5914

Doeller CF, King JA, Burgess N (2008) Parallel striatal and hippocampal systems for landmarks and boundaries in spatial memory. Proc Natl Acad Sci 105(15):5915–5920

Gianni E, Lee SA (in prep.). The role of visual boundary-structures in early spatial mapping

Gouteux S, Thinus-Blanc C, Vauclair J (2001) Rhesus monkeys use geometric and nongeometric information during a reorientation task. J Exp Psychol Gen 130(3):505

Hartley T, Lever C, Burgess N, O'Keefe J (2014) Space in the brain: how the hippocampal formation supports spatial cognition. Phil. Trans. R. Soc. B 369(1635):20120510

Hermer L, Spelke ES (1994) A geometric process for spatial reorientation in young children. Nature 370(6484):57

Hermer-Vazquez L, Moffet A, Munkholm P (2001) Language, space, and the development of cognitive flexibility in humans: the case of two spatial memory tasks. Cognition 79(3):263–299

Kosslyn SM, Pick HL Jr, Fariello GR (1974). Cognitive maps in children and men. Child Dev 707–716

Lee SA, Spelke ES (2008) Children's use of geometry for reorientation. Dev Sci 11(5):743–749

Lee SA, Spelke ES (2010) Two systems of spatial representation underlying navigation. Exp Brain Res 206(2):179–188

Lee SA, Spelke ES (2011) Young children reorient by computing layout geometry, not by matching images of the environment. Psychon Bull Rev 18(1):192–198

Lee SA, Sovrano VA, Spelke ES (2012) Navigation as a source of geometric knowledge: young children's use of length, angle, distance, and direction in a reorientation task. Cognition 123 (1):144–161

O'Keefe J, Nadel L (1978). The hippocampus as a cognitive map. Clarendon Press, Oxford

Artificial Cognitive Maps: Selecting Heterogeneous Sets of Geographic Objects and Relations to Drive Highly Contextual Task-Oriented Map Views

Lucas Godfrey and William Mackaness

Abstract We present work from an on-going project to develop techniques of auto-mated cartography. We introduce Artificial Cognitive Maps as an approach to integrating insights from spatial cognition with geographic data. The ultimate goal is to drive highly contextual map views that more effectively support navigation tasks such as travelling across large, complex cities. With a focus on our now ubiquitous small screen mobile devices, we propose that distortions on the traditional metric cartographic representation may support a reduction in cognitive load for the user, but that the logic and parameters of these distortions should be founded on the natural distortions present in our cognitive representations of geographic objects and their relation.

Keywords Spatial cognition · Automated cartography · Navigation · Multi-scale data

1 Project Background

The motivation for the work presented here is a broad hypothesis that the graphical display of spatial information delivered to small screen mobile devices for the purposes of navigation may benefit from a relaxation of our rigid, metric representation of geographic space (Godfrey and Mackaness 2017). To traverse complex journeys, particularly journeys that require the transfer between varying modes of transport, the map user may make use of a number of conceptual scales, as well as a number of forms of spatial representation—for example topographic views of the city, in contrast to more highly schematised views of the public transport network. The mobile map user must transition between these varying views, for example to

L. Godfrey (✉) · W. Mackaness
School of GeoSciences, University of Edinburgh, Edinburgh, UK
e-mail: lucas.godfrey@ed.ac.uk

W. Mackaness
e-mail: william.mackaness@ed.ac.uk

© Springer International Publishing AG 2018
P. Fogliaroni et al. (eds.), *Proceedings of Workshops and Posters at the 13th International Conference on Spatial Information Theory (COSIT 2017)*, Lecture Notes in Geoinformation and Cartography, https://doi.org/10.1007/978-3-319-63946-8_13

reconcile the need to see a high level of detail around key decision-points with the need to get an overview of the overall journey.

We propose that introducing geometric distortions to the display of spatial information may support representations of journeys that better reflect travel as a continuous experience, and act to reduce the cognitive load of using a small screen device as a means to navigate. Before we are able to test hypotheses around the impact of these 'continuous' map views on the user, we must first specify some underlying distortions of metric space that may give rise to these experimental views. The core approach we employ to tackle this issue is based on a proposal that if we are to purposefully distort the visual representation of geographic relations, these distortion should reflect the kind of distortions that exist naturally within people's cognitive representations of the environment. It is acknowledged that there has been much work to date within this area, and it is the ambition of the work presented here to move some of the key insights from the spatial cognition research into a more explicitly cartographic context—namely using our understanding of natural perceptions and representations of geographic space as the logic by which we surface 'continuous' map views to the user through cognitively ergonomic geometric distortions.

Artificial Cognitive Maps (ACM) refer to the data structures that serves as input to these visualisation methods, as opposed to the map view itself—this is similar to the distinction between 'cognitive maps' and 'cognitive configurations' (Golledge and Stimson 1997). Here we propose that a relative, task-centric view of the geography may be inferred given the particular context of the mobile map user, and that this inferred set of geographic objects and relations may be used as a basis for surfacing a heterogeneous graphical representation. These views are heterogeneous as they are dependent not just on location but also the direction of travel and the particular modes of transport available to the traveler.

2   Functional Representation: A Task-Oriented Approach

We use functional representation as a way of forming logical connections between decisions and geographic entities and relationships. A primitive route object is defined as a sequence of decisions joining the start and destination. Here 'task' and 'context' are closely linked, in the simplest case being defined by the journey start, destination, and possible modes of travel. The selection and display of geographic entities is functional when its logic is based on the purpose that entities serve relative to the current navigation task (Freksa et al. 2005), rather than displaying a generic representation of the structure of the geography based on a uniform map scale. It should be noted that in our vocabulary a road intersection is itself considered an entity. In the example in Fig. 1 we see three representations of the same geographic area. In (a) we see a simple topographic illustration that is provided for some context. In (b) and (c) we see the two routes through the network with arrows indicating the route and direction of travel. In (b), although the roads (shown in

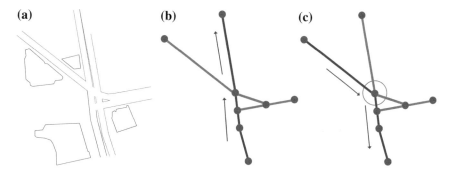

Fig. 1 Functional representation of two routes passing through the same network node

grey) that join on both the right and left provide some context, they are not decision-points in the journey. In (c), the traveler must pass through a number of the same network nodes as (b), but there is a decision point where the traveler must turn right (denoted by the grey circle).

While a cognitive map reflects aspects of the environment that are particularly salient to the traveler, an ACM reflects aspects that are inferred to be of functional relevance to the map user. In other words, an ACM can be thought of as an a priori approximation of an a posteriori cognitive representation of a successfully completed journey. Before the geometric structure of the cartographic visualisation is processed, a set of decision nodes are selected. This set includes the start, destination, the various turning points and the transfer points (nodes that represent a transfer to a different mode of travel). This simple topological structure then serves as the foundation for the broader set of nodes and edges that form the overall structure and content of the map view.

3   From Cognitive Graphs to Cognitive Cartography

While clearly from a cartographic perspective a simplified graph view is insufficient as a graphical aid for navigational decision-making, our approach is to consider the kinds of distortions that may be appropriate on the overall network structure as a first step. In fact, it has been argued that our cognitive representations of the structure of geographic space are appropriately conceptualised as graphs.

The cognitive graph approach (Warren et al. 2017) proposes that human spatial knowledge may be best thought of as a 'labelled graph' (i.e. a property graph) in that we encode local metric information in a 'noisy', 'biased' and 'geometrically inconsistent' manner. The argument is that we make use of representations of topological structures with 'loose' geometric attributes assigned to nodes and edges, but with varying accuracy and in a context specific manner. This view supports the ACM approach in that relative distances and angles are contingent on the current task and context (e.g. mode of travel), and therefore rather than the structural representation of the geography being specified in a uniform manner at each scale, individual nodes and edges can be weighted, with their final value being chosen from a potential range based on current requirements (e.g. a maximum rotation of 90°). This approach has strong links with the classic anchor-point hypothesis (Couclelis et al. 1987) in that while topological relations are preserved, the exact spatial distribution of geographic objects is dependent on the experience (or predicted experience) of the traveler. The ultimate aim of the approach is to show graphical navigational information in a way that is better suited to the physical affordances of a small screen.

In Fig. 2 we see the basic ACM model for a multi-modal route with all angle and edge weight distortions set at '0' (i.e. the metric topographic view). While the model is biased toward a route orientation strategy, geographic context is included by way of edges (streets) that are likely to be visible to the traveler, even if they are not traversed (shown in grey). There are three types of nodes within the basic ACM model—decision nodes, structural nodes that represent intersections and the start and end of polylines, and context nodes that represent the end of the first polyline segment for roads that intersect the journey but are not themselves traversed (i.e. the start of intersecting roads).

Fig. 2 Graph of multi-model route across the city of Edinburgh—*blue edges* denote travel by foot, *green edges* denote travel by tram

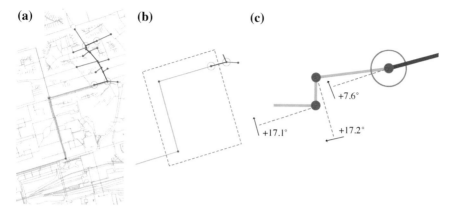

Fig. 3 Transition from a topographic to a more highly schematised view with 'smoothed' transfer to the second phase of walking from the tram phase (*green* to *blue*)

In Fig. 3 we see the next step of processing on the property graph, illustrated by a disambiguated view of the route. In Fig. 3c the angularRange and edgeWeight variables are updated based on user tasks. The angularRange variable denotes the amount and direction of distortion in terms of the angular rotation of an edge, and edgeWeight denotes the distortion of relative edge length. This example reflects the requirement to show route sections with low frequency of decision-making in a more highly schematised form, with the geometric representation of the tram phase conforming to a 0°/45°/90° network schematic, but with a 'smoothed' transition

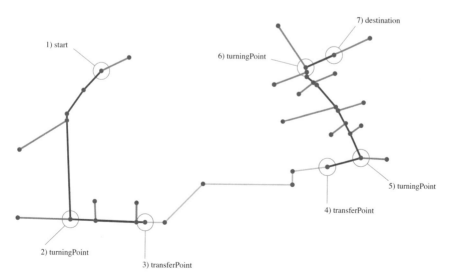

Fig. 4 Basic ACM for a multi-modal route based on seven key decisions and two transfers (changes of transport mode)

back to the 'street-level' topographic view. The aim is to reconcile differing forms of spatial representation such that the user is able to more easily get a sense of the overall structure of the journey, as well as the detail required around key decision points (Mackaness et al. 2011; Quigley 2001). In Fig. 4 this output is illustrated based on our test route.

4 Summary and Next Steps

In summary, decision-action pairs that can be inferred along the route serve as the underlying topology of the graph. The 'locations' of these actions in geographic space essentially serve as anchor-points, but rather than anchors based on a person's experience, as proposed by Couclelis et al. in the original paper (Couclelis et al. 1987), they are automatically inferred based on the likely functional character of the geography given the current task. Geometric distortions are then applied by updating angularRange and edgeWeight given the characteristics of the current journey.

By defining the basic logic of an ACM we lay the foundation for further work to generate mixed-scale and mixed-space map views that integrate varying forms of spatial information. The next step is to investigate the user impact of these types of distortion on network views in the context of the mobile device form factor.

Acknowledgements We would like to thank the Ordnance Survey and the Engineering and Physical Sciences Research Council for supporting this research.

References

Couclelis H, Gale N, Golledge R, Tobler W (1987) Exploring the anchor-point hypothesis in spatial cognition. J Env Psychol 7:99–122

Freksa C, Barkowsky T, Klippel A, Richter K (2005). The cognitive reality of schematic maps. In: Meng L, Zipf A, Reichenbacher T (eds) Map-based mobile services. Springer

Godfrey L, Mackaness W (2017) The bounds of distortion: truth, meaning and efficacy in digital geographic representation. Int J Cartogr 3(1):31–44

Golledge RG, Stimson RJ (1997). Spatial behaviour: a geographic perspective. Guilford Press

Mackaness W, Tanasescu V, Quigley A (2011). Hierarchical structures in support of dynamic presentation of multi resolution geographic information for navigation in urban environments. In: Proceedings of the 19th GIS Research UK Annual Conference University of Portsmouth

Quigley A (2001). Large scale relational information visualization, clustering and abstraction [online]. https://aquigley.host.cs.st-andrews.ac.uk/aquigley-thesis-mar-02.pdf

Warren WH, Rothmans DB, Schnapps BH, Ericsson JD (2017) Wormholes in virtual space: from cognitive maps to cognitive graphs. Cognition 166:152–163

Categorizing Cognitive Scales of Spatial Information

Thomas Hervey, Daniel W. Phillips and Werner Kuhn

Abstract We investigate the relations between human cognitive scales and spatial information. To help organize spatial information, particularly around how humans perceive and interact with spaces around them, we explore the intersection of Kuhn's (2012) spatial information taxonomy, and Montello (1993) spatial scale taxonomy. We discuss results and challenges while using this intersection to categorize phenomena from an earthquake case study.

Keywords Scale · Cognition · Taxonomy · GIS

1 Transitions Between Our Environment and Our Cognition

Humans categorize to simplify their understanding of the world. One way of categorizing is based on content (rather than, for example, form or structure). GIS and maps often depict phenomena from a scale we cannot directly apprehend (e.g., the country of Italy) (Kuhn 2012). Therefore, to use these tools effectively, humans must cognitively relate their environmental experiences to different-sized spaces. We posit that such tools will be more intuitive when spatial content representations closely reflect our cognitive scales of real world spatial phenomena.

However, it is unclear how spatial content and scale taxonomies are related to each other. Do spatial phenomena naturally afford particular content representations, and at what scale do we associate these phenomena? This work attempts to answer these questions by orthogonally categorizing real world spatial phenomena by a content and a scale taxonomy. This raises an additional question, of whether the spatial

T. Hervey (✉) · D.W. Phillips · W. Kuhn
University of California, Santa Barbara, CA 93106, USA
e-mail: thomas.hervey@geog.ucsb.edu

© Springer International Publishing AG 2018
P. Fogliaroni et al. (eds.), *Proceedings of Workshops and Posters at the 13th International Conference on Spatial Information Theory (COSIT 2017)*, Lecture Notes in Geoinformation and Cartography, https://doi.org/10.1007/978-3-319-63946-8_14

content categories are scale neutral, i.e. capable of being represented at all scale levels. In the next section we present examples of both a content and scale taxonomy, after which we attempt to cross-tabulate them using phenomena from a case study.

2 Taxonomies of Spatial Content and Spatial Scales

The Core Concepts of Spatial Information (Kuhn 2012) act as conceptual lenses on environments. Each core concept comes with an evolving set of core operations to construct instances (for example, to construct a field representation) and to observe properties (for example, to observe the value of a field at a certain position and time). The concepts are currently being tested for information about geographic, astronomic, brain, and minuscule spaces (e.g., those of molecules), to see if they are valid at all scales. Four concepts can serve as classes in a content taxonomy: objects (spatially-bounded entities), networks (the system of links between objects), fields (continuous surfaces), and events (time-bounded entities).

A number of different spatial scale taxonomies have been proposed, and (Freundschuh and Egenhofer 1997) give a good overview of their criteria. Montello's taxonomy (Montello 1993) (updated in Montello 1998) has been widely adopted in the literature. Each class is defined by the size of a space relative to the human body, as well as the means of apprehending that space, whether technological aid, manipulation, and/or locomotion is required. These scale classes are: minuscule (much smaller than the body; requires microscopes), figural (smaller than the body; manipulable), vista (larger than the body), environmental (much larger than the body; requires locomotion), and gigantic (very much larger than the body; requires figural scaled representations).

3 2016 Amatrice Earthquake: A Case Study

On August 24, 2016, a magnitude 6.2 earthquake struck Central Italy, causing almost 300 deaths and billions USD in damage. Many historic buildings in Amatrice were destroyed, including the prominent Church of Sant'Agostino. In order to respond to and rebuild from such a disaster, public officials and stakeholders can assess the situation through the lenses of both spatial content and spatial scale. They can determine which scale level certain spatial phenomena and problems fall into, and then consider what type of phenomenon that content represents.

We use phenomena involved in this earthquake, particularly various damage to the Church of Sant'Agostino, to see what the relation is between these taxonomies, and how they hold together. We hypothesize that these taxonomies meaningfully overlap as a matrix of cells, and appropriately group phenomena. Each cell in Table 1 features one earthquake-related phenomenon within a particular content and scale class.

Table 1 2016 Amatrice earthquake phenomena categorized by concept rows and scale columns

	Minuscule	Figural	Vista	Environmental	Gigantic
Object	Grain making up the clay brick	Clay brick in the church facade	facade of the Church of Sant'Agostino	Amatrice	Apennine Mountains
Network	Bonds of molecules within that grain	Chemistry of the clay	Mortar between bricks in the facade	Streets of Amatrice	Roads across the mountains
Field	Pressure on those bonds	Stress within the clay brick	Structural integrity of the facade	Slope of the terrain	Shaking intensity
Event	Breaking of those bonds	Cracking in the clay brick	Collapse of the facade	Aftershock	Main earthquake

4 Challenges and Observations

Constructing Table 1 proved challenging in a number of respects. First, it is difficult to derive useful spatial examples for several of the cells in Table 1, such as a minuscule field. Second, we had varying success in striving for vertical coherence (having each column refer to the same object), horizontal coherence (making each row a series of nesting scale levels), and topical coherence (designing Table 1 to treat the same topic). A third challenge is that all GIS and map representations are ultimately figural because people interact with them on a tabletop space or a computer. Future work will entail identifying difficulties when applying this tabulation to other topics. Furthermore, we aim to determine if scale neutrality is possible for the Core Concepts taxonomy. Despite these challenges, we feel that we advance basic scientific understanding by convincingly showing that a content taxonomy can orthogonally interact with a scale taxonomy in a meaningful and potentially useful way.

Acknowledgements We acknowledge Emmanuel Papadakis, Fenja Kollasch for their editing contributions, and Sara Lafia, and Crystal Bae for their discussion. This work is a result of a seminar course, and is an extension of the 2017 COSIT Cognitive Scales of Spatial Information (CoSSI) workshop.

References

Freundschuh SM, Egenhofer MJ (1997) Human conceptions of spaces: implications for geographic information systems. Trans GIS 2(4):361–375. doi:10.1111/j.1467-9671.1997.tb00063.x
Kuhn W (2012) Core concepts of spatial information for transdisciplinary research. Int J Geogr Inf Sci 26(12):2267–2276. doi:10.1080/13658816.2012.722637

Montello DR (1993) Scale and multiple psychologies of space. In: Frank AU, Campari I (eds) Spatial information theory: a theoretical basis for GIS. Proceedings of COSIT '93. Lecture notes in computer science 716. Springer, Berlin, pp 312–321

Montello DR (1998) Thinking of scale; the scale of thought. In: Montello DR, Golledge RG (eds) Scale and detail in the cognition of geographic information: report on the Varenius specialist meeting, 14–16 May, Santa Barbara, CA, pp 11–12

The Concept of Location in Astronomic Spaces

Fenja Kollasch and Werner Kuhn

Abstract Spatial computing occurs at multiple scales, ranging from the minuscule to the gigantic. This raises the question whether the concepts involved in computing are scale-invariant. The Core Concepts of Spatial Information have originally been proposed for geographic spaces, though with the intent to make them applicable across scales. We have started to systematically validate them for astronomic spaces, starting with the concept of location.

Keywords Astronomic spaces · Astronomy · Core concepts · Location

1 Core Concepts of Spatial Information

In 2015, Kuhn and Ballatore (2015) presented a revised set of *core concepts of spatial information*, offering a language for spatial computing, and serving as conceptual lenses on environments. There are seven core concepts which can be divided in *content concepts* and *quality concepts*. The five content concepts are defined as *location*, *field*, *object*, *network*, and *event*. The two quality concepts are *granularity* and *accuracy*, a third one, *provenance*, is currently being added. They give knowledge about the level of detail in spatial information and the correctness of this information, respectively. The core concepts of spatial information have so far been tested with case studies from geographic spaces only. In astronomy, space comes at completely different scales. The distances between objects of interests are considerably larger and spaces are not limited to finite boundaries. Scale, in turn, has a significant

F. Kollasch (✉)
University of Bremen, Bibliothekstrasse 1, 28359 Bremen, Germany
e-mail: kollasch@uni-bremen.de

W. Kuhn
Center for Spatial Studies and Department of Geography, University of California at Santa Barbara (UCSB), Santa Barbara, USA
e-mail: werner@spatial.ucsb.edu

© Springer International Publishing AG 2018
P. Fogliaroni et al. (eds.), *Proceedings of Workshops and Posters at the 13th International Conference on Spatial Information Theory (COSIT 2017)*, Lecture Notes in Geoinformation and Cartography, https://doi.org/10.1007/978-3-319-63946-8_15

influence on how spatial information is produced and understood (Montello 1993). However, we posit that the core concepts are scale-neutral. As a first attempt to show this, we test whether the concept of location is applicable to astronomic spaces.

2   Location in Astronomy

Location is the most fundamental concept of spatial information (Vahedi et al. 2016). It can be seen as a spatial relation between a figure and a ground. Applying a relation r to a figure f and a ground g locates f relative to g. For instance, if we observe Sirius from the earth and note its declination and right ascension, we locate it relative to the earth. The partial application of r to g is considered *a location*:

$$r : g \rightarrow a\ location \tag{1}$$

By regarding r, f, and g in respect to each other, we can find out if f is in relationship r with g. This question is equivalent to "Is f located by $r(g)$?" The answer is either true or false:

$$located : (r, g, f) \rightarrow Bool \tag{2}$$

$$located(r, g, f) \leftrightarrow located(r(g), f) \tag{3}$$

The location for a figure also depends on how the ground is located. If we now want to obtain the angular distance between the stars Bellatrix and Sirius as observed from the Earth, we are locating Bellatrix (figure) in relation ("distance") to *Sirius relative to the earth* (ground), which is also a location. However, this is not the same as locating Bellatrix in distance to *Sirius without any reference to the earth*. In this scenario we would obtain an euclidean distance, not an angular distance. We can see here that a ground can also be a location and that the way it is located has impact on the location for figures relative to this ground.

   In geography one can always assume that a figure is located on the earth, and figure and ground are of finite sizes. In astronomy, however, we are often dealing with infinitely large spaces. Also, the earth is not always an ideal point of reference.

2.1   Celestial Reference Frames

A common way to describe locations is modeling them as a tuple of coordinates. Even though they seem like an absolute description of a location, coordinates are actually relative to the system they are specified in. Astronomic locations presented as coordinates can appear in many different coordinate systems with different centers, representation forms, and origins (Kovalevsky et al. 2012), in particular some without any reference to the earth. Common celestial reference systems are the *equatorial system* relative to the celestial poles and the celestial equator, the *horizontal system* relative to the observer's horizon, the *ecliptic system* aligning along the ecliptic plane, and the *galactic system* aligning along the galactic plane. Reference systems can be seen as standardized ground relation combinations and thus are *a location*.

2.2 Cartesian Representations

Locating an object by a celestial reference frame delivers two dimensional spherical coordinates. Extending them to a three dimensional representation is achieved by using the objective distance to a system-dependent reference object. With this information, the coordinates can be transformed into a Cartesian or cylindrical representation. Using simple geometric laws, the Cartesian coordinate system can easily be shifted to another origin. For instance, if a space probe is observing an asteroid, it might be interesting to know the position of the asteroid respectively to the probe to plan further maneuvers.

2.3 Distances

Coordinates are a quantitative way to describe locations. By using prepositions such as "near", "next to", or "in", figures can be located qualitatively to grounds. The most common qualitative relation to locate astronomic entities is *distance*. This form of location is especially interesting when they come along with fields or events. The values of a field, for instance the velocity field of a galaxy, often depend only on the distance from the field source. During events such as supernovae or gamma ray bursts, the main focus may lay on the question "are we at our current location involved?", which leads to the question "how far away did this event happen?"

2.4 Extents

Locations are not only single positions but can also be regions that contain other locations and thus allow relations as "is in" or "part of". Extended locations usually have a footprint position which is used to resolve questions about quantitative relations. In astronomy, extents are often three-dimensional if one is not only referring to a certain region at the night sky. However, for most astronomic locations it is hard to obtain crisp boundaries. Solar systems, galaxies, or star clusters can be treated as regions but although they are finite, no one can say where exactly such a region ends. Thus, they can not only appear as a three dimensional space but also as a collection of locations that are part of the extent.

3   Modeling Astronomic Locations

In order to test whether such a conceptualization is adequate for modeling astronomic locations, we used the `AstroPy coordinates` (Astropy Collaboration et al. 2013) Python package to implement them. This package provides various astronomic coordinate representations, reference frames, and conversion tools. By using these components with built-in operations, we can ensure correct computations on locations described with coordinates.

Although coordinates are a common way to describe locations, they are not the only one. Locating an object "five light years away from earth" is also a valid location. Astronomic locations can be rather complex since they may depend on further properties like the observer's location, the distance, or the observation time.

To represent this, we wrapped the high-level class `SkyCoord` which represents a footprint position and capsules all location-relevant informations. Based on this class, we have implemented four types of locations:

- `SphericalCoord`, consisting of a longitude, a latitude, a distance (optional) and further arguments, dependent on a celestial reference system
- `CartesianCoord` providing an x, y, and z coordinate tuple, relative to an origin
- `Distance` as a combination of the distance and the object the distance is measured to
- `AstroExtent` describing both a three dimensional ellipsoid-shaped space, and a collection of locations

It is possible to transform these representations into other representations if the instance has the necessary information. A Cartesian coordinate is a distance, too. An extent can be broken down to its footprint. The four types also allow a shifting of their reference objects. In this way, we created a flexible conceptualization of astronomic locations based on the core concepts.

4   Conclusion and Future Work

It is clear that locating astronomic phenomena requires different representations than locating geographical ones. However, we can see that astronomic as well as geographic locations depend on their reference system, i.e. on a relation to a reference object. This means that Eqs. 1 and 2 are valid for astronomic spaces as well. The basic concept of location remains the same. Our implementation is able to provide answers for questions like "Where is Bellatrix seen from Sirius?" or "Can I see Vega from Santa Barbara this year at July 04, 11pm?" Thus, it follows the approach of question-based spatial computing that is proposed through the core concepts.

Our next step will be to extend these tests to the other core concepts of spatial information. If we can successfully show that the other concepts can be used to represent astronomic phenomena and compute with these representations, then we will be able to state that they are indeed not limited to geographic spaces. Furthermore, such a conceptualization could help facilitate software development for astronomical spatial computing, as it provides a simple design pattern for spatial computing in astronomy.

References

Astropy Collaboration, Robitaille TP, Tollerud EJ, Greenfield P, Droettboom M, Bray E, Aldcroft T, Davis M, Ginsburg A, Price-Whelan AM, Kerzendorf WE, Conley A, Crighton N, Barbary K, Muna D, Ferguson H, Grollier F, Parikh MM, Nair PH, Unther HM, Deil C, Woillez J, Conseil S, Kramer R, Turner JEH, Singer L, Fox R, Weaver BA, Zabalza V, Edwards ZI, Azalee Bostroem K, Burke DJ, Casey AR, Crawford SM, Dencheva N, Ely J, Jenness T, Labrie K, Lim PL, Pierfederici F, Pontzen A, Ptak A, Refsdal B, Servillat M, Streicher O (2013) Astropy: a community python package for astronomy. A&A 558:A33. doi:10.1051/0004-6361/201322068

Kovalevsky J, Mueller I, Kolaczek B (2012) Reference frames: in astronomy and geophysics. Astrophysics and Space Science Library. Springer Netherlands. https://books.google.com/books?id=P7nrCAAAQBAJ

Kuhn W, Ballatore A (2015) Designing a language for spatial computing. Springer International Publishing, Cham, pp 309–326. doi:10.1007/978-3-319-16787-9_18

Montello DR (1993) Scale and multiple psychologies of space. Springer, Berlin, Heidelberg, pp 312–321. doi:10.1007/3-540-57207-4_21

Vahedi B, Kuhn W, Ballatore A (2016) Question-based spatial computing—a case study. Springer International Publishing, Cham, pp 37–50. doi:10.1007/978-3-319-33783-8_3

Is Wireless Functional Near-Infrared Spectroscopy (fNIRS) 3D Neuroimaging Feasible to Map Human Navigation in the Real-World?

Stefania Lancia, Silvia Mammarella, Denise Bianco and Valentina Quaresima

Abstract Real-time maps (with temporal and spatial resolution: 1–10 Hz and ~1 cm, respectively) of cortical activation can be obtained by functional near-infrared spectroscopy (fNIRS), which noninvasively measures cortical hemodynamic changes (as oxygenated and deoxygenated hemoglobin changes). The very recent launch in the market of commercial wireless/wearable fNIRS systems encourages their application in the field of human navigational studies to be carried out in the real-life situations.

Keywords fNIRS · Spatial navigation · Human behavior · Real-life monitoring · Cerebral cortex

1 Introduction

Functional near-infrared spectroscopy (fNIRS) is a non-invasive vascular-based functional neuroimaging technology which measures, simultaneously from multiple measurement sites, concentration changes of oxygenated-hemoglobin (O_2Hb) and deoxygenated-hemoglobin (HHb) at the level of the cortical microcirculation blood vessels. fNIRS uses near-infrared light (700–1000 nm) which passes easily through the scalp/skull/brain surface (Quaresima and Ferrari 2017). Real-time maps of cortical O_2Hb/HHb changes (hemodynamic response to a given stimulus) can be obtained applying an array of sources/detectors over the scalp (temporal resolution: 1–10 Hz; spatial resolution about 1 cm). The increase of cerebral oxygenation, observable by fNIRS, is secondary to the cortical neuronal activation (neurovascular coupling) in response to specific stimuli (cognitive, motor, etc.). Unlike the well-known neuroimaging technique functional magnetic resonance imaging (fMRI), fNIRS (thanks to its intrinsic advantageous features) represents an optimal

S. Lancia (✉) · S. Mammarella · D. Bianco · V. Quaresima
Department of Life, Health and Environmental Sciences, University of L'Aquila, Via Vetoio, L'Aquila, Italy
e-mail: stefania.lancia@graduate.univaq.it

© Springer International Publishing AG 2018
P. Fogliaroni et al. (eds.), *Proceedings of Workshops and Posters at the 13th International Conference on Spatial Information Theory (COSIT 2017)*, Lecture Notes in Geoinformation and Cartography, https://doi.org/10.1007/978-3-319-63946-8_16

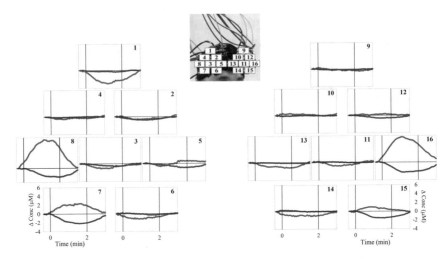

Fig. 1 Example of cortical hemodynamic changes over the monitored prefrontal cortex during a 2-min cognitive task. The cortical activation is indicated from an increase in O₂Hb (*red line*) and a concomitant decrease in HHb (*blue line*). The numbers *1–16* of the panels refer to the cerebral projections of the measurement points. A significant cortical activation is clearly observable in the measurement points: *7, 8* and *16* corresponding to the ventrolateral prefrontal cortex

tool for successfully exploring functional activation of the human cerebral cortex during cognitive tasks execution (Fig. 1). For instance, it allows the monitoring of brain activation of people while interacting with other people and/or while executing a cognitive task in a quiet environment without the noise which characterizes the fMRI scanner environment. Moreover, fNIRS allows the cortical investigation on subjects while performing even single trial tests which involve a component of originality and require a precise strategy planning, as it happens in some human navigation tasks. fNIRS signals are less susceptible to the participants' movements while executing their task in as natural a way as possible. Therefore, fNIRS seems to meet most of the requirements so that it can be termed a helpful tool for objectively evaluating the human behavior in real-life situations.

2 fNIRS to Map Human Navigation in the Real World

The impact of spatial navigation on everyday life is unquestionable. Spatial navigation is a complex cognitive ability to find one's way around the environment maintaining a sense of direction and position while moving around. It can be supported by external representations, such as maps, or internal mental representations based on sensory experience (Wolbers and Hegarty 2010). In everyday life, people can implement different strategies to deal with spatial navigation (route, survey, landmarks modalities). Most of the times, spatial navigation is goal-directed (e.g., when we need to reach a target) and, consequently, it requires to mentally evaluate alternative sequences of actions to infer optimal trajectories for reaching the goal (i.e. it requires a navigation strategy planning before starting the navigation itself). Planning is a complex skill that implies several cognitive processes such as working memory, mental imagery, visuo-spatial attention, and problem solving (Boccia et al. 2014).

Several previous neuroimaging studies (mainly using fMRI), focused on the identification and evaluation of the brain regions implicated in several aspects of spatial navigation. Interestingly, in the past it was reported that a fronto-parietal network (including dorsolateral prefrontal cortex, ventrolateral prefrontal cortex, posterior parietal cortex, precuneus, insula, right parahippocampal gyrus, occipital regions, thalamus, and cerebellum) is recruited during planning a route (Viard et al. 2011). Unfortunately, fMRI technology makes it able to explore only cerebral areas involved in some aspects of spatial navigation such as planning and/or spatial mental imagery. It is well known that spatial orientation and navigation rely to a large extent on locomotion and its accompanying activation of motor, vestibular, and proprioceptive systems. Therefore, although virtual navigation has become a popular tool for understanding navigational processes, it is important to recognize its limitations because participants are lying motionless in a supine position while viewing a monitor (Taube et al. 2013).

To the best of our knowledge in the past only a bulky fNIRS laboratory system was utilized to explore the cortical areas involved during human spatial navigation in a virtual environment (Kober et al. 2013). An increased oxygenation was found over the parietal cortical areas. In addition, they performed their study combining fNIRS device with an EEG system, advising the best-combined approach for identifying cerebral areas during human spatial navigation.

In order to explore the brain behavior in real-life situations, different battery operated multi-channel wearable/wireless fNIRS systems have been recently commercialized (Table 1). These fNIRS imagers have different interesting performances. For instance, NIRSIT has the feature of showing even "on line" high density activation maps on a tablet located at 5–7 m from the subject. The use of these systems in the real-world situations can give a promising contribution in several fields such as neuroergonomics, social sciences, neurosciences, etc. (Quaresima and Ferrari 2017). Some interesting studies in these fields have been carried out (also in the laboratory where the authors operate) over the last 5 years, using different fNIRS systems (Carrieri et al. 2017). The interest for mapping human brain by fNIRS during real-life situations is growing up. Nevertheless, it must be recognized that the spatial resolution offered by fNIRS is smaller than that one offered by fMRI.

Recent technological developments have led to the advancement of the portability of traditional brain imaging (Balardin et al. 2017; Ladouce et al. 2017). The portable and wearable neuroimaging devices, such as electroencephalography (EEG) and fNIRS, represent valuable tools for assessing multiple aspects of spatial navigation.

Table 1 Commercial wireless/wearable fNIRS Systems

Device	Company, country	Time-resolution (Hz)	Number of channels	Web site
Brite[23]*°+	Artinis, The Netherlands	100	23	www.artinis.com
fNIR Imager 1200 W°	fNIRS Devices LLC, USA	4	16	www.fnirdevices.com
WOT-HS	Hitachi, Japan	10	34	www.hitachihightech.com
NIRSport	NIRx Medical Technologies, LLC, USA	7.8	64	www.nirx.net
NIRSIT*°	OBELAB, Republic of Korea	8.1	204	www.obelab.com
LIGHTNIRS	Shimadzu, Japan	13.3	20–22	www.an.shimadzu.co.jp/bio/nirs/light_top.htm
OEG-17APD*°	Spectratech Inc., Japan	6.1	16–57	www.spectratech.co.jp/En

Hz Hertz. *Control unit included in the imager; °Bluetooth/wireless; +Integrated 9 axis movement sensor

Fig. 2 Setting for real-life mapping of the prefrontal cortex activity by a wearable fNIRS system equipped with 22 channels (WOT-220, Hitachi High-Technologies Corporation, Japan). The flexible probe unit covers the head in correspondence of the underlying dorsolateral and the rostral prefrontal cortex. The processing unit is worn on the waist. Photo courtesy of Dr. Pinti (University College London). For details of the study (Pinti et al. 2015)

Fig. 3 Setting for real-life fNIRS measurements of the prefrontal cortex activity by a wearable fNIRS system equipped with 204 measurement points, 48 channels at 3 cm depth (NIRSIT, OBELAB, Republic of Korea). Medical device approval in Republic of Korea. For details of the instrument (Choi et al. 2016). Photo taken at the Fountain of the 99 spouts (year 1273) in L'Aquila, Italy

The very recent introduction in the market of relatively low-cost wireless fNIRS devices could revolutionize the approach of human navigation studies. One of the first examples of the use of a mobile fNIRS in a real environment (Central London) was performed using the instrument shown in Fig. 2. In addition, the last generation of wireless fNIRS, represented by high-density 3D diffuse optical tomography system like NIRSport and NIRSIT (Table 1), utilizes multiple source-detector spacing to build 3D tomographic maps of the human brain cortex. This summer, in the laboratory where we operate the use of the NIRSIT system (Choi et al. 2016) is scheduled for performing human navigation studies in real-life environments (Fig. 3).

3 Conclusions

The development of fNIRS technology has strongly gained from the advances in microelectronics, computer technology, and optical engineering. With the advent of further miniaturization and integration such as integrated optics, wearable and even disposable fNIRS technology can be envisioned. The fNIRS systems that will emerge from these developments would further enlarge the number of fNIRS applications. The prediction of the future directions of fNIRS in human navigational studies is not easy to be done. However, fNIRS seems to have the requirements for giving a potential outstanding contribution in this research field. Unfortunately it must be taken into account that fNIRS instruments suffer some limitations such as: (1) the detection of brain activation is limited to the restricted monitored superficial cortical areas according to the size of the optical probe and the deeper areas, supposed to be involved in human navigation, cannot be investigated; (2) the duration of the measurements is restricted to less than 60 min; and (3) at the present time, the data analysis, performed only off-line, is quite complex and is lengthy.

Acknowledgements The Authors wish to thank OBELAB, Inc. (Seoul, Republic of Korea) for the loan of the NIRSIT instrument.

References

Balardin JB, Zimeo Morais GA, Furucho RA, Trambaiolli L, Vanzella P, Biazoli C Jr, Sato JR (2017) Imaging brain function with functional near-infrared spectroscopy in unconstrained environments. Front Hum Neurosci 11:258

Boccia M, Nemmi F, Guariglia C (2014) Neuropsychology of environmental navigation in humans: review and meta-analysis of FMRI studies in healthy participants. Neuropsychol Rev 24(2):236–251

Carrieri M, Lancia S, Bocchi A, Ferrari M, Piccardi L, Quaresima V (2017) Does ventrolateral prefrontal cortex help in searching for the lost key? Evidence from an fNIRS study. Brain Imaging Behav. doi:10.1007/s11682-017-9734-7

Choi JK, Kim JM, Hwang G, Yang J, Choi M, Bae M (2016) Time-divided spread-spectrum code-based 400 fW-detectable multichannel fNIRS IC for portable functional brain imaging. IEEE J Solid-State Circuits 51:484–495

Kober SE, Wood G, Neuper C (2013) Measuring brain activation during spatial navigation in virtual reality: a combined EEG-NIRS study. In: Trautman S, Julien F (eds) Virtual environments: developments. Applications and challenges. NOVA Publisher, Hauppauge, pp 1–24

Ladouce S, Donaldson DI, Dudchenko PA, Ietswaart M (2017) Understanding minds in real-world environments: toward a mobile cognition approach. Front Hum Neurosci 10:694

Pinti P, Aichelburg C, Lind F, Power S, Swingler E, Merla A, Hamilton A, Gilbert S, Burgess P, Tachtsidis I (2015) Using fiberless, wearable fNIRS to monitor brain activity in real-world cognitive tasks. JoVE 106:53336. doi:10.3791/53336

Quaresima V, Ferrari M (2017) Functional near-infrared spectroscopy (fNIRS) for assessing cerebral cortex function during human behavior in natural/social situations. Organ Res Methods. doi:10.1177/1094428116658959

Taube JS, Valerio S, Yoder RM (2013) Is navigation in virtual reality with fMRI really navigation? J Cogn Neurosci 25:1008–1019

Viard A, Doeller CF, Hartley T, Bird CM, Burgess N (2011) Anterior hippocampus and goal-directed spatial decision making. J Neurosci 31:4613–4621

Wolbers T, Hegarty M (2010) What determines our navigational abilities? Trends Cogn Sci 14 (3):138–146

The Influence of the Web Mercator Projection on the Global-Scale Cognitive Map of Web Map Users

Lieselot Lapon, Kristien Ooms and Philippe De Maeyer

Abstract For decades, cartographers and cognitive scientists have been speculating about the influence of map projections on mental representations of the world. We investigate if the mental map of young people is influenced by the increasing availability of web maps and its Web Mercator projection. An application is developed to let participants scale the area of some regions compared to Europe. The outcome gives insight into implications of using a—potentially misleading—map projection on the global-scale cognitive map.

Keywords Cognitive map · Web Mercator projection · Cartography

1 Background

Extensive technological improvements in the development of faster processors, more bandwidth and cheaper large storage units gave rise to the evolution of digital mapping (Gale 2013). Since the beginning of the 21st century, maps are available for internet users due to the development of web mapping services such as Google Maps. Social media websites (Twitter, Flickr or Facebook) incorporate these mashups in their platforms, providing users a way to produce and publish personalized web maps. Consequently, the personalization of web maps and the development of Web 2.0 are causing an enormous spread of online maps and cartographic data. Nevertheless, remarks about the development of these maps can be made.

L. Lapon (✉) · K. Ooms · P. De Maeyer
Department of Geography, Ghent University, Krijgslaan 281, 9000 Ghent, Belgium
e-mail: lieselot.lapon@ugent.be

K. Ooms
e-mail: kristien.ooms@ugent.be

P. De Maeyer
e-mail: philippe.demaeyer@ugent.be

© Springer International Publishing AG 2018
P. Fogliaroni et al. (eds.), *Proceedings of Workshops and Posters at the 13th International Conference on Spatial Information Theory (COSIT 2017)*, Lecture Notes in Geoinformation and Cartography, https://doi.org/10.1007/978-3-319-63946-8_17

Cases of misusing geographic data by representing cartographic data in a misleading way are reported in the era of paper maps (Monmonier 1996). Although this is not different in the digital age, the impact of these cartographic inaccuracies is more due to the way digital information is spread. In the interconnected world that we live in today, the velocity of circulating information is increasing drastically. One of the cartographic characteristics that can be misleading is the choice of map projection. The Web Mercator projection, for example, was originally introduced by Google Maps in 2005, and is adopted by several other web mapping services (Microsoft Bing Maps, OpenStreetMap, Yahoo Maps).

Battersby et al. (2014) investigated the properties of the Web Mercator projection, such as conformality. This implies a preservation of local angles around points on the map. This property, that is useful for nautical purposes, causes area deformation. This lack of area preservation can lead to interpretation problems, since the area of countries towards the poles is represented proportionally larger than these near the equator.

In 2006, Battersby and Montello (2009) conducted experiments about the area estimation of world regions and the projection of the global-scale cognitive map. Their research is embedded in the discipline of cognitive research that tries to understand how humans create and utilize mental representations of the Earth's environment (Slocum et al. 2001). The goal of their research was to unravel if map projections, especially the Mercator projection, had an influence on the global-scale cognitive map of students. The authors discovered that the participants can estimate the area fairly accurate relative to the actual area of the regions.

2 Research Question

The research of Battersby and Montello (2009) was conducted in 2006; the early years of the development of web maps. Consequently, the study group consisted of students that grew up with (printed and analogue) atlases and wall maps in the class room, which were not systematically projected with the Mercator projection. In contrast to the web maps which use the Web Mercator projection since 2005. Therefore, it is interesting to focus on a study group that grew up with these internet applications. Accordingly, the experiment was repeated in 2017.

The aim of our research is thus to investigate the influence of the use of Web Mercator projection on young people's cognitive map. Most of these youngsters have the technical abilities to use navigational and location-based services. They also use regularly social media, which are a source of information and cartographic data. It is expected that these capabilities make them regular users of web mapping services and consequently, prone to the influence of the Mercator projection.

3 Methodology

A web application was developed in which participants could estimate the size (area) of the selected region (country or continent) relative to the fixed reference region (Europe or USA). Students of Belgium and the United States, with a wide variety of study backgrounds, are asked to participate. The test regions are visualized one after the other on the reference region. Participants can graphically scale the area of these countries and the reference region relatively to one another using a slider until they estimate a size, which is considered correct compared with the reference region Europe or USA. An additional questionnaire about their technical and cartographical skills provides a good instrument to detect if the influence is a result of educational materials or of the use of web maps.

In order to register the certainty of the answers or to reveal speculation, the participant also has to answer two questions for each region. First, the participants have to rate on a Likert scale how sure they are about their area estimation. Second, the participants are asked whether they know where to position this country or region on a world map.

The second part of the test consists of two questions "Which representation of the world is the most familiar to you?" and "Which representation of the world do you think is not area distorted?". For both questions four projections of the world are shown: 'Gall-Peters projection', 'Mercator projection', 'Robinson projection' and 'Equirectangular projection'. The participants can choose one of these projections or the button 'No idea'.

At the beginning of the test, the participants' personal characteristics—age, gender, nationality, education level, cartographical background, use of social media and web maps—are gathered using a questionnaire.

4 Further Work

In a first phase, students of Ghent University, Belgium, with a wide variety of study backgrounds were asked to participate. In a second phase students of the United States will be asked to do the same study, but with the United States as reference area. Thus, possibilities arise to compare the data of the US students of 2006 with these of 2017. Furthermore, data collected in 2017 in the United States can be compared with these of Belgium. An additional questionnaire about their technical and cartographical skills provides a good instrument to detect if the influence is a result of educational materials or of the use of web maps.

At the end of the research it will be possible to answer the main questions: "Are the cognitive maps of people that are confronted regularly with web maps influenced by the representation of the world, and more specifically with the map projection that is used?" "Is there a difference between Belgian and US students?"

"Is there a difference between groups with a different cartographic knowledge level?".

In the future, the study will be extended to other nationalities and educational backgrounds. Is the nationality or place of residence of any influence on how people perceive the world map? For example: "Are there differences between the mental maps of habitants of countries that are represented smaller on the Mercator world map (areas around the equator) and these of habitants of Europe or the United States?".

Apart from answering the research questions, this study will also contribute to the development of an awareness of distortions on world maps amongst the participants. That people are not fully aware of these distortions is something that cannot be denied, surely when you are confronted with so many surprising reactions while talking about this topic to novices. This awareness will grow while they complete the test. However, without any further explanation the information will get lost. Therefore, a tool will be developed to communicate to the participants their results compared with the real relative areas. Besides, the participants will receive more basic background information about the cartographic processes, map projections and its distortions. The collection and providing of this information will, hopefully, make the users of maps more critical for the construction of maps and its content.

References

Battersby SE, Finn MP, Usery EL, Yamamoto KH (2014) Implications of Web Mercator and its use in online mapping. Cartographica 49(2):85–101

Battersby SE, Montello DR (2009) Area estimation of world regions and the projection of the global-scale cognitive map. Ann Am Geogr 99(2):273–291

Gale G (2013) Push pins, dots, customization, brands and services: the three waves of making digital maps. Cartogr J 50(2):155–160

Monmonier M (1996) How to lie with maps. The University of Chicago Press, Chicago

Slocum TA, Blok C, Jiang B, Koussoulakou A, Montello DR, Fuhrmann S, Hedley NR (2001) Cognitive and usability issues in geovisualization. Cartogr Geogr Inf Sci 28:1–28

Spatial Navigation by Boundaries and Landmarks in Williams Syndrome in a Virtual Environment

Marilina Mastrogiuseppe, Victor Chukwuemeka Umeh and Sang Ah Lee

Abstract Spatial navigation by humans and other animals engages representations of environmental boundaries as well as featural landmarks. In this paper we examine the use of these two types of information using a virtual reality (VR) task in patients with Williams syndrome (WS), a genetic disorder involving abnormalities in brain regions that support spatial cognition. Our preliminary data suggests that boundary-based navigation is more fragile in this syndrome, compared to control subjects matched for mental and chronological age.

Keywords Navigation · Williams syndrome · Geometry · Features · Path integration · Spatial mapping

1 Introduction

Our daily life requires us to represent and plan our movement through complex environments. How the brain accomplishes such computations is a topic of wide scientific interest. Both behavioral and neural evidence from studies of spatial navigation in both humans and nonhuman animals suggest that there are several distinctions that should be made in spatial representation. One major distinction is between environmental geometry and landmark use (Lee 2017). The neural basis of this dissociation in humans has been examined in several functional magnetic resonance imaging (fMRI) studies (e.g., Doeller et al. 2008; Park et al. 2015;

M. Mastrogiuseppe (✉) · V.C. Umeh · S.A. Lee
Center for Mind/Brain Sciences, University of Trento, Corso Bettini 31,
38068 Rovereto, Italy
e-mail: m.mastrogiuseppe-1@unitn.it

S.A. Lee
e-mail: sangah.lee@kaist.ac.kr

S.A. Lee
Department of Bio and Brain Engineering, Korea Advanced Institute of Science
and Technology, Daejeon, Korea

© Springer International Publishing AG 2018
P. Fogliaroni et al. (eds.), *Proceedings of Workshops and Posters at the 13th
International Conference on Spatial Information Theory (COSIT 2017)*, Lecture Notes
in Geoinformation and Cartography, https://doi.org/10.1007/978-3-319-63946-8_18

Ferrara and Park 2016). Doeller et al. (2008) used a non-immersive virtual environment task with feedback, in which participants had to learn the locations of two targets relative to a landmark or the boundary of the environment. Activation of the right dorsal striatum correlated with the improvement in performance for the landmark-related targets, whereas activation of the right posterior hippocampus correlated with the improvement in performance for the boundary-related targets.

Given the specificity of spatial learning using boundaries and landmarks, we set out to understand whether neurological conditions that impair hippocampal function can selectively impair the use of boundaries (and not landmarks). Moreover, if such a condition can be caused by specific genetic modifications, we will be one step closer to understanding the genetic basis of spatial cognition. Williams Syndrome (WS) is a particularly interesting case for such inquiry because it is caused by the microdeletion of about 25 genes in region q11.23 of chromosome 7. Besides characteristic physical traits and health issues, WS causes abnormalities of functions and metabolism of the hippocampal areas of the brain known to be involved in navigation (Meyer-Lindenberg et al. 2004) and is characterized by a unique cognitive profile that includes severe impairment in a range of spatial functions (Meyer-Lindenberg et al. 2005). Because of its distinctive cognitive profile and well-defined genetic modifications, this syndrome has come under increased interest by cognitive neuroscientists as a model for investigating the correlation between a specific genetic deletion and its cognitive and behavioral expression (Martens et al. 2008). Lakusta et al. (2010) previously tested whether people with WS could also use geometry to reorient themselves after they became disoriented in a laboratory environment. The authors showed that WS individuals are impaired in boundary-based navigation while relatively good at navigating by a featural landmark. However, a recent replication and extension by Ferrara and Landau (2015) suggests that this conclusion is too strong. By modifying the layout slightly to enhance the salience of the room's corners and object's hiding locations, they found that the majority of WS participants also used geometry above chance. Nevertheless, because there were very few trials tested in each session (four trials with two testing conditions) and because the environment for practical reasons had to be very small (small chamber: 1.2×1.8 m; large chamber: 2.4×3.7 m), we felt it necessary to test spatial navigation in WS using a different approach.

In this study, we developed a VR task to get a better characterization for the spatial profile in WS. There are several advantages of using VR when studying human navigation. VR tasks provide a versatile, engaging and safe alternative for conducting human spatial navigation research. The flexibility allows researchers to control and manage the environmental cues present in a scene easily and quickly, enabling the creation of virtual environments perfectly designed for studying cue specificity in navigation. Moreover, the acquisition of data using VR can be more fast, precise, and automatic respect to a real-environment task, allowing researchers to collect data more quickly and more precisely.

2   Methods

A computer game was created using Unity (Unity Technologies SF) in which participants initially found themselves in the middle of a five by three meters space. Using a joystick, they were told to find and collect a gem (one per trial) in the environment after which they were disoriented by a 'sandstorm' and placed at a random position. Then, they were instructed to go back to the position where they had collected the gem and press a button on the joystick. To make this more interesting especially for younger participants, the game was introduced with a storyline depicting the player as a pirate on a treasure island whose assignment was to gather gems for his/her crew's treasure chest. When the response position corresponded with the position of the hidden gem (± 50 cm), the player was rewarded with virtual gold coins or toys (for children). The gem disappeared immediately after collection and during the 'sandstorm' the screen was obfuscated. There were two environments that subjects were tested in: A boundary-only environment and a boundary+landmarks condition (See Fig. 1). The gem positions were randomized and could be closer to the center (Inner) or closer to the boundaries (Outer) of the arena.

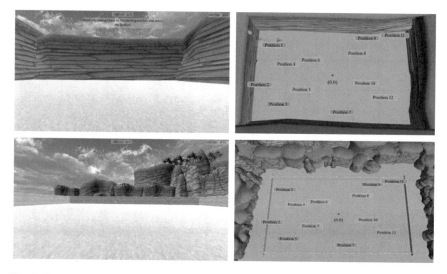

Fig. 1 Scenes from the two virtual environments: Boundary-only environmental condition with high boundaries and no landmarks (*above, left*), and Landmark environmental condition with low walls and various distal landmarks (*below, left*). Overhead views of Boundary-only (*above, right*) and Landmark (*below, right*), marked with Inner (*green*) and Outer (*orange*) gem positions

2.1   Participants

108 subjects were tested of which 29 were WS and 79 were typically developing (TD) subjects matched for mental (MA) and chronological age (CA). For the preliminary analyses, we considered 14 WS individuals (mean MA = 6.83, mean CA = 23.91), TD subjects matched for mental age (MA, mean age = 7.86) and those matched for chronological age (CA, mean age = 23.65).

3   Results

Analyses were performed considering the distance error which represents the Euclidean distance between the response position and the actual gem position in that trial. Distance error was calculated in all three groups in both game environments based on the average of 10 trials. Our results showed that all the groups performed better in the landmark compared to the boundary condition (Fig. 2a). The mean distance error was higher in the WS group compared with the two control groups in both the landmark and boundary condition ($F_{(2,27)} = 9.78$, $p = 0.001$;

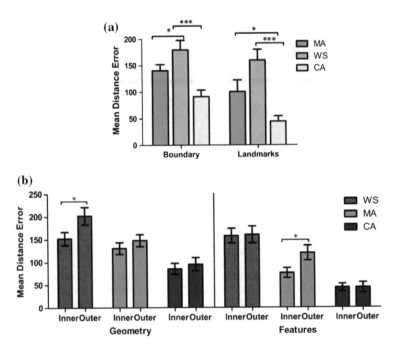

Fig. 2 Mean distance errors (in cm) for: **a** WS, MA and CA groups; **b** Inner and outer positions in WS, MA and CA groups. Error *bars* represent the standard errors of the mean. * $p \leq 0.05$, ** $p \leq 0.01$, *** $p \leq 0.001$

$F_{(2,27)} = 7.89$, $p = 0.002$). Interestingly, pairwise comparisons showed that WS performed significantly worse with respect to MA controls only in the boundary environment condition ($p = 0.023$) but there was no significant difference between WS and MA groups in the landmark environment condition ($p = 0.091$). In both conditions WS perform significantly worse with respect to CA controls ($p < 0.001$). Only in the landmark condition is there a difference between the MA and CA groups ($p = 0.036$). Interestingly, the distance error analysis considering the inner and outer positions showed that WS perform worse with gems in outer positions only in the geometry condition ($p = 0.048$). Conversely, MA participants show a lower performance with gems in outer positions only in the features condition ($p = 0.039$).

4 Discussion

Our results show a clear pattern of impairment among people with WS in a VR task, with failures to use both boundary-based and landmark-based representations of the environment. Consistent with the deficit in spatial navigation, previous studies have shown an unusual morphology of the hippocampus in WS patients, with a local volume reduction at the posterior apex (the tail-end), and expansion at the anterior base (extending from the midsection to the dorsal hippocampal head) (Meyer-Lindenberg et al. 2005). The converse of this pattern is observed in humans proficient in spatial navigation (i.e., London taxi drivers), who show increased posterior and decreased anterior hippocampal volume (Maguire et al. 2000).

We have replicated and extended Lakusta et al.'s (2010) finding of robust feature use, showing that this is also true in a VR environment. In the landmark condition, WS participants may have been relying on an alternative mechanism, possibly using view-dependent representations to identify the correct position (Lakusta et al. 2010; Wang and Spelke 2002). Moreover, our study extend previous results from Ferrara and Landau (2015), showing that also in a VR environment WS navigation mechanism is not best characterized by a complete absence of geometric sensitivity, but rather that it may be fragile, if compared to both mental and chronological age controls.

In WS, the use of features is relatively spared, whereas the use of boundary geometry appears to be impaired, in particular when the target locations were closer to the boundaries. In fact, we have found that only in the boundary condition WS patients perform worse when the gem to be collected and remembered is farther from the center of the arena (outer vs. inner position). The fact that MA controls showed the opposite pattern, with a worse performance in outer positions only in the landmark condition, is consistent with the idea that WS pattern differs qualitatively from that of TD individuals. This pattern is consistent with theories and evidence suggesting that the two navigational systems are separable in terms of both behavioral mechanism (Keinath et al. 2017; Lee and Spelke 2010; Lee et al. 2006) and underlying neural instantiation (Doeller and Burgess 2008; Doeller et al.

2008). Given recent findings that demonstrate the importance of distance and direction from the target (Lee et al. 2012; Lee et al. 2013)—we speculate that the WS deficit might be due to a reduced sensitivity to metric variables in particular when the task is heavily boundary-dependent. This would be consistent with other findings about the WS profile, which includes reduced sensitivity to estimated numerosity (Libertus et al. 2014).

There is growing evidence that navigation systems could serve as the basis for other types of representations, such episodic memory and various types of non-navigational metric information such as time and viewed space (Deuker et al. 2016). A better understanding of the navigation and spatial coding in neurodevelopmental disorders like WS is important in illuminating the connection between gene-brain-behavior and in clarifying the role of hippocampal maturation and function for spatial navigation and memory.

References

Deuker L, Bellmund JL, Navarro Schröder T, Doeller CF (2016) An event map of memory space in the hippocampus. Elife 5

Doeller CF, Burgess N (2008) Distinct error-correcting and incidental learning of location relative to landmarks and boundaries. Proc Natl Acad Sci USA 105:5909–5914

Doeller CF, King JA, Burgess N (2008) Parallel striatal and hippocampal systems for landmarks and boundaries in spatial memory. Proc Natl Acad Sci USA 105:5915–5920

Ferrara K, Landau B (2015) Geometric and featural systems, separable and combined: evidence from reorientation in people with Williams syndrome. Cognition 144:123–133

Ferrara K, Park S (2016) Neural representation of scene boundaries. Neuropsychologia 89:180–190

Keinath AT, Julian JB, Epstein RA, Muzzio IA (2017) Environmental geometry aligns the hippocampal map during spatial reorientation. Curr Biol 27(3):309–317

Lakusta L, Dessalegn B, Landau B (2010) Impaired geometric reorientation caused by a genetic defect. Proc. Natl. Acad. Sci. USA 107:2813–2817

Lee SA (2017) Curr Opin Behav Sci 58–65

Lee SA, Spelke ES (2010) Two systems of spatial representation underlying navigation. Exp Brain Res 206(2):179–188

Lee SA, Shusterman A, Spelke ES (2006) Reorientation and landmark-guided search by young children: evidence for two systems. Psychol Sci 17(7):577–582

Lee SA, Sovrano VA, Spelke ES (2012) Navigation as a source of geometric knowledge: young children's use of length, angle, distance, and direction in a reorientation task. Cognition 123(1):144–161

Lee SA, Vallortigara G, Flore M, Spelke ES, Sovrano VA (2013) Navigation by environmental geometry: the use of zebrafish as a model. J Exp Biol 216(Pt 19):3693–3699

Libertus ME, Feigenson L, Halberda J, Landau B (2014) Understanding the mapping between numerical approximation and number words: evidence from Williams syndrome and typical development. Dev Sci 17(6):905–919

Maguire EA, Gadian DG, Johnsrude IS, Good CD, Ashburner J, Frackowiak RSJ, Firth CD (2000) Navigation-related structural change in the hippocampi of taxi drivers. Proc Natl Acad Sci USA 97:4398–4403

Martens MA, Wilson SJ, Reutens DC. (2008). Research Review: Williams syndrome: a critical review of the cognitive, behavioral, and neuroanatomical phenotype. J Child Psychol Psychiatry 49(6):576–608

Meyer-Lindenberg A, Kohn P, Mervis CB, Kippenhan JS, Olsen RK, Morris CA, Berman KF (2004) Neural basis of genetically determined visuospatial construction deficit in Williams syndrome. Neuron 43(5):623–631

Meyer-Lindenberg A, Mervis CB, Sarpal D, Koch P, Steele S, Kohn P, Marenco S, Morris CA, Das S, Kippenhan S (2005) Functional, structural, and metabolic abnormalities of the hippocampal formation in Williams syndrome. J Clin Invest 115:1888–1895

Park S, Konkle T, Oliva A (2015) Parametric coding of the size and clutter of natural scenes in the human brain. Cereb Cortex 25(7):1792–1805

Wang RF, Spelke ES (2002) Human spatial representation: insights from animals. Trends Cogn Sci 6(9):376

Linked Data for a Digital Earth: Spatial Forecasting with Next Generation Geographical Data

Marvin Mc Cutchan

Abstract This document outlines potential research on integrating heterogeneous geographical data for forecasting purposes within the context of Digital Earth. The approach presented in this document relies on Linked Data principles which provide advantages for data integration but also data access. A structure for embedding geographic data into Linked Data is proposed. This structure is then utilized for forecasting spatial phenomena by establishing spatial association rules. The expected outcome of this proposed work is a framework capable of extracting association rules from the Linked Open Data web and an investigation of these rules as well as their potential.

Keywords Linked data · Spatial prediction · Machine learning

1 Introduction

The purpose of this document is to delineate potential research on the utilization of Linked Data for the concept of Digital Earth and therefore focuses on (1) the integration of heterogeneous geographical data using principles of Linked Data and (2) forecasting spatial phenomena on a digital Earth by creating spatial association rules (Koperski and Han 1995) based on this integrated geographical data. Linked Data refers to a paradigm for publishing Data on the internet and aims at creating a web of machine readable data by managing it in a structured manner (Wood et al. 2014). Digital Earth is conceptualized as a platform utilizing geoinformation from a series of different data sources, such as satellite imagery or governmental vector data for visualization and analysis purposes (Craglia et al. 2012; Goodchild 2008). This document is structured as follows: Section two discusses related research and

M. Mc Cutchan (✉)
TU Wien, Gusshausstr. 27-29, Vienna, Austria
e-mail: marvin.mccutchan@geo.tuwien.ac.at

© Springer International Publishing AG 2018
P. Fogliaroni et al. (eds.), *Proceedings of Workshops and Posters at the 13th International Conference on Spatial Information Theory (COSIT 2017)*, Lecture Notes in Geoinformation and Cartography, https://doi.org/10.1007/978-3-319-63946-8_19

finalizes with a conclusion that identifies research questions for the proposed research. Section three outlines the research methodology. Section four describes the expected outcome. The last section discusses potential future work, based on the research proposed.

2 Related Research

Within this chapter a review of existing literature is provided. First, relevant literature within the domain of Linked Data is outlined, then Digital Earth. The literature review on Digital Earth is based on position papers. Finally, a summary is made which is followed by the identification of a research gap.

2.1 Existing Literature

Kuhn et al. (2014) provide an overview of the impact of Linked Data in Geoinformation Science (GI) and argue that an augmentation of spatial data with non-spatial data is provided. Additionally they argue that the synergy allows to tackle common problems in GI in a new way. In Usery and Varanka (2012) and Scharrenbach et al. (2012), an attempt to integrate raster data and vector data is provided. Scharrenbach et al. (2012) link vector data to objects created by a threshold images segmentation. Usery and Varanka (2012) transform vector data into raster data in order to ensure interoperability. Janowicz et al. (2013) outline a concept of a RESTful proxy and data model using linked sensor data for the OGC Sensor Observation Service as well as a strategy for creating Uniform Resource Identifier. Grütter et al. (2016) investigate the quality of the topological representation of geographical data within the Linked Open Data web and conclude that the provided quality depends on how data is interlinked. Zhu et al. (2017) demonstrate how vector data can be interlinked, considering characteristics such as spatiotemporal topology or the category of a feature.

Goodchild (2008) provides a requirement analysis for the vision of Digital Earth and concludes that interoperability as well as the simulation of geographical data for forecasting purposes are key aspects which remain almost completely unrealized. Craglia et al. (2012) outline additional requirements for Digital Earth and conclude that semantically rich multi-disciplinary models for forecasting have to be developed.

2.2 Research Gap

Considering current literature on Digital Earth, it can be said that the aspect of forecasting as well as data integration are still far from being completely realized and are therefore expected to be fruitful for future research. It is argued that Linked Data principles are suitable for championing these challenges as they excel traditional GIS methods in three key aspects: (1) They enable to dynamically add semantic content to existing geographical data and do not require to change the original data (Zhu et al. 2017), which would be the case with data formats such as shape files. (2) Heterogeneous geographical data can be integrated by establishing predicates interlinking the datasets (Wiemann and Bernard 2015; Usery and Varanka 2012; Scharrenbach et al. 2012). (3) Linked Data enables the coexistence of different models as it is based on the open world assumption (Wood et al. 2014). This is not possible with current solutions which rely on the closed world assumption. Based on this research gap and the potential synergy of Linked Data and Digital Earth following research questions are raised:

1. How can Linked Data be used in order to perform forecasting based on heterogeneous geographical data?

    a. How can techniques such as inference be used for this purpose?
    b. How does geographical data have to be interconnected in order to enable forecasting?
    c. Is it possible to model temporal aspects for forecasting and how can they be utilized?
    d. How do different types of geographical data (e.g. raster or vector) contribute to the forecasting result?
    e. Which interoperability issues arise when different data types are being fused for the procedure of forecasting and how can they be solved?

2. Which spatial phenomena can be detected using forecasting?

    a. How well can spatial phenomena be forecast?
    b. Are there regional or even global phenomena which can be observed (such as urbanization)?
    c. Are there spatial phenomena which are significantly related to each other?

These research questions aim to answer and canvas following hypothesis: Can Linked Data improve spatial forecasting?

3 Methodology

The methodology is based on two major steps:

1. Gather and integrate geographical data in a common structure.
2. Based on the integration, find patterns which allow forecasting.

Step 1 integrates heterogeneous geographical data in a common hierarchical structure over a defined region of interest (ROI) (see Fig. 1). Every level of the structure is set up by a set of squares. For integration, geo-objects are connected with

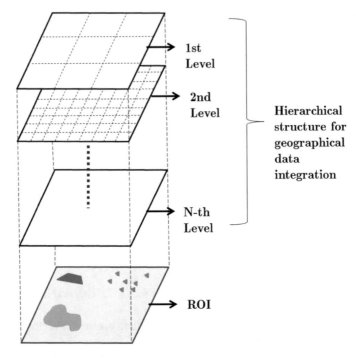

Fig. 1 Illustration of the hierarchical structure covering the ROI

predicates to the best fitting square according to the geo-objects minimum bounding rectangle (see Fig. 2).

Step 2 establishes spatial association rules (Koperski and Han 1995) for classes of geo-objects. Each spatial association rule is associated with a confidence C:

$$X \longrightarrow Y \tag{1}$$

$$\{X_1, X_2, \ldots, X_n\} \longrightarrow \{Y_1, Y_2, \ldots, Y_m\}) \tag{2}$$

$$C(X, Y) = \frac{\sum_{i=1, j=1}^{n,m} O(X_i, Y_j)}{\sum_{i=1}^{n} A(X_i)} , (C \in \mathbb{R}, 0 \leq C \leq 1) \tag{3}$$

X and Y are two sets of classes of geo-objects. The area associated with a geo-object is the area of the square it is connected with. $O(X_i, Y_j)$ computes the area of the spatial overlap of all instances of a class X_i to another class Y_j of geo-objects. $A(X_i)$ computes the area for all instances of X_i. Ultimately, spatial association rules with a significant confidence connect the corresponding classes using the Web Ontology Language (Wood et al. 2014). Future predictions are then performed using inference (Wood et al. 2014) on the established class structure.

Fig. 2 Geo-objects are linked (*black arrows*) to the squares according to their extend. Every square has an index

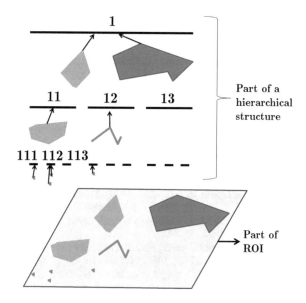

4 Expected Outcome

The work finalizes with a framework gathering spatial association rules from the Linked Open Data web (Wood et al. 2014). This framework is used to answer the proposed research questions and hypothesis.

5 Further Work

Incorporating the temporal dimension in a explicit manner as well as machine learning strategies such deep learning are expected to be potential extensions of the proposed research.

References

Craglia M, de Bie K, Jackson D, Pesaresi M, Remetey-Fülöpp G, Wang C, Annoni A, Bian L, Campbell F, Ehlers M, van Genderen J, Goodchild M, Guo H, Lewis A, Simpson R, Skidmore A, Woodgate P (2012) Digital Earth 2020: towards the vision for the next decade. Int J Digital Earth 5(1):4–21. ISSN 1753-8947. doi:10.1080/17538947.2011.638500

David W, Marsha Z, Luke R, Michael H (2014) Linked data, 1st edn. Manning Publications Co., Greenwich, CT, USA. ISBN 1617290394, 9781617290398

Goodchild M.F. (2008) The use cases of digital Earth. Int J Digital Earth 1:31–42. ISSN 1753-8947. doi:10.1080/17538940701782528

Grütter R, Purves RS, Wotruba L (2016) Evaluating Topological Queries in Linked Data Using DBpedia and GeoNames in Switzerland and Scotland. Trans GIS. ISSN 14679671: doi:10.1111/tgis.12196

Janowicz K, Bröring A, Stasch C, Schade S, Everding T, Llaves A (2013) A RESTful proxy and data model for linked sensor data. Int J Digital Earth 6(3):233–254. ISSN 1753-8947. doi:10.1080/17538947.2011.614698

Koperski K, Han Jiawei (1995) Discovery of Spatial Association Rules in Geographic Information Systems. Adv Spat Databases 951:47–66

Kuhn W, Kauppinen T, Janowicz K (2014) Linked data—a paradigm shift for geographic information science. In: Geographic information science: 8th international conference, GIScience 2014, Vienna, Austria, September 24–26, 2014, pp 173–186. ISSN 16113349. doi:10.1007/978-3-319-11593-1_12

Lynn Usery E, Varanka D (2012) Design and development of linked data from the National Map. In: Semantic web, 3(4):371–384. ISSN 15700844.10.3233/SW-2011-0054

Scharrenbach T, Bischof S, Fleischli S, Weibel R (2012) Linked raster data. In: GIScience 2012: seventh international conference on geographic information science, 2012. doi:10.5167/uzh-74705

Wiemann S, Bernard L (2015) Spatial data fusion in spatial data infrastructures using linked data. Int J Geogr Inf Sci 1–24. ISSN 1365-8816. doi:10.1080/13658816.2015.1084420

Zhu Y, Zhu A-X, Song J, Yang J, Feng M, Sun K, Zhang J, Hou Z, Zhao H (2017) Multidimensional and quantitative interlinking approach for Linked Geospatial Data. Int J Digital Earth 1–21. ISSN 1753-8947. doi:10.1080/17538947.2016.1266041

Route Learning from Maps or Navigation in Aging: The Role of Visuo-Spatial Abilities and Self-assessed Visuo-Spatial Inclinations

Chiara Meneghetti, Veronica Muffato and Rossana De Beni

Abstract It is well known that older adults have an impaired spatial performance than young; but the role of age and visuo-spatial factors in learning a route from a map or navigation has not been investigating throughout yet. After performing visuo-spatial tasks and questionnaires, young, young-old and old-old adults learned a map and a video and then performed map drawing and route repetition tasks. Age, visuo-spatial abilities and self-assessed visuo-spatial inclinations combine together to influence route learning.

Keywords Route learning · Older adults · Visuo-spatial abilities · Visuo-spatial inclinations

1 Introduction

The literature on spatial cognition and aging has demonstrated a decline in spatial learning performance with age (Klencklen et al. 2012), whether the environment is learned from a map (Borella et al. 2015) or by navigating within it (Taillade et al. 2016). There is still a paucity of evidence, however, regarding the difference between the two learning modalities (Yamamoto and De Girolamo 2012). Apart from external factors (such as of source of learning and type of recall tasks), individual visuo-spatial factors, such as visuo-spatial abilities and self-assessed visuo-spatial inclinations, can predict spatial performance too (Hegarty et al. 2006), even in aging (Meneghetti et al. 2014). The present study thus explores

C. Meneghetti (✉) · V. Muffato · R. De Beni
Department of General Psychology, University of Padova, via Venezia 8, Padova, Italy
e-mail: chiara.meneghetti@unipd.it

V. Muffato
e-mail: veronica.muffato@gmail.com

R. De Beni
e-mail: rossana.debeni@unipd.it

© Springer International Publishing AG 2018
P. Fogliaroni et al. (eds.), *Proceedings of Workshops and Posters at the 13th International Conference on Spatial Information Theory (COSIT 2017)*, Lecture Notes in Geoinformation and Cartography, https://doi.org/10.1007/978-3-319-63946-8_20

how age and individual visuo-spatial factors influence spatial mental representations (Wolbers and Hegarty 2010) derived from learning an environment from a map and from a video (navigation).

2 Method

2.1 Participants

The study involved: 46 young adults (aged 25–34; 23 females); 43 young-old adults (aged 65–74; 23 females); and 38 old-old adults (aged 75–84; 21 females). The older participants had no cognitive impairment.

2.2 Materials and Procedure

Participants individually attended two sessions. In the first, they completed tasks to test their visuo-spatial abilities, and self-assessed visuo-spatial inclination questionnaires (De Beni et al. 2014). In the second, they learned a route from either a map or a video, then they drew a sketch map and completed a route repetition task, in a balanced order across participants. They were then asked to learn a route through another environment using the learning condition not used before (video or map), and they completed the same two spatial recall tasks.

3 Result

Several mixed-effect models were used to examine sketch map drawing accuracy and route repetition accuracy. In all models, as random effects we input the participants, and the points involving a change of direction in the route repetition task. As fixed effects, we considered group (young vs. young-old vs. old-old), learning condition (map vs. video), visuo-spatial measures (visuo-spatial tasks and questionnaires), and all their interactions. Regression analyses were run in steps: random effect (step 0), gender, and years of education (step 1), group and learning condition (step 2), visuo-spatial factors (step 3), two-way interactions (step 4), and three-way interactions (step 5). Models were then compared using the Akaike Information Criterion (AIC). Group had an effect on both the sketch map and the route repetition tasks, with young adults performing better than either young-old or old-old participants. Learning condition only had an effect on the sketch map task (performance was better after learning from a map than after learning from a video), not on the route repetition task. Visuo-spatial factors (tasks and self-assessed measures) had an effect on the sketch map and route repetition accuracy. In the sketch map task, two- and three-way interactions emerged, while in the route repetition task only the group visuo-spatial abilities interaction emerged. This suggests that to some extent age group, learning condition and visuo-spatial factors combine together in influencing spatial performance.

4 Discussion and Conclusion

Age and individual visuo-spatial factors (such as visuo-spatial abilities and self-assessed visuo-spatial inclinations) combine together to influence mental spatial representation processes after learning an environment from a map or by navigating within in. These findings expand our understanding in the domain of spatial cognition and aging.

References

Borella E, Meneghetti C, Muffato V, De Beni R (2015) Map learning and the alignment effect in young and older adults: how do they gain from having a map available while performing pointing tasks? Psychol Res 79:104–119

De Beni R, Meneghetti C, Fiore F, Gava L, Borella E (2014) Batteria Visuo-spaziale. Strumenti per la valutazione delle abilit visuo-spaziali nellarco di vita adulta [Visuo-spatial battery: instrument for assessing visuo-spatial abilities across adult life span. Hogrefe, Firenze, Italy

Hegarty M, Montello DR, Richardson AE, Ishikawa T, Lovelace K (2006) Spatial abilities at different scales: individual differences in aptitude-test performance and spatial-layout learning. Intelligence 34:151–176

Klencklen G, Despres O, Dufour A (2012) What do we know about aging and spatial cognition? Reviews and perspectives. Ageing Res Rev 11:123–135

Meneghetti C, Borella E, Pastore M, De Beni R (2014) The role of spatial abilities and self-assessments in cardinal point orientation across the lifespan. Learn Individ Differ 35:113–121

Taillade M, N'Kaoua B, Sauzeon H (2016) Age-related differences and cognitive correlates of self-reported and direct navigation performance: the effect of real and virtual test conditions manipulation. Front Psychol 6:1–12

Wolbers T, Hegarty M (2010) What determines our navigational abilities? Trends Cogn Sci 14:138–146

Yamamoto N, De Girolamo GJ (2012) Differential effects of aging on spatial learning through exploratory navigation and map reading. Front Aging Neurosci 4:17

Towards Personalized Landmarks

Eva Nuhn and Sabine Timpf

Abstract Recent studies state that people familiar with an environment prefer landmarks with semantic salience, while people unfamiliar with an environment prefer landmarks with outstanding visual and structural characteristics. Current landmark salience models are not able to provide such personalized landmarks. To tackle this problem we proposed a multidimensional model for personalized landmarks and identified attributes of the personal dimension. This poster focuses on the next step, the calculation of potential values of the attributes and the discussion of their relationships.

Keywords Spatial cognition · Landmarks · Pedestrian · Navigation · Familiarity · Personalization

1 A Multidimensional Model for Personalized Landmarks

Landmarks are references for wayfinding, which are external to the traveller (Lynch 1960). The choice of landmarks for route instructions is based on several criteria, such as the mode of travel or the desired route characteristics (Lovelace et al. 1999). Recent studies have shown that the level of spatial knowledge of the recipient is also an important factor in landmark selection. Hamburger and Röser (2014) for example, showed that familiar buildings are easily more recognized than unfamiliar ones. Based on these findings Quesnot and Roche (2015) confirmed that people familiar with an environment clearly prefer local semantic landmarks. In contrast, people unfamiliar with an environment prefer landmarks with outstanding visual and structural characteristics. Quesnot and Roche (2015) also stated that semantic

E. Nuhn (✉) · S. Timpf
Geoinformatics Group, Institute for Geography, University of Augsburg,
Alter Postweg 118, 86159 Augsburg, Germany
e-mail: eva.nuhn@geo.uni-augsburg.de

S. Timpf
e-mail: sabine.timpf@geo.uni-augsburg.de

© Springer International Publishing AG 2018
P. Fogliaroni et al. (eds.), *Proceedings of Workshops and Posters at the 13th International Conference on Spatial Information Theory (COSIT 2017)*, Lecture Notes in Geoinformation and Cartography, https://doi.org/10.1007/978-3-319-63946-8_21

landmarks should be included in route instructions according to the traveller's spatial knowledge of the environment. The assessment of the salience of landmarks is a challenging task (Caduff and Timpf 2008). Providing personalized landmarks makes it even more challenging and current landmark salience models are not able to provide personalized information.

In our research we investigate how landmarks may be personalized for route instructions. We proposed a multidimensional model for personalized landmarks (Nuhn and Timpf 2016), which considers a personal dimension taking into account the traveller's familiarity with an environment in addition to the known visual, semantic and structural dimensions of landmarks. In a follow-up work (Nuhn and Timpf 2017) we identified attributes of the personal dimension for inclusion within such a multidimensional model: *spatial knowledge, personal interests, personal goals* and *personal background.* Four categories of *spatial knowledge* were determined dependent on the amount of knowledge, i.e. the familiarity with the environment: *no knowledge, landmark knowledge, route knowledge* and *survey knowledge.*

2 Calculating Salience of the Attributes of the Personal Dimension

As a next step we need to calculate the values of all attributes of all dimensions within the multidimensional model. The focus of this poster is thus on calculating the potential values and discussing their interrelationships. If a landmark candidate is part of landmark knowledge it is already familiar to the traveller. Therefore, its personal salience is set to 100% (see Table 1). If the landmark is located next to an already navigated route (i.e. part of route knowledge) it is investigated if landmarks were part of former route instructions or not. If they were part of former route instructions the traveller should also know them and therefore their personal salience is equal to the salience of landmark knowledge. Landmarks, which were not yet used for navigating may not attract the traveller's attention, therefore their personal salience (pertaining to spatial knowledge) is set to 75%. In case of a landmark being part of survey knowledge its personal salience is set to 50%. This salience is lower than the others because it can be assumed that landmarks located in survey knowledge areas are less consciously perceived than landmarks that are

Table 1 Personal saliences

Spatial knowledge	Personal salience in (%)
Landmark knowledge	100
Route knowledge (part of former route instructions)	100
Route knowledge (not part of former route instructions)	75
Survey knowledge	50
No knowledge	0

part of landmark or route knowledge. Landmarks located in areas with no spatial knowledge are assigned a personal salience of zero.

Within the multidimensional model the attribute values of the dimensions and the determination of their effect on the landmarkness or salience of a landmark candidate are calculated. In the case of a landmark candidate with a personal salience of 100% we propose to use it for the route instruction. If the personal salience is 50% or greater the semantic salience needs to be assessed. If the semantic salience exceeds a predefined threshold the landmark can be included in route instructions for people familiar with the environment. If the personal salience is zero then a landmark with a high visual or structural salience has to be selected.

In future work we will test the applicability and usefulness of our multidimensional model with a case study. The results will be compared with landmarks obtained from conventional landmark salience models, to see if the consideration of the personal dimension is valuable in landmark salience assessment for the provision of personalised landmarks.

References

Caduff D, Timpf S (2008) In the assessment of landmark salience for human navigation. Cogn Process 9:249–267

Hamburger K, Röser F (2014) The role of landmark modality and familiarity in human wayfinding. Swiss J Psychol 73:205–213

Lovelace KL, Hegarty M, Montello DR (1999) Elements of good route directions in familiar and unfamiliar environments. In: Freksa C, Mark DM (eds) Spatial information theory. Cognitive and computational foundations of geographic information science. COSIT'99, Stade, Springer, Germany, pp 65–68

Lynch K (1960) The image of the city. MIT, Boston

Nuhn E, Timpf S (2016) A multidimensional model for personalized landmarks. In: Gartner G, Huang H (eds) 13th international conference on location based services. Austria, Vienna, pp 4–6

Nuhn E, Timpf S (2017) Personal Dimensions of Landmarks. In: Bregt A, Sarjakoski T, Van Lammeren R, Rip F (eds) Societal Geo-innovation. Springer, pp 129 –143

Quesnot T, Roche S (2015) Quantifying the significance of semantic landmarks in familiar and unfamiliar environments. In: Fabrikant SI, Raubal M, Bertolotto M, Davies C, Freundschuh S, Bell S (eds) Spatial information theory. COSIT 2015, Santa Fe, Springer, USA, pp 468–489

Developing and Evaluating VR Field Trips

Danielle Oprean, Jan Oliver Wallgrün, Jose Manuel Pinto Duarte, Debora Verniz, Jiayan Zhao and Alexander Klippel

Abstract We present our work on creating and assessing virtual field trip experiences using different VR and AR setups. In comparative studies, we address the question of how different settings and technologies compare regarding their ability to convey different kinds of spatial information and to foster spatial learning. We focus on a case study on an informal settlement in Rio, Brazil, in which we used an informal assessment to help inform and improve the design of different VR site experiences.

Keywords Virtual reality · Augmented reality · Spatial learning

1 Introduction

Immersive virtual reality (VR) and augmented reality (AR) technologies are seeing a resurgence in popularity thanks to massively improved and more cost-effective products, a trend that can be expected to lead to an increased usage of VR approaches in the education of fields that are inherently spatial, such as architecture, archeology, or the geosciences. They enable field trip-like classroom experiences of places that are inaccessible or too expensive to visit. However, while empirical studies have shown the potential of VR in the teaching-learning process (Barilli et al. 2011; Roussou 2004), there still exist many challenges for designing effective learning environments for different available devices as well as many open questions on how approaches and technologies affect the conveyance of spatial information and spatial learning in general.

D. Oprean (✉) · J.M. Pinto Duarte · D. Verniz
Architecture, The Pennsylvania State University, State College, USA
e-mail: dxo12@psu.edu

J.O. Wallgrün · J. Zhao · A. Klippel
ChoroPhronesis, Department of Geography, The Pennsylvania State University,
State College, USA

© Springer International Publishing AG 2018
P. Fogliaroni et al. (eds.), *Proceedings of Workshops and Posters at the 13th
International Conference on Spatial Information Theory (COSIT 2017)*, Lecture Notes
in Geoinformation and Cartography, https://doi.org/10.1007/978-3-319-63946-8_22

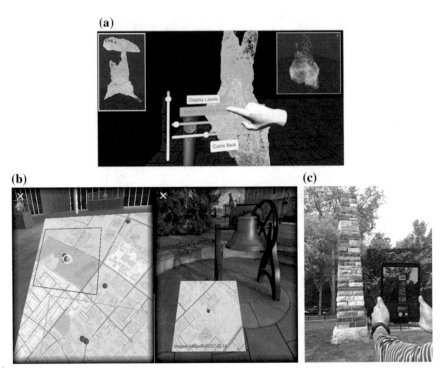

Fig. 1 Examples of VR site experiences created for different spatial science fields: **a** Experience of the Icelandic volcano Thrihnukar for geoscientist. **b** A VR tour of the Penn State main campus (overview map *left*, image view with zoom-in map *right*). **c** First prototype of an AR application displaying historic information about objects on the Penn State campus

In our work, we have been designing virtual experiences for different devices, (including consumer-level VR platforms such as the HTC Vive as well as mobile VR solutions such as the Google Cardboard/Daydream and Samsung GearVR) and in different domains. Figure 1 shows some examples of this work: a VR experience of the Icelandic volcano Thrihnukar that provides interactive tools for geoscientists, for example, to perform measurements (Zhao et al. 2017) (Fig. 1a), and a VR tour for the Penn State main campus (Fig. 1b). We are also currently designing a VR field trip for the ancient Mayan site of Cahal Pech in Belize for archeologists and the general public, and we have started to extend these projects to also create AR-based in-situ experiences of the different sites (see Fig. 1c).

In the following, we outline an informal study from the area of (landscape) architecture about the virtual experience of the informal settlement Santa Marta in Rio, Brazil. We summarize how the results of the study were used to improve the design of other environments used for exploring topics in the geosciences.

2 VR Experiences and Study

In over a year's worth of development and testing, we designed three different VR systems for site experiences of a single location for the Santa Marta informal settlement in Rio, Brazil.

Figure 2 shows the three different site experiences we developed for the HTC Vive, Android based smartphones in combination with the Google Cardboard (both developed in Unity3D[1]), and a WebVR based web site using A-Frame.[2] While the general setup is the same on all three platforms, allowing the user to select different 360° images and videos through an overview map and then viewing the selected image in VR or on the screen, the details of the display and, in particular, the interface and interactions needed to be designed specifically for each of the three setups.

In the study conducted during the development stage, students, participating in a joint architecture and landscape architecture studio course, remotely visited the settlement using all of the platforms. An informal exploration of students perceptions towards the platforms was used to gain insights into use in future architecture studios as well as design improvements for future development and implementation. The results demonstrate success of the platforms in making students aware of relevant visual site features such as configuration of public spaces, texture of building surfaces, and interplay of light and shadow. They also show some limitations related to failing to engage other senses and the social dimension, which are essential to convey more complex dimensions of the space. We are currently working on more formal evaluations to assess different experiences. However, some results pertaining to the development specifically were used to make improvements to not only the Santa Marta VR experience but our other projects that followed in development afterwards.

3 Improvements to VR Experiences

Through the use of our informal evaluations of the Santa Marta VR experience, we have made improvements to the process for generating VR experiences for different scenarios. The evaluations provided user-based insight to help resolve and inform development of features as well as helping to identify different concerns. We identified a number of issues and received different suggestions that we were able to use to improve the VR experience for Santa Marta and for other VR experiences, as summarized below.

---

[1]https://unity3d.com/.
[2]https://aframe.io/.

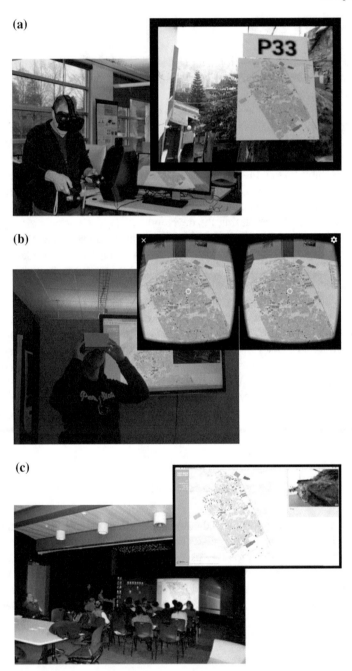

Fig. 2 Different versions of the settlement experience with the view of the user shown at the top right of each of the images: **a** VR setup in the HTC Vive, **b** VR setup for Android-based smartphone in combination with the Google Cardboard, **c** web site version

Issues/Suggestion identified	Implementation
Image quality	Added image labels for reference to currently viewed image
	Testing for highest quality image that would be compatible across multiple devices
	Correcting inverted image mapping to sphere
	Added workbench feature for allowing information recording within VR experiences
Accuracy	Incorporating geo-referencing of content
Spatial perceptions	Adjusting the camera settings in the VR camera to match the actual camera
Other modalities	Adding 360 sound to the experiences
	Adding controller support to the mobile version
Video features	Incorporating a Pause/Play ability for videos

4 Conclusions and Outlook

Immersive technologies, from AR to VR and everything in between (Milgram and Kishino 1994), have finally weeded out issues that were seen as barriers for them to become a phenomenon of mass communication, like maps. Just a decade ago, (Fisher and Unwin 2002) identified three major issues of VR systems:

- Ergonomic issues such as feeling disorientated and nauseous, also referred to as motion sickness (Hettinger and Riccio 1992);
- a rather complex technology for getting content to users in comparison with other media such as the World Wide Web;
- substantial costs for everyone who wants to use a VR system.

It is fair to say that none of these issues, at a broader scale, exist anymore (Slater and Sanchez-Vives 2016). It is important to note that the second barrier identified by Fisher and Unwin (2002) has been largely addressed by advances in software technology but specifically by advances in the spatial sciences through developments in environmental sensing (Khorram et al. 2012). This is the time for the spatial sciences to seriously deconstruct the opportunities afforded by immersive technologies as a communication paradigm. Our group is embracing these opportunities in two ways. First, through a thorough investigation of characteristics of immersive technologies with a goal to identify those that matter for improving communication, understanding, and learning (an approach also termed the foundational approach (Oprean 2014)). Within this approach we are currently comparing, more systematically, differences induced by, for example, different fields of view (Oprean et al. 2017). Second, by developing efficient workflows that allow bringing content into an immersive environment we can further explore how individuals interact and analyze data from climate change (Simpson et al. 2016), to volcanos (Zhao et al. 2017), to less accessible places such as informal settlements and archeological sites. The joint future of spatial sciences and immersive technologies is bright. As we continue with development of different VR experiences, we hope to be able to use more formal assessment to refine and improve our workflows and end products.

Acknowledgements Support for this research by the Penn State Center for Online Innovation and Learning (COIL), the Stuckeman Center for Design Computing (SCDC), the National ScienceFoundation under Grant Number NSF #1526520, and by the Brazilian National Council for Scientific and Technological Development (CNPq) is gratefully acknowledged.

References

Barilli ECVC, Ebecken NFF, Cunha GG (2011) The technology of virtual reality resource for information in public health in the distance: an application for the learning of anthropometric procedures. Ciênc. saúde coletiva 16:1247–1256

Fisher P, Unwin D (2002) Virtual reality in geography: an introduction. In: Fisher PF, Unwin D (eds) Virtual reality in geography. Taylor & Francis, London, New York, pp 1–4

Hettinger LJ, Riccio GE (1992) Visually induced motion sickness in virtual environments. Presence: Teleoper Virtual Environ 1(3):306–310

Khorram S, Koch FH, van der Wiele CF, Nelson SAC (2012) Future trends in remote sensing. In: Remote sensing. Springer US, Boston, MA, pp 125–129

Milgram P, Kishino F (1994) A taxonomy of mixed reality visual displays. IEICE Trans Inf Syst E77-D(12):1321–1329

Oprean D (2014) Understanding the immersive experience: Examining the influence of visual immersiveness and interactivity on spatial experiences and understanding. Doctoral dissertation. University of Missouri

Oprean D, Simpson M, Klippel A (submitted) Remote immersive collaboration: an evaluation of immersive capabilities on spatial experiences and team membership. J Digit Earth

Roussou M (2004) Learning by doing and learning through play. An exploration of interactivity in virtual environments for children. Comput Entertain 2(1)

Simpson M, Wallgrün JO, Klippel A, Yang L, Garner G, Keller K, Bansal S (2016) Immersive analytics for multi-objective dynamic integrated climate-economy (DICE) models. In: Hancock M, Marquardt N, Schöning J, Tory M (eds) Proceedings of the 2016 ACM companion on interactive surfaces and spaces—ISS Companion '16. ACM Press, New York, USA, pp 99–105

Slater M, Sanchez-Vives MV (2016) Enhancing our lives with immersive virtual reality. Front Robot AI 3:74

Zhao J, Wallgrün JO, LaFemina P, Oprean D, Klippel A (2017) iVR for geosciences. In: Second workshop on K-12 embodied learning through virtual and augmented reality (KELVAR). IEEE Digital Library

A Reference Landform Ontology for Automated Delineation of Depression Landforms from DEMs

Gaurav Sinha, Samantha T. Arundel, Kathleen Stewart, David Mark,
Torsten Hahmann, Boleslo Romero, Alexandre Sorokine, Lynn Usery
and Grant McKenzie

Abstract This reference landform ontology is intended to guide automated delineation of landforms from digital elevation models (DEMs) and semantic information retrieval about landforms. Since only form related information is available from DEMs, the categories of this reference ontology are defined based only on morphological criteria. The choice of the landform categories is informed by ethnophysiographic and spatial cognition research. The proposed taxonomy is work in progress and reflects the current focus on automated delineation and mapping of depression landforms (e.g., basins, valleys and canyons).

Keywords Landform · Topography · Depression · Delineation · Reference ontology

G. Sinha (✉)
Department of Geography, Ohio University, Athens, OH, USA
e-mail: sinhag@ohio.edu

S.T. Arundel · L. Usery
Center of Excellence for Geographic Information Science (CEGIS),
USGS, Rolla, MO, USA

K. Stewart · G. McKenzie
Department of Geographical Sciences, University of Maryland, College Park,
MD, USA

D. Mark
Department of Geography, University at Buffalo SUNY, Buffalo, NY, USA

T. Hahmann
National Center for Geographic Information and Analysis (NCGIA),
School of Computing and Information Sciences, University of Maine, Orono, ME, USA

B. Romero
Department of Geography, University of California, Santa Barbara, CA, USA

A. Sorokine
Oak-Ridge National Laboratory, Oak Ridge, TN, USA

© Springer International Publishing AG 2018
P. Fogliaroni et al. (eds.), *Proceedings of Workshops and Posters at the 13th International Conference on Spatial Information Theory (COSIT 2017)*, Lecture Notes in Geoinformation and Cartography, https://doi.org/10.1007/978-3-319-63946-8_23

1 Background

GeoVoCamps provide a forum for building ontologies or controlled vocabularies for tractable knowledge domains. This paper reports initial findings from a Geo-VoCamp meeting held in College Park, MD in November 2016 to guide the US Geological Survey with its design of a conceptual reference ontology. The primary purpose of this reference ontology is to support natural language topographic information retrieval and context-sensitive algorithms for user-controlled automated delineation of cognitively salient landforms (e.g., hill, mountain, valley). The first step at this GeoVoCamp meeting was to refer and reuse concepts from the *surface network* (SN) and *surface water feature* (SWF) terrain ontologies that resulted from previous GeoVoCamps. The SN ontology (Sinha et al. 2017) formalizes the minimum mereotopological semantics for describing the shape of a surface. The SWF ontology (Sinha et al. 2014) formally distinguishes between terrain features that act as containers (channel, depression, and interface) and the contained bodies of water (stream segment, water body, and fluence). This work further combines these ideas with the formal ontological approach to geo-physical objects and the negative parts —so-called holes or voids—they may host (Casati and Varzi 1994; Hahmann and Brodaric 2012).

2 Conceptualization of the Reference Landform Ontology

The primary category of this reference ontology is *landform*, which represents entities that are three-dimensional features located on the solid surface of the Earth or similar planetary bodies. Landforms may be material (e.g., mountains) or have both material and immaterial parts, such as a water body that consists of a river bed, the depression it hosts, and the water therein (Brodaric et al. 2017). Landforms may be assigned some characteristic geometric, topological, mereotopological, temporal and material properties. Knowledge of the agents and types of processes that create landforms or in which landforms participate should also be specified when possible to support geoscientific conceptualizations. However, because this is a reference ontology intended for broad usage in both scientific and naïve geographic contexts, only a few landform properties that people can intuitively sense and cognize are used to define the top-level categories.

Figure 1 is a graphical representation of the taxonomic relationships identified (thus far) for this reference landform ontology. The choice of these fundamental categories is based on landform categories and descriptions from multiple languages and cultures, especially as revealed from ethnophysiographic research of one of the authors (Mark and Turk 2003; Mark et al. 2007). Since delineating landforms from DEMs is an important motivation for this ontology, only the fundamental criterion of form (shape) is used to define the categories of this reference ontology. Other important categories that people define using material,

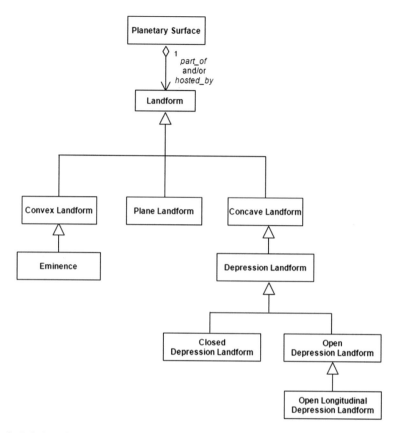

Fig. 1 Relationships between categories of the reference landform ontology

color, and cultural qualities are, therefore, not currently included in this reference landform ontology.

As shown in Fig. 1, there are three fundamental types of landforms that people have been found to conceptualize across cultures and languages: *convex landform*, *concave landform*, and *plane landform*. These categories are covert in the sense that most languages do not specifically have terms recognizing these abstract, top-level landform categories. However, the various landform categories and related concepts encountered in ethnophysiographic research provide strong evidence for explicit modeling of these categories in a reference landform ontology. Convex and concave landforms cover an overwhelming majority of landform categories, with plane landforms covering remaining (perceptually) flat areas. This categorization also suggests the need for designing different methods for searching and delineating concave, convex, and plane landforms.

2.1 Convex and Concave Landforms

Convex landforms protrude outward or upward from the surface. *Eminence* covers an important sub-group of convex landforms (e.g., mountain, hill, butte) that stand above their surroundings. Eminence ontology is still being developed as part of another project. The reference landform ontology described here reflects this group's focus on clarifying concepts that are critical for designing methods for delineating different kinds of *depression landforms* (defined below).

In contrast to convex landforms that protrude out, concave landforms are indented, necessarily hosting holes and giving rise to a sense of material missing from the surrounding host surface. The most significant subcategory of concave landforms is *depression landform*, which covers low-lying landforms surrounded by higher land (e.g., basin, river-bed, valley). Examples of concave landforms that are not depressions would be caves or tunnels.

Depression landforms can be further classified as *closed* or *open*. Note that the *closed depression landform* category specializes the *SN: Basin* category and is conceptually identical to the *SWF: Depression* category. For terminological simplicity and specificity, in this reference ontology, the term *depression landform* refers to the superordinate category of *all* depressions.

A *closed depression landform* is surrounded by higher ground, and has, as part, one level rim (represented by a *SN: contour*) marking the depression's upper edge, one pour point at the level of the rim, and a wall or basal surface that is impermeable enough to allow water storage. In sufficiently wet conditions, closed depression landforms store water and may be perceived to form still water bodies (e.g., puddles, lakes, ponds) with the water they contain, consistent with the view proposed in a recent ontology of water features and their parts (Brodaric et al. 2017). All other depressions are examples of *open depression landform* because they lack either an enclosing rim or their basal surface is sufficiently permeable. Thus, open depression landforms cannot store water for long periods of time (e.g., sink-holes).

Both open and closed depression landforms can be further categorized by their planimetric shape to distinguish landforms that are elongated, that is, have a single primary "length axis" from those that are not elongated. Elongated open depression landforms (e.g., valley, canyon, ravine, canal, trench, fissure, fault) are frequently referenced in natural language and are captured in this reference ontology by a named subcategory of open depression landforms: *open-longitudinal depression landform*. These depression landforms have a primary, sloping longitudinal axis bounded by upward sloping sides, and are generally open at both longitudinal ends. In wet conditions, they contain or host flowing bodies of water (e.g., rivers, streams). Most instances of the *SWF: Channel* category are examples of and also parts of instances of *open-longitudinal depression landform*.

3 Ontology Extension, Alignment and Application for Delineation of Depression Landforms

How can such a conceptual reference ontology help in landform delineation? Its usefulness lies in specifying a controlled vocabulary and categorical relationships and properties that can help in reasoning about which automated delineation tools must be chosen or which type of landforms can be delineated. For example, delineation of closed and open depression landforms requires different methods. Valley, canyon, gorge, ravine, gully, hollow, gulch, chasm, rill, canal, and trench in the English language are possible sub-categories of *open-longitudinal depressions*. Differentiating between instances of these types maybe quite challenging, requiring detailed three-dimensional morphometric measurements, and possibly knowledge of geomorphological agents and processes.

If such specific information is not available, as an alternative, based on semantic similarities between landform categories (e.g., valley/canyon/gorge, gully/gulch), delineation requests for one category could return instances of all related categories, or recommend searching for only the superordinate category (i.e., all types of *open longitudinal depression landforms*, instead specifically search for only valleys, canyons, or gorges). On the other hand, geoscientific classifications may be more precisely specified and supported for delineation.

An important next step is to formalize this ontology with specification of all classes and properties and linking it to previously developed ontologies such as the SN and SWF reference terrain ontologies, and more generally applicable reference ontologies of voids (Hahmann and Brodaric 2012) and water features (Brodaric et al. 2017). These ontology alignments will not only capture the detailed semantics of the categories but also extend inferencing capabilities to provide deeper insights about when, where and how to delineate landforms based on semantic queries. For example, a search for lake boundaries can be automatically inferred as also requiring delineation of a closed depression landform; or a query for a valley floor can be simplified as delineation of an area within a certain distance and/or depth of a *SN: CourseLine*.

It is also anticipated that this group (and others) will continue to add more specialized landform categories to extend the scope of this reference ontology. Some of the categories will necessitate inclusion of non-morphological criteria such as size, material (e.g., sand dune, drumlin), color, geomorphological origin, or culturally significant factors, and therefore, require supplementary data in addition to DEM datasets. On the other hand, ethnophysiographic research has shown that diverging ontological assumptions about landforms must be contended with when recognizing such specialized landform categories. Thus, diverse multilinguistic, multicultural and multimodal (e.g., using text, maps, photos, videos) human subject experiments must also be conducted to validate the contexts in which delineation methods might be guided by this reference ontology.

Acknowledgements Any use of trade, firm, or product names is for descriptive purposes only and does not imply endorsement by the U.S. Government.

References

Brodaric B, Hahmann T, Gruninger M (2017) Water features and their parts. Manuscript submitted to applied ontology

Casati R, Varzi AC (1994) Holes and other superficialities. MIT Press

Hahmann T, Brodaric B (2012) The void in hydro ontology.In: Proceedings of formal ontology in information systems (FOIS 2012). IOS Press, pp 45–58

Mark DM, Turk AG (2003) Landscape categories in yindjibarndi: ontology, environment, and language. LNCS 2825. Springer, Berlin, Heidelberg, pp 28–45

Mark DM, Turk AG, Stea D (2007) Progress on yindjibarndi ethnophysiography. In: Proceedings of the 8th international conference on spatial information theory (COSIT'07) LNCS 4736. Springer,Verlag, pp 1–19

Sinha G, Mark DM, Kolas D, Varanka D, Romero BE, Feng C.C, Usery LE, Liebermann J, Sorokine A (2014) An ontology design pattern for surface water features. In: Proceedings of the 8th international conference on geographic information science GIScience (LNCS 8728). International Publishing, Springer, Switzerland, pp 187–203

Sinha G, Kolas D, Mark DM, Romero BE, Usery LE, Berg-Cross G, Padmanabhan A (2017) Surface network ontology design patterns for linked topographic data. Revised manuscript to be submitted to Semantic Web

Correspondence Between PLCA and Maptree: Representations of a Space Configuration

Kazuko Takahashi

Abstract We discuss the correspondence of two qualitative spatial representations: PLCA and maptree. They can provide a topological configuration of a space with finer granularity by depicting the construction of a figure using points and lines. We define conversions between these two representations to show that they have the same granularity level of expression. We also investigate preservation of planarity on the conversions.

Keywords Qualitative sptial representation · Incidence relation · PLCA · Maptree

1 Introduction

In Qualitative Spatial Reasoning (QSR) (Cohn and Renz 2007; Ligozat 2013), image data are often represented using spatial relationships between objects projected in a two-dimensional (2D) Euclidean space. This means that the region occupied by one object may overlap that of another object. It is usually represented in the form of binary or ternary relations (e.g., Randell et al. 1992). There are other methods of representation that use incidence relations of elements, such as points or lines and their inclusions. *PLCA* was designed to represent the connection patterns of regions using incidence relations (Takahashi and Sumitomo 2007). In PLCA, an area is defined as an element that does not overlap with each other. A *maptree* is a representation that is also based on an incidence relation (Worboys 2012, 2013). Its data structure is an extension of a combinatorial map corresponding to the embedding of disconnected graphs. A few studies have compared QSR systems using incidence relations, which can give a more granular level of representation than systems using binary relations.

The entire space is regarded as being divided by edges in the representations based on incidence relations. The treatment of a scene including disconnected components

K. Takahashi (✉)
Kwansei Gakuin University, 2-1, Gakuen, Sanda 669-1337, Japan
e-mail: ktaka@kwansei.ac.jp

© Springer International Publishing AG 2018

P. Fogliaroni et al. (eds.), *Proceedings of Workshops and Posters at the 13th International Conference on Spatial Information Theory (COSIT 2017)*, Lecture Notes in Geoinformation and Cartography, https://doi.org/10.1007/978-3-319-63946-8_24

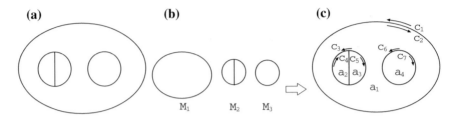

Fig. 1 Treatment of disconnected components

is crucial. Specifically, it is regarded as a composition of connected components, in which the locations of the components are expressed.

Consider the representations of the figure in Fig. 1a. There are three components: M_1, M_2, and M_3 (Fig. 1b). In PLCA, these components are considered to be the areas with the interior. After embedding these on a closed surface, four disjoint areas are generated: a_1, a_2, a_3, and a_4. Of these, a_1 has two holes (Fig. 1c). The border of an area may not be a Jordan curve, and one area may have multiple borders. Each edge in an embedding has two sides that confront the opposite areas, such as c_1 and c_2 or c_3 and c_4. By contrast, in maptree, these components are considered to be connected graphs, consisting only of strings. After embedding these on a closed surface, the complement in the surface of a connected graph is a collection of faces. Each edge in an embedding has two sides that confront the opposite areas, such as c_1 and c_2 or c_3 and c_4. The borders of the graphs are these edges.

This shows that the two representations reflect different recognitions of a scene, and it is interesting to consider their convertibility. Additionally, PLCA has several advantages. First, Coq proof assistant gives a constructive definition with its formal proof, and several properties of planarity are also proved formally (Takahashi et al. 2016). Second, a transformation method from a PLCA expression to RCC, a representation using a binary relation, is constructed by adding an attribute to each area (Takahashi and Sumitomo 2007). This transformation method can be extended to another fine-grained expression using a binary relation (Lewis et al. 2013). On the other hand, although maptree inherits the theoretical background of a combinatorial map, its properties have not been clarified sufficiently. It follows that, if we can convert maptree into PLCA, then the above advantages of PLCA can be applicable to maptree.

In this paper, we give a specific definition of the planar maptree, then define conversion rules between PLCA and maptree, and investigate preservation of planarity on these conversions.

This paper is organized as follows. We describe PLCA and maptree in Sects. 2 and 3, respectively. Then, we describe conversions between these two representations in Sect. 4. Finally, we present our conclusions in Sect. 5.

2 PLCA

A PLCA expression is defined as a five-tuple, $\langle P, L, C, A, om \rangle$. In PLCA, there are four kinds of object: points, lines, circuits, and areas. A *point* is the most primitive object and points are distinguishable from each other. A *line* represents segments between two points and is defined as a pair of points. Each line has a direction from the first to the second element of the pair. The inverse direction of a line l is denoted by $\bar{l}$,[1] and $\bar{\bar{l}} = l$. A *circuit* represents a closed outline and is defined as a list of lines. Each circuit is closed, that is, the first element of the first element l_0 and the second element of the last element l_n are the same. An *area* represents a region enclosed with circuits and is defined as a set of circuits. Additionally, we use a specific circuit in the outermost side of the figure, denoted by *outermost (om)*.

For example, a PLCA expression for Fig. 1c is shown.

$$\langle P, L, C, A, c_1 \rangle \qquad l_1 = (p_1, p_1) \; c_1 = [l_1] \quad a_1 = \{c_2, c_3, c_6\}$$
$$P = \{p_1, p_2, p_3, p_4\} \qquad l_2 = (p_2, p_3) \; c_2 = [\bar{l_1}] \quad a_2 = \{c_4\}$$
$$C = \{c_1, c_2, c_3, c_4, c_5, c_6, c_7\} \; l_3 = (p_3, p_2) \; c_3 = [l_2, l_4] \; a_3 = \{c_5\}$$
$$L = \{l_1, l_2, l_3, l_4, l_5\} \qquad l_4 = (p_2, p_3) \; c_4 = [\bar{l_2}, \bar{l_3}] \; a_4 = \{c_7\}$$
$$A = \{a_1, a_2, a_3, a_4\} \qquad l_5 = (p_4, p_4) \; c_5 = [l_3, \bar{l_4}]$$
$$c_6 = [\bar{l_5}] \quad c_7 = [\bar{\bar{l_5}}]$$

A PLCA expression is too permissive to find a corresponding topological space. For example, if there exists more than one area that contains the same circuit, such an expression does not make sense. Thus, we set a restriction on this data structure and define a consistent PLCA. We consider planarity only for a consistent PLCA.

Definition 1 (*planar PLCA*) If a consistent PLCA satisfies PLCA-constraints, PLCA-connectedness, and PLCA-euler, then it is said to be *a planar PLCA*.

PLCA-constraint is a condition stipulating that only straight lines are allowed and that there is no isolated point, no bridge between points nor isolated lines. PLCA-connectedness guarantees that no objects are separated, including the *outermost*. That is, each object is traceable from the *outermost*. Without this condition, some elements may be independent from the others, and we would not know where to embed them. PLCA-euler guarantees that a PLCA expression can be embedded in a 2D space so that the orientation of each circuit can be defined correctly.

A planar PLCA expression provides a surface subdivision of a 2D space. Here, we consider a surface subdivision as a configuration in which both sides of each line always belong to distinct areas.

---

[1] We use this notation to coincide with the one in maptree. Although, in PLCA, l^+ and l^- are used to show the directions of a line.

3 Maptree

A combinatorial map is a representation of an embedding of a connected graph. In a combinatorial map, *a dart (or a half-line)*, is defined as a primitive, and other elements are defined as a permutation of a set of darts.

Let A be a finite set. We call any bijective function $\phi : A \to A$ *a permutation* of A. For $a_1, \ldots, a_n \in A$ and a permutation ϕ, if $a_2 = \phi a_1$, $a_3 = \phi a_2$, ... $a_1 = \phi a_n$, we call $(a_1 \ldots a_n)$ *a cycle*. Then, any permutation is written as a collection of cycles. Let Φ be a collection of permutations of A. Φ is *transitive* if for any elements $x, y \in A$, we can transform x to y by a sequence of permutations from Φ. For a dart δ, we call a pair of δ and $\bar{\delta}$ *a complementary pair*, and $\bar{\bar{\delta}} = \delta$.

Definition 2 (*combinatorial map*) *A combinatorial map* $M\langle S, \alpha, \tau \rangle$ consists of:

1. A finite set $S = \{\delta_1, \ldots, \delta_n, \bar{\delta_1}, \ldots, \bar{\delta_n}\}$ of complementary pairs of darts.
2. A permutation α of S where $\alpha = \phi_1 \ldots \phi_m$. ($\phi_1, \ldots, \phi_m$ are called α-*cycles*.)
3. A permutation τ of S where $\tau = (\delta_1 \bar{\delta_1})(\delta_2 \bar{\delta_2}) \ldots (\delta_n \bar{\delta_n})$.

Subject to the constraint that the collection of permutations $\{\tau, \alpha\}$ is transitive.

Given a combinatorial map M with α-cycles $\phi_1, \ldots, \phi_m$, *p-star associated with M* is an edge-labelled tree with a central black node from which edges connect to white nodes, the i-th edge being labelled with ϕ_i ($1 \le i \le m$). A *bw-tree* is a colored tree with the nodes colored black or white, subject to the condition that no two adjacent nodes have the same color. White nodes correspond to faces (regions).

Definition 3 (*maptree*) Let M be a finite collection of combinatorial maps. A *map-tree* T_M is an edge-labelled bw-tree formed by the merging of p-stars at white nodes.

Here, we introduce a formal definition of a planar maptree, which is not explicitly described in (Worboys 2013).

Let T_M be a maptree for $M = (M_1, \ldots, M_k)$ ($1 \le i \le k$) where $M_i \langle S_i, \alpha_i, \tau_i \rangle$. We consider the region assignment function *reg* from a set of α-cycles in M to a set of regions. That is, for each α-cycle ϕ, we associate the region Δ denoted by $reg(\phi) = \Delta$. For α-cycles ϕ_i of α_i and ϕ_j of α_j, if $reg(\phi_i) = reg(\phi_j)$, then it is said that M_i *and* M_j *share a region*. Henceforth, the term 'maptree' denotes the maptree associated with a region assignment function.

Definition 4 (*planar maptree*) A maptree T_M for $M = (M_1, \ldots, M_k)$ is said to be *a planar maptree* if the following conditions are satisfied:

1. Exactly one root node Δ_{root} for T_M exists.
2. For ϕ, ϕ' ($\phi \ne \phi'$), if they are in the same permutation α, then $reg(\phi) \ne reg(\phi')$.
3. When $k \ge 2$, for each M_i ($1 \le i \le k$), there exists an α-cycle ϕ such that $reg(\phi) = \Delta(\ne \Delta_{root})$ and that Δ is shared with M_j ($1 \le j \le k, i \ne j$).
4. When $k \ge 2$, for any pair of M_i and M_j ($1 \le i, j \le k, i \ne j$), they share at most one region.

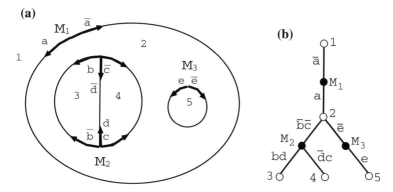

Fig. 2 An embedding of a disconnected graph and the corresponding maptree

For example, the planar maptree corresponding to an embedding of a disconnected graph in Fig. 2a is represented as follows and depicted in Fig. 2b.

It consists of three combinatorial maps M_1, M_2, and M_3:

$$M_1\langle\{a,\bar{a}\},(a)(\bar{a}),(a\bar{a})\rangle$$
$$M_2\langle\{b,c,d,\bar{b},\bar{c},\bar{d}\},(\bar{b}\bar{c})(bd)(\bar{d}c),(b\bar{b})(c\bar{c})(d\bar{d})\rangle$$
$$M_3\langle\{e,\bar{e}\},(e)(\bar{e}),(e\bar{e})\rangle$$

The region assignment function is defined as follows: $reg((\bar{a})) = 1 = \Delta_{root}$, $reg((a)) = reg((\bar{b}\bar{c})) = reg((\bar{e})) = 2$, $reg((bd)) = 3$, $reg((\bar{d}c)) = 4$ and $reg((e)) = 5$.

4   Conversions Between PLCA and Maptree

We show the conversion from a planar maptree to a PLCA expression.

Let T_M be a planar maptree for $M = (M_1, \ldots, M_k)$ $(1 \le i \le k)$, where $M_i\langle S_i, \alpha_i, \tau_i\rangle$ is a combinatorial map, and let reg be a region assignment function. First, we calculate $\beta_i = \tau_i\alpha_i^{-1}$ for each i. We bijectively map each dart, α-cycle, β-cycle, and region in T_M to elements in PLCA, and make a set of P, L, C, and A. Then, we generate their incidence relations.

Let $\mathscr{S}_S, \mathscr{S}_\alpha, \mathscr{S}_\beta$, and $\mathscr{S}_R$ be sets of darts, α-cycles, β-cycles, and regions appearing in T_M, respectively. The notation $x \mapsto y$ denotes that an element x is mapped to y.

1. Make sets of P, L, C, and A. We bijectively map each β-cycle to a point, dart to a line, and α-cycle to a circuit in PLCA. For a dart $\delta \in \mathscr{S}_S$, if $\delta \mapsto l$, then $\bar{\delta} \mapsto \bar{l}$. For a region, $\Delta_{root} \mapsto \bot$, because the external region of the *outermost* does not exist in PLCA; otherwise, it is mapped to an area in PLCA.
2. Generate incidence relations.

   • For $\delta \in \mathscr{S}_S$, $\delta \mapsto l$; if δ is in ρ, $\bar{\delta}$ is in ρ', and $\rho \mapsto p, \rho' \mapsto p'$, then $l = (p, p')$.

- For $\phi \in \mathscr{S}_\alpha$, $\phi \mapsto c$; if $\phi = (\delta_1 \ldots \delta_s)$ and $\delta_j \mapsto l_j$ $(1 \leq j \leq s)$, then $c = \{l_1, \ldots, l_s\}$.
- For $\Delta \in \mathscr{S}_R$, $\Delta \mapsto a$; if $reg(\phi_j) = \Delta$, $\phi_j \mapsto c_j$, then $a = \{c_1, \ldots, c_t\}$, where t is the number of ϕ_j that satisfies $reg(\phi_j) = \Delta$.
- For Δ_{root}, if $reg(\phi) = \Delta_{root}$, $\phi \mapsto c$, then $om = c$.

For the obtained PLCA expression, PLCA-consistency, PLCA-connectedness and PLCA-euler clearly hold from the planarity of a maptree. However, PLCA-constraints is not satisfied, because a planar maptree admits isolated lines and bridges as well as multiple edges connecting the same pair of points. Therefore, we need a condition so that it provides a surface subdivision of a 2D space.

Proposition 1 *Let T_M be a planar maptree. If each α-cycle ϕ in T_M satisfies the following conditions, then T_M provides a surface subdivision of a 2D space: (i) there does not exist δ such that δ and $\bar{\delta}$ are both included in ϕ, and (ii) $|\phi| \geq 3$.*

We can similarly define a conversion rule from a planar PLCA expression to a maptree. In this case, the crucial point is that the mapped data are correctly divided into a set of combinatorial maps. We have proved that the obtained one is a planar maptree.

5   Conclusion

We have discussed the correspondence of two qualitative spatial representations based on incidence relations, PLCA and maptree. We have defined conversions between these two representations and clarified the condition that a planar maptree provides a surface subdivision of a 2D space. The main contribution of this paper is to relate the two representations that reflect different recognitions of a scene: area-based and string-based. We have implemented in Prolog prototypes of the conversion programs in both directions.

The proofs are done manually, and we will provide a strict proof using proof assistants in future. We also want to show the relationship between representations using incidence relations and binary relations.

References

Cohn A, Renz J (2007) Handbook of knowledge representation, chapter 13. In: Qualitative spatial reasoning, pp 551–596. Elsevier

Lewis JA, Dube MP, Egenhofer MJ (2013) The topology of spatial scenes in $\mathbb{R}$. In: COSIT13, pp 495–515. LNCS 8116, Springer

Ligozat G (2013) Qualitative spatial and temporal reasoning, Wiley

Randell D, Cui C, Cohn A (1992) A spatial logic based on regions and "Connection". KR1992 165–176

Takahashi K, Moriguchi S, Goto M (2016) Formalization of a surface subdivision allowing a region with holes without coordinates. ADG2016 190–207

Takahashi K, Sumitomo T (2007) The qualitative treatment of spatial data. Int J Artif Intel Tools 16(4):661–682

Worboys M (2012) The maptree: a fine-grained representation of space. GIScience 2012, pp 298–310. LNCS 7478, Springer

Worboys M (2013) Using maptrees to characterize topological change. COSIT13, pp 74–90. LNCS 8116, Springer

Building Social Networks in Volunteered Geographic Information Communities: What Contributor Behaviours Reveal About Crowdsourced Data Quality

Quy Thy Truong, Guillaume Touya and Cyril de Runz

Abstract Modelling social interactions in volunteered geographic information projects requires defining what binds contributors together. In order to be as realistic as possible, instead of choosing one social aspect to study, we choose to build a multi-layered social network that contains several types of interaction between VGI contributors. The analysis of such a multigraph should allow the detection of communities and the definition of typical profiles of contributors. A use case on Open-StreetMap illustrates what inferences can be made about contributions based on their authors.

Keywords Social networks · Volunteered geographic information · Data quality

1 Introduction

Among the objections raised against crowdsourcing, data quality is still a major debated issue, especially in Volunteered Geographic Information (VGI). In order to ensure VGI quality, Goodchild and Li (2012) have expressed a social approach that "relies on a hierarchy of trusted individuals who act as moderators or gate-keepers". This statement implies analysing contributors' actions in order to identify those who are trustworthy. Moreover, it also requires qualifying contributors' reactions to one another's contribution. The key question is how to model contributors' interactions based on VGI contributions? This paper suggests modelling contributors' interactions using a social multiplex network. It depicts initial results on an OpenStreetMap

Q.T. Truong (✉) · G. Touya
University of Paris-Est, LASTIG COGIT, IGN, ENSG, F-94160 Saint-Mande, France
e-mail: quy-thy.truong@ign.fr

C. de Runz
Modeco, CReSTIC, University of Reims Champagne-Ardenne,
CS 30012 Reims Cedex 2, France

P. Fogliaroni et al. (eds.), *Proceedings of Workshops and Posters at the 13th International Conference on Spatial Information Theory (COSIT 2017)*, Lecture Notes in Geoinformation and Cartography, https://doi.org/10.1007/978-3-319-63946-8_25

(OSM) use case, showing that, despite the typical heterogeneity of VGI contributions/contributors, some contributors are trusted by the community and so are their contributions.

2 Related Works

Viewing VGI communities as social networks is not a breakthrough. Mooney and Corcoran (2014) and Stein et al. (2015) have proposed social graph models to account for collaboration between OSM contributors. More generally, crowdsourcing communities are based on contributors reacting to one another's contributions. However, only one aspect at a time was used to model collaboration.

According to Kivelä et al. (2014), "reducing a social system to a network in which actors are connected (. . .) by only a single type of relationship is often an extremely crude approximation of reality. As a result, sociologists recognized decades ago that it is crucial to study social systems by constructing multiple social networks using different types of ties among the same set of individuals". This principle was successfully used by Jankowski-Lorek et al. (2016) in order to understand the impact of teamwork in the quality of crowdsourced data through a Multidimensional Behavioural Social Network of Wikipedia community. Stein et al. (2015) also confronted two types of collaboration (width and depth of collaboration) in order to identify contributor types. Following the same idea, we propose to model contributors' interactions using a social multiplex network. This graph provides a more complete portrait of VGI contributors when it comes to collaboration inside a community.

3 Modelling VGI Contributors in a Multiplex Social Network

We build a specific type of multi-layered network, called a multiplex network, which is a sequence of graphs $\{G_\alpha\}_{\alpha=1}^{b} = \{(V_\alpha, E_\alpha)\}$ where E_α is the set of edges, V_α is the set of nodes such as $V_\alpha = V_\beta = V$ for all α, β, and b is the number of graphs (Kivelä et al. 2014). In our case, V is the set of contributors. The challenge here is to define the sets of edges E_α. The following propositions are inspired by literature's models of social interactions.

In VGI literature, collaborations have been mainly considered from an "object edition" point of view. One can define $E_{co-edition} = \{(A, B, e)\}$ such as contributor A created a version $v + 1$ of contributor B's version v of the very same object, and e the intensity of co-edition i.e. the number of times A directly responded to B's contributions (Mooney and Corcoran 2014).

Besides, co-edition can be extended to the definition of interlocking collaboration that enables to characterize the intensity of collaboration (Stein et al. 2015). In this case, interlocking collaboration is not restricted to successive edits between two contributors. Thus, collaboration width is described in $E_{width} = \{(A, B, w)\}$ where w is the number of objects contributors A and B commonly edited; collaboration depth is represented in the set $E_{depth} = \{(A, B, d)\}$ where d is the maximum number of times contributor A responded (directly or not) to contributor B on one object. Despite subtle differences, all those graphs enable to evaluate whether a contributor reaches agreement with the community.

Furthermore, based Wikipedia's norm network (Heaberlin and DeDeo 2016) where two pages are connected when one refers to the other via a hyperlink, reusing someone else's contribution to make one's own contribution is also another type of interaction. The set $E_{depth} = \{(A, B, d)\}$ approximates trust relationships between two users A and B, especially in the case $d > 0$, for a contributor uses the contribution of a peer if the former trusts in the latter.

Eventually, connecting contributors who contributed in the same area may highlight existing partnerships among the community, especially if they contribute at the same time. Thus, we can define the set $E_{co-location} = \{(A, B, l)\}$ that connects contributors depending on the overlapping l of their respective areas of contribution (Neis and Zipf 2012; Zielstra et al. 2014) and the set of edges $E_{co-temporal} = \{(A, B, t)\}$ that describes the temporal link that separates a pair of contributors. The aggregation of the co-location and the co-temporal graphs corresponds to the model of spatiotemporal co-occurrence proposed in (Crandall et al. 2010). This model was defined in order to infer social ties; therefore the simultaneous analysis of these graphs ought to deal with the same issue.

4 Use Case: Analysis of OpenStreetMap Contributors

Three graphs are created based on OSM contributions in Paris between 2010 and 2015: a width collaboration graph, a co-edition graph and a use graph. These graphs are shown in Figs. 1, 2 and 3. Visualisation parameters are given in Table 1. According to the width collaboration graph (Fig. 1), user #17397 is a very active contributor whose editions were generally not questioned by the others. Therefore, user #17397s profile is relatively close to the moderator type. Besides, on the co-edition graph (Fig. 2), major contributors who have been directly edited by user #17397 can be identified. Similarly, user #18855 edited less and yet, his contributions have widely been corrected afterwards by many contributors. Hence, we can infer that the contributions of user #18855 could not reach agreement among OpenStreetMap community. Eventually, in the use graph (Fig. 3) user #17397 can again be noticed as a contributor who is keen on reusing peoples contributions to map his/her own

Fig. 1 Width collaboration graph of OSM contributors of Paris (IDs are anonymized). A directed edge from user A to user B means that user A has modified an object that was previously edited by user B. Edge thickness reveals the number of unique objects user A has edited after user B

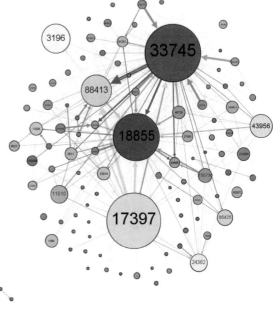

Fig. 2 Co-edition graph of OSM contributors of Paris (IDs are anonymized). An edge directed from user A to user B means that user A has directly edited user B's object version. Edge thickness reveals the number of times user A modified user B's contributions

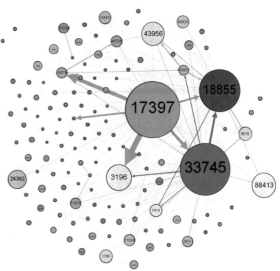

objects. Reciprocally many contributors rely on user #17397s contributions to enrich OSM database. Consequently, user #17397 does not only behave like a moderator but also seems to be recognised as trustworthy by the community.

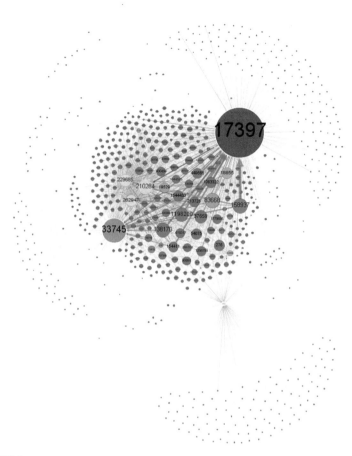

Fig. 3 OSM use graph in Paris. An edge directed from user A to user B means that user A created a new object based on user B's contribution. For instance, contributor A may use contributor B's nodes to map his own way, or contributor A can create a new relation based on contributor B's ways or nodes. Edge thickness reveals the number of user A's objects that are composed of user B's contributions

5 Conclusions and Outlook

In the same line as Social Network Analysis, this paper put forward a multiplex social graph to model the behaviours of VGI contributors. Interactions between contributors are captured on several types of VGI contribution tasks; thus, social ties are more thoroughly retraced. Admittedly, a social multiplex network highlights how contributors view one another within a VGI community, but the analysis enables to dig deeper and thus tackle the issue of VGI quality. Indeed, the application of this model on an OSM use case brought to light information on data quality. Unsurprisingly, the diversity of behaviours explains the heterogeneity of VGI quality. Nevertheless,

Table 1 Graphs visual parameters. According to the multiplex network's definition, each graph shares the same set of nodes but for clarity, only nodes that are part of an edge are displayed

Parameter	Definition
Node colour	Proportional to indegree as follows: blue = low; white = intermediate; red= high
Node size	Proportional to outdegree
Edge colour	Identical to node source colour
Edge size	Proportional to weight i.e. intensity of the collaboration

contributors who were reckoned as trusted individuals by the community suggest the reliability of their contributions. By analysing additional graphs in a more automated way, we hope to capture most of contributors' modes of operation. The main goal of these graphs is to make emerge a typology of contributors in which VGI contributors will be automatically classified. To that end, future research will deal, on the one hand, with community discovery in multiplex graphs including clustering methods (Battiston et al. 2014). On the other hand, the resulting typology should be compared to existing qualitative surveys like in Coleman et al. (2010), Duféal et al. (2016).

References

Battiston F, Nicosia V, Latora V (2014) Structural measures for multiplex networks. Phys Rev E 89:032, 804. doi:10.1103/PhysRevE.89.032804. https://link.aps.org/doi/10.1103/PhysRevE.89.032804

Coleman DJ, Geogiadou Y, Labonte J (2010) Volunteered geographic information: the nature and motivation of producers. Int J Spatial Data Infrastruct Res 4

Crandall DJ, Backstrom L, Cosley D, Suri S, Huttenlocher D, Kleinberg J (2010) Inferring social ties from geographic coincidences. In: Proceedings of the National Academy of Sciences 107(52):22,436–22,441. doi:10.1073/pnas.1006155107. http://dx.doi.org/10.1073/pnas.1006155107

Duféal M, Jonchères C, Noucher M (2016) ECCE CARTO—DES ESPACES DE LA CONTRIBUTION A LA CONTRIBUTION SUR L'ESPACE—Profils, pratiques et valeurs d'engagement des contributeurs d'OpenStreetMap (OSM). Research report, UMR 5319. https://halshs.archives-ouvertes.fr/halshs-01371544

Goodchild MF, Li L (2012) Assuring the quality of volunteered geographic information. Spatial Stat

Heaberlin B, DeDeo S (2016) The evolution of wikipedia's norm network. Future Internet 8(2):14+. doi:10.3390/fi8020014. http://dx.doi.org/10.3390/fi8020014

Jankowski-Lorek M, Jaroszewicz S, Ostrowski Ł, Wierzbicki A (2016) Verifying social network models of wikipedia knowledge community. Inf Sci 339:158–174. doi:10.1016/j.ins.2015.12.015. http://dx.doi.org/10.1016/j.ins.2015.12.015

Kivelä M, Arenas A, Barthelemy M, Gleeson JP, Moreno Y, Porter MA (2014) Multilayer networks. J Complex Netw 2(3):203–271. doi:10.1093/comnet/cnu016. http://dx.doi.org/10.1093/comnet/cnu016

Mooney P, Corcoran P (2014) Analysis of interaction and co-editing patterns amongst openstreetmap contributors. Trans GIS 18(5):633–659. doi:10.1111/tgis.12051. http://dx.doi.org/10.1111/tgis.12051

Neis P, Zipf A (2012) Analyzing the contributor activity of a volunteered geographic information project the case of openstreetmap. Int J Geo-Inf pp 146–165

Stein K, Kremer D, Schlieder C (2015) Spatial collaboration networks of OpenStreetMap. In: Jokar Arsanjani J, Zipf A, Mooney P, Helbich M (eds) Openstreetmap in GIScience. Lecture notes in geoinformation and cartography, pp 167–186. Springer International Publishing. doi:10.1007/978-3-319-14280-7_9. http://dx.doi.org/10.1007/978-3-319-14280-7_9

Zielstra D, Hochmair H, Neis P, Tonini F (2014) Areal delineation of home regions from contribution and editing patterns in openstreetmap. ISPRS Int J Geo-Inf 3(4):1211–1233. doi:10.3390/ijgi3041211. http://dx.doi.org/10.3390/ijgi3041211

Guiding People Along More Intuitive Indoor Paths

Nina Vanhaeren, Kristien Ooms and Philippe de Maeyer

Abstract Route planning algorithms in indoor navigation systems are currently limited to shortest path algorithms or derivatives. The development of a cognitive route planning algorithm, providing cognitive more comfortable paths to navigators, could improve these systems. The first phase of the development of such an algorithm entails the identification of relevant path characteristics of the indoor environment. Therefore, a user study is enrolled: an in-depth discussion with a focus group of experts is followed by an international online survey.

Keywords Indoor navigation · Route planning algorithm · Cognition

1 Background

As long as people need to decide where to go and how to reach a destination, navigation will remain one of the fundamental problems in human cognition, wayfinding and geospatial research. Wayfinding is the goal-directed part of navigation based on decision making and planning, and influenced by both personal and environmental factors (Montello 2005). The improvement of indoor navigation systems could ease the wayfinding task in the indoor environment.

Route planning is a key element of navigation guidance applications as it aims at computing an optimal route between a starting and a destination point (Montello 2005). Route planning algorithms in existing (indoor) navigation applications are limited to the shortest or fastest path (Vanclooster et al. 2014a). However, studies have proven that people do not always prefer the shortest or fastest route to reach

N. Vanhaeren (✉) · K. Ooms · P. de Maeyer
Ghent University, Krijgslaan 281, Ghent, Belgium
e-mail: nina.vanhaeren@ugent.be

K. Ooms
e-mail: kristien.ooms@ugent.be

P. de Maeyer
e-mail: philippe.demaeyer@ugent.be

© Springer International Publishing AG 2018
P. Fogliaroni et al. (eds.), *Proceedings of Workshops and Posters at the 13th International Conference on Spatial Information Theory (COSIT 2017)*, Lecture Notes in Geoinformation and Cartography, https://doi.org/10.1007/978-3-319-63946-8_26

133

their destination (Golledge 1999). More intuitive and easier-to-follow routes reduce the risk of getting lost, require a smaller wayfinding effort, guide in recalling routes and are overall perceived as more comfortable (Vanclooster et al. 2014a).

This research focuses on developing a cognitive route planning algorithm to support indoor navigation systems. The development of such an algorithm will improve indoor navigation systems by guiding people along cognitive more comfortable paths. Since these paths are in line with the user's mental structure and thus adhere better to natural human wayfinding behaviour, the cognitive load for the wayfinder will be reduced to the minimum (Vanclooster et al. 2013).

2 Research Question

Until now, little research has been devoted to the definition of indoor path characteristics that differentiate a more intuitive path from the common indoor shortest or fastest path. In order to create this cognitive indoor routing algorithm, it is crucial to understand the determining characteristics of path selection and to interpret how wayfinders make route choices in indoor environments. In other words, this research wants to focus on identifying the path characteristics people use during indoor wayfinding. The outcome of this research provides essential insights into users' natural route planning behaviour in indoor environments.

3 Approach

To determine path characteristics of cognitive routes in indoor environments, a research design was developed in line with previous studies about outdoor navigation, route choice criteria and their accompanying path characteristics (such as intersection complexity, visibility, turning points) (e.g. Dalton 2003; Golledge 1995; Hillier and Iida 2005; Hochmair 2005). Furthermore, the research design was based on existing research on indoor wayfinding behaviour (Hölscher et al. 2006), usability engineering and user-centred design (UCD) (e.g. Haklay and Nivala 2010; Nielsen 1994; van Elzakker and Wealands 2007).

To obtain a well-founded and coherent selection of relevant cognitive path characteristics and to incorporate the definition of the user requirements into the design process, a focus group and an online survey are employed.

The focus group is composed of diverse academic researchers and experts experienced with indoor environments, navigation and human behaviour studies. This focus group helps to define and formulate, through multiple discussions, possible cognitive path characteristics in indoor environments. Hereafter, the results of the focus group discussions are scrutinised through an online survey, in which a large group and diverse range of participants can be reached.

In the online survey, different routes in various indoor environments are recorded and displayed to the participants. Subsequently, in a questionnaire participants are asked to answer questions about these routes, their characteristics and preferences.

Integrating the results of the focus group and the online survey leads to a coherent selection of relevant cognitive path characteristics and provides complementary information on the main path characteristics in the indoor environment. Through this combination of qualitative and quantitative research, the path characteristics that differentiate a more intuitive path from the currently used indoor shortest or fastest paths (e.g. Kwan and Lee 2005; Thill et al. 2011) are defined.

4 Outcome

In general, the results of this research will provide essential knowledge on which characteristics are determinative in human path selection in indoor environments and how humans interpret these path characteristics. It will not only provide information on how to make navigation aid more comfortable, but it will also contribute to the overall understanding of indoor wayfinding and navigation.

5 Further Work

The obtained path characteristics will be incorporated in the cognitive route planning algorithm. The development of such an algorithm requires a theoretical conceptualisation of the underlying spatial concepts of each of those path characteristics, which have to model the users' perception on these path characteristics. The underlying indoor spatial model has to be taken into account in this process, as this determines the structure of the algorithm and could influence the results and accuracy of the algorithmic implementation (Vanclooster et al. 2014b).

References

Dalton RC (2003) The secret is to follow your nose. Environ Behav 35(1):107–131

Golledge RG (1995) Path selection and route preference in human navigation: a progress report. In: Frank AU, Kuhn W (ed) Spatial information theory a theoretical basis for GIS, vol 988, pp 207–222. Springer, Berlin

Golledge RG (ed) (1999) Wayfinding behavior: cognitive mapping and other spatial processes. Johns Hopkins University Press, Baltimore and London

Haklay, MM, Nivala, AM (2010). User-centred design. In: MM. Haklay (ed), Interacting with geospatial technologies, pp 89–106. Wiley

Hillier B, Iida S (2005) Network and psychological effects: a theory of urban movement. In: Cohn AG, Mark DM (eds) Spatial information theory. Springer, Berlin, pp 475–490

Hochmair H (2005) Towards a classification of route selection criteria for route planning tools. In: Fisher PF (ed) developments in spatial data handling. Springer, Berlin, pp 481–492

Hölscher C, Vrachliotis G, Meilinger, T (2006). The floor strategy: wayfinding cognition in a multi-level building. In: Proceedings of 5th international space syntax symposium. Delft, The Netherlands: Techne Press

Kwan MP, Lee J (2005) Emergency response after 9/11: the potential of real-time 3D GIS for quick emergency response in micro-spatial environments. Comput Environ Urban Syst 29 (2):93–113

Montello DR (2005) Navigation. In: Shah P, Miyake A (eds) The cambridge handbook of visuospatial thinking. Cambridge University Press, Cambridge, pp 257–294

Nielsen J (1994) Usability engineering. Morgan Kaufmann Publishers Inc, San Francisco, CA, USA

Thill JC, Dao THD, Zhou Y (2011) Traveling in the three-dimensional city: applications in route planning, accessibility assessment, location analysis and beyond. J Transp Geogr 19(3):405–421

van Elzakker CPJM, Wealands K (2007) Use and users of multimedia cartography. In: Cartwright W, Peterson MP, Gartner DG (eds) multimedia cartography. Springer, Berlin Heidelberg, pp 487–504

Vanclooster A, Viaene P, Van de Weghe N, Fack V, De Maeyer P (2013). Analyzing the applicability of the least risk path algorithm in indoor space. In: ISPRS Annals of photogrammetry, remote sensing and spatial information sciences II-4/W1, pp 19–26, December

Vanclooster A, Ooms K, Viaene P, Fack V, Van De Weghe N, De Maeyer P (2014a) Evaluating suitability of the least risk path algorithm to support cognitive wayfinding in indoor spaces: an empirical evaluating suitab. Appl Geogr 53:128–140

Vanclooster A, Van de Weghe N, Fack V, De Maeyer P (2014b) Comparing indoor and outdoor network models for automatically calculating turns. J Loc Based Serv 8(3):1–18

Context and Vagueness in Automated Interpretation of Place Description: A Computational Model

Diedrich Wolter and Madiha Yousaf

Abstract We investigate automated interpretation of human-generated place descriptions. This paper presents work in progress towards a computational model that can represent spatial knowledge occurring in place descriptions and fosters efficient querying of a spatial database. In this paper we analyse how context information shapes the meaning of a place description and we outline a computational model to represent vague spatial knowledge and context occurring in human-generated place descriptions.

Keywords NL place descriptions · Context · Spatial reasoning

1 Problem Statement

Several works address interpretation of place descriptions, motivated by either questions regarding semantics of spatial language or by prospects on applications in human-centered computing, for example exploitation of volunteered geographic information (VGI) (e.g., Adams and McKenzie 2013). Our work is aimed at a computational model for the latter, that is a formal knowledge representation and a reasoning algorithm, which allows natural language place descriptions to be interpreted by identifying the described place within a spatial database such as OpenStreetMap (OSM). Our overall aim is to develop an automatic system applicable to a range of settings. We consider natural language descriptions aiming to communicate location information to an unknown recipient. In particular, we are interested in two research questions:

D. Wolter (✉) · M. Yousaf
University of Bamberg, An der Weberei 5, 96047 Bamberg, Germany
e-mail: diedrich.wolter@uni-bamberg.de

M. Yousaf
e-mail: madiha.youssaf@uni-bamberg.de

© Springer International Publishing AG 2018
P. Fogliaroni et al. (eds.), *Proceedings of Workshops and Posters at the 13th International Conference on Spatial Information Theory (COSIT 2017)*, Lecture Notes in Geoinformation and Cartography, https://doi.org/10.1007/978-3-319-63946-8_27

1. How can a spatial relation expression ("north of", "near", "in front of") be interpreted with an appropriate spatial relation? What is an appropriate semantics for relations to represent vague knowledge inherent to relations (e.g., near) or entities themselves (e.g., extent of a mountain)?
2. How can a set of spatial relations describing a location be efficiently matched against a spatial knowledge base to identify location candidates?

We restrict our consideration to descriptions that refer to objects represented in the spatial database using spatial language, either directly ("the *campground* near mount Foo"), or as sub-region of larger entities ("*northern beach* of the lake near mount Foo"). Also, we disregard interpretation of place names, for example we do not aim to interpret "students' district" which may stand for a city district close to university which contains affordable restaurants. The contribution of this paper is to discuss contextual influences and to propose a computational framework which can capture vague and context-sensitive information about places. First we present computationally motivated models of place and context, then we outline our overall model.

2   What Is a Place?

Before embarking on interpretation of place descriptions we have to commit to a definition of place. The notion of place and how it can be defined has been discussed comprehensively (Bennett and Agarwal 2007; Davies et al. 2009; Winter and Truelove 2010; Vasardani et al. 2013), yet there exists no definition capturing human intuition which leads to a fully specified computational data model. Discrepancies thus arise between how human conceptualise space and how entities are represented in a geographic information system (GIS). A single entity in a GIS may correspond to several places (for example several barbecue places in a single park), or vice versa (for example adjacent meadows which constitute a place for playing soccer). Vasardani et al. (2013) contrasts definitions of place which involve elements place name, spatial extent, and—in case of gazetteers—a type of place, often defined using some form of ontology. We argue that places can obtain their meaning from actions that can be performed within their spatial extents and therefore a definition of space should include the conceptualisation, i.e., the process which associates a particular concept

Fig. 1 Computational model of place based on name, spatial extent, and conceptualisation

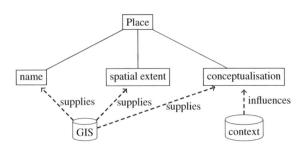

with a spatial area. Our application-specific definition which is illustrated in Fig. 1 comprises three elements:

1. Places are designated by names, either a unique label (e.g., "Eiffel tower") or a circumscription (e.g., "picnic place"). We assume unique labels to by available from a GIS and we treat any circumscription as a free-ranging variable that can represent any type of place.
2. Places are defined by their spatial extent. Although some places have indeterminate boundaries we assume location information to be provided from a GIS. In case spatial extents are not directly provided by the GIS, either a spatial extent of some GIS entity is used as upper approximation or a single location representing a prototypical location of the place is computed.
3. Places are defined by their conceptualisation. While a GIS provides information about type of objects, it may require a context-sensitive interpretation of type information. We are particularly concerned with handling spatial context variables. For example, if a specific entity is referred to as a *large lake*, then entities of the same type and similar extent are reasonable instantiations of *large lake*, too.

3 Towards an Operational Definition of Context

Context shapes the meaning in all communications and presents a multidisciplinary topic. A lot of definitions of context exist which all refine the umbrella term *conditions and circumstances*, often differently. Bazire and Brézillon (2005) argue that context is inherently domain- or task-specific and thus no generic definition of context can be formulated. In order to arrive at an operational definition that can be realised as a computational model we employ a bottom-up approach of identifying *context variables* which are elements influencing what the most reasonable interpretation of a place description is. We have chosen the term context variable since variables often represent unknowns: Context is not explicitly given but must be inferred from the pieces of information available.

We distinguish three classes of context variables in interpretation of place description: the *environment*, the *human* who generates a place description, and the *place description* as a linguistic phrase. This classification adapts a characterisation of context in performing map-based tasks by Freksa et al. (2007) to a text-based task. Our linguistic context implied by the place description corresponds to their map context, yet we separate objective environmental factors from all cognitive factors. Let's have a detailed look on our classification which is depicted in Fig. 2:

Environment The environment defines the physical domain in which place descriptions are interpreted. It provides input to human conceptualisation (link A in the figure), that is, what mental representation is built which then serves as basis for generating the place description. Considering actions a particular entity allows a human to perform contributes to interpretation, for example a river can

Fig. 2 Contextual influence
in interpretation of place
descriptions

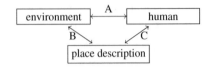

be followed, a hill enables going further up or down, etc. Second, entities in the environment can define reference frames which need to be identified in order to interpret a spatial relation occurring in a place description (link B in the figure). For example, "in front of X" may refer to an intrinsic front if provided by entity X provides it (e.g., a building with a designated front), yet it may also refer to a route described, i.e., it refers to a location just before reaching X. Moreover, the environment shapes the meaning of relations by providing information about scale. For example, "near Egypt" implies a different interpretation of nearness in absolute distance terms than "near the central station".

Human Cognitive principles shape mental representation of the environment (link A), they also influence how places are described verbally (link B). Moreover, the intention of the human to generate the place description and her model of the recipient are important. However, we have to disregard these aspects for time being, since we assume only the place description and environment to be known.

Place Description The place description provides a linguistic aspect which allows us to draw conclusion about cognitive influences (link B). For example, a term such as "north-northwest" introduces a finer level of granularity in direction than "north". Also, the descriptions provides (qualitative) information about entities in the environment (link A).

4 A Context-Sensitive Model of Place Descriptions

We first review the requirements a computational model needs to satisfy, both general properties and those arising from context variables. With respect to general properties of the spatial knowledge model we first observe that *different aspects of spatial knowledge* (cardinal direction, distance, etc.) can occur in a single place description and must be jointly expressible in our model. Second, several *qualitative relations and concepts need to be considered as being vague*, that is, there exists no determinate boundary of acceptability. Our computational model must thus be able to handle partial satisfiability of a relation or concept and it should aim to determine the most plausible interpretation my *maximising satisfiability*. For example, if one is referring to a park near the city centre, a reasonable strategy would be to look for the park closest to a location most certainly inside the city centre.

With respect to context-dependent modelling, context variables influence the *mapping of GIS entity types to concepts*, for example, reference to a park for picnic or for playing football poses different space requirements on the park. Technically speaking, environmental features establish conceptual entities referred to in a place description. Our model thus needs to include variables that represent conceptual entities and represent rules that determine the mapping. The same technique is also applicable to identify *frames of reference* environmental features offer. Use of words may influence the *granularity* of concepts within a given hierarchy, that is the relation semantics of "north" depend on whether it is contrasted to either "north-north-west" or "west". Our model thus has to provide means for interpreting relation terms with spatial relations. Last but not least, environmental features and actions referred to in the place description introduce *scale* information, for example "near" depends on scale information implied by mode of transportation (e.g., driving by car vs. walking) or a reference object (e.g., "near Egypt" vs. "near the signpost").

In consideration of the requirements discussed, we propose to tackle interpretation of place description as a constrained optimisation problem using discrete and continuous variables. To circumvent the long-standing (qualitative) spatial reasoning problem to reason about several types of relations in a joint manner, we extend an approach based on Boolean combinations of linear inequalities (Kreutzmann and Wolter 2014), called AND/OR LPs,[1] which already can capture diverse relations within a single formal framework. This model allows us to represent variables that are either Booleans (to compose alternative interpretations) or numerical, for example to represent scale and spatial locations. Moreover, we are able to state dependencies among the distinct valuations of context variables by modelling a constraint which is part of an overall conjunctive formula representing the place description. We propose to employ linear inequalities to model partial satisfiability of a vague relation. Reasoning with AND/OR LPs employs linear programming (LP) techniques, but does not make use of the optimisation criterion LPs have originally been designed for. By modelling vagueness as optimisation criterion using linear inequalities, one can optimise for least overall vagueness in a decision taken during reasoning.

The overall model we propose is based on the approach by Vasardani et al. (2013) which starts by considering a place description as a set of prepositional phrases (PPs) obtained by means of parsing natural language. We extend the model by introducing an explicit step to interpret a natural language relation as an AND/OR LP spatial relation, denoted by function φ. Also, we introduce context variables as non-spatial variables. The objective to identify a place p, given a set of PPs can then be approached as a constraint reasoning problem which is involved with a single constraint formula

$$\bigwedge_{i=1}^{N} \varphi(\mathrm{PP}_i)(o_{i,1}, \dots, o_{i,n_i}).$$

Variables $o_{i,j}, j = 1, \dots, n_i$ stands for entities referred to in PP_i. In this formula we use $\bigwedge$ as a conjunction operator that combines distinct pieces of information obtained

---

[1] LP stands for Linear Programming, a system of linear inequalities.

from a single PP. The operator is related to logic "and" but differs in terms of aligning contextual interpretation among distinct PPs and being a conjunct of vague relations that are partially satisfied.

5 Summary and Conclusions

We review contextual influence to interpretation of place descriptions and propose a computational model based on constrained optimisation to match place descriptions to information provided by a GIS in an automated manner. To arrive at an implementation we have to commit ourselves to several working definitions of ill-defined concepts such as concept or place. We have chosen a bottom-up approach of identifying a set of context variables. In future work we will specify details of the model and study the contribution of context variables to obtain reasonable interpretations. Last but not least, efficiency of the computational model will be investigated.

Acknowledgements This work is supported by the German Research Foundation (DFG) in context of the priority program 1894 "Volunteered Geographic Information". Financial support is gratefully acknowledged.

References

Adams B, McKenzie G (2013) Inferring thematic places from spatially referenced natural language descriptions. In: Sui D, Elwood S, Goodchild M (eds) Crowdsourcing geographic knowledge, Volunteered geographic information (VGI) in theory and practice. Springer, pp 201–221

Bazire M, Brézillon P (2005) Understanding context before using it. Proc Context 2005:29–40

Bennett B, Agarwal P (2007) Semantic categories underlying the meaning of 'place'. In: Proceedings of the 8th international conference on spatial information theory, pp 78–95

Davies C, Holt I, Green J, Harding J, Diamond L (2009) User needs and implications for modelling vague named places. Spat Cogn Comput 9(3):174–194. Special Issue on Computational Models of Place

Freksa C, Klippel A, Winter S (2007) A cognitive perspective on spatial context. In: Cohn AG, Freksa C, Nebel B (eds) Spatial cognition: specialization and integration. Internationales Begegnungs- und Forschungszentrum für Informatik (IBFI), Schloss Dagstuhl, Germany, Dagstuhl, Germany, no. 05491 in Dagstuhl Seminar Proceedings. http://drops.dagstuhl.de/opus/volltexte/2007/980

Kreutzmann A, Wolter D (2014) Qualitative spatial and temporal reasoning with AND/OR linear programming. In: Proceedings of 21st European conference on artificial intelligence (ECAI), pp 495–500

Vasardani M, Winter S, Richter KF (2013) Locating place names from place descriptions. Int J Geogr Inf Sci 27(12):2509–2532

Winter S, Truelove M (2010) Talking about place where it matters. In: Frank A, Mark D (eds) Las Navas 20th anniversary meeting on cognitive and linguistic aspects of geographic space. Springer, Las Navas, Spain, pp 121–139

Modeling Spatio-Temporal Variations for the Language-Driven Development of Simulated Environment Generators

Liqun Wu, Thomas Brinkhoff and Axel Hahn

Abstract We present in this contribution a language to describe environmental changes in spatial-aware simulations from the requirement view. For this language, we model the expression of spatio-temporal change patterns named *variation* with explicit components. Then, we analyze three categories of *variations* based on the expression model. This step determines possible types of *variations* that may appear in descriptions of environments. After that, mapping principles for mapping these types to software artifacts are introduced. Our work provides support to prototyping simulated environment generators from conceptual descriptions. By processing a description in our language, a skeleton of software which can produce the described *variations* is generated.

Keywords Executable description language · Language-driven development · Simulated spatial-temporal environment

1 Introduction

In explanatory simulations, environmental phenomena that have high impacts on systems of interest are generated as controlled conditions. They provide appropriate stimuli in various simulation scenarios. (Klügl et al. 2005). System models need to run in different simulated environments. An example is the verification and the val-

L. Wu(✉) · A. Hahn
Department für Informatik, Carl von Ossietzky Universitt Oldenburg,
Oldenburg, Germany
e-mail: liqun.wu@uni-oldenburg.de

A. Hahn
e-mail: axel@uni-oldenburg.de; hahn@uni-oldenburg.de

T. Brinkhoff
Institut für Angewandte Photogrammetrie und Geoinformatik,
Jade Hochschule, Germany
e-mail: thomas.brinkhoff@jade-hs.de

© Springer International Publishing AG 2018
P. Fogliaroni et al. (eds.), *Proceedings of Workshops and Posters at the 13th International Conference on Spatial Information Theory (COSIT 2017)*, Lecture Notes in Geoinformation and Cartography, https://doi.org/10.1007/978-3-319-63946-8_28

idation of a controller for autonomous vessels via a simulator. The controller has to be tested in various scenarios with different tidal, current and weather conditions. Spatio-temporal change patterns of these phenomena generated by software components have to match the requirements of scenario modelers.

However, a gap exists between desired change patterns in the view of modelers, and capabilities of software that generate such environments in the view of developers. A description of a required environment usually has less details than an implementable software model. Besides, scenario modelers express change patterns from an overview perspective of an observer. The result description can exist at one time instant. Yet in a software, some of these patterns may be embedded in static software structures and some other emerge in execution processes.

Aiming at bridging this gap, we model a change pattern named as *variation* at first. Based on the model, we analyze the types of *variations* that human may perceive and express in spatio-temporal environments. We then continue by analyzing the nature of these variations per type to determine their necessary software counterparts that can produce them. Our work provides a description language of conceptual spatio-temporal variations in simulated environments from the requirement view, which are explicitly mapped to artifacts of generator software. It raises the abstraction level of domain models for simulated environment generators and thus enables active participation of scenario modelers in the environment generator software development, following the language-driven development paradigm (Sierra 2014).

2   Modeling Spatio-Temporal Variation Types

A spatio-temporal phenomenon is known to have spatial, temporal and attributive properties (Guttag and Horning 1978) which can be represented by values in these three types of domain spaces. A *variation* is defined as a relation between the values in two domain spaces of a phenomenal representation, since the value change in one domain space (as the *variant*) has to be observed by varying the values in another (as the *variable*) in a controlled way. The *variations* are associated to a structural description that describes standalone members of the environment. This strcutural description is based on widely accepted spatial conceptualizations (Couclelis 1992; Worboys 1995; Galton 2004; Yuan 1997). It uses a zoom view to enable the expression of the *variations* in different scale of perception: level 1 captures the occurrence of phenomena in an environment, level 2 views a phenomenon as a distinct object and level 3 captures spatial heterogeneity within the object extent that implies a spatial field view.

The relation forms of *variations* that can be viewed as functions are individual-specific. However, *variations* of common natures can be produced by the same types of artifacts in corresponding software and can be viewed as the same type. These *variations* types are captured by analyzing the rationale of all possible combinations of following aspects to construct a *variation*: the domain space type of the *variant*, the domain space type of the *variable* and the level of description. Further, we identify three categories of *variations* listed as follows.

- **Independent variations**: this category of *variations* are relations between values of the same phenomenon in two domain spaces. They describe self-evolving change patterns that can be computed purely per initial setting. In other words, these *variations* are controlled stimuli that can be produced without knowing the current state of the system or of other phenomenon. An example is a defined moving trajectory of an environmental entity. These *variations* are the basic conceptual variability of a spatial phenomenon in simulated environments. *Variation* types that a specific phenomenon may have, depend on the required dimensions of the simulated environment and required scale of the phenomenon in the simulation.
- **Inter-phenomenon variations**: *variants* of this category are spatial or attributive values of one phenomenon. *Variables* are spatial or attributive values of another phenomenon. A type of *variation* in this category is meaningful, only when the type of its *variant* is a valid type as the *variant* of the independent *variation* types at the corresponding description level. A simplified example is that the spatial occurence (i.e., the spatial values) of ice depends on the temperature of the air. Such *variations* lead to links between computations of two phenomena in result software programs.
- **Derivative variations**: *variants* of this category are the relation form of another *variation*, i.e., the mode of the behavior. Such *variation* could be the result of inter-phenomena influences or reactive influences to an environmental phenomenon from the system of interest. An example is the moving speed of an iceberg which is altered by the force of current. Different from the previous two categories, phenomena with such *variations* are not controlled stimuli, but rather should be modeled in the same manner as the system of interest, e.g., as agent models in agent-based simulations. Thus, our language only provides vocabularies that create access points to such model from environment generators.

3   Prototyping Environment Generators by Description

We use UML terminology for the language specification to provide an implementation-independent mapping from description items to software artifacts. The mappings are defined at M2 meta level of the Meta Object Facility (MOF) (Object Management Group 2016). Concepts in the language metamodel are defined as UML stereotypes. This metamodel provides the abstract syntax of our language. The corresponding software artifacts are also specified in UML. Some of the mapping principles are listed as follows.

- *Variations* are mapped to *Actions* in *Activity* flows for computing attributive values at given spatio-temporal locations.

- *Activity* flows derived from *Variations* are associated to corresponding *Class* derived from the structural description, one *Class* for each phenomemon as well as for each attributive field of this phenomenon.
- An *Activity* flow starts from *variations* at higher description levels, and *Actions* derived from *variations* with time as their type of *variables* are placed before others in the flow.
- *Variations* with no temporal domain space involved are additionally mapped to member *Classes* in their associated *Classes*.

A description of an environment conforming to the description metamodel corresponds to the M1 model level of MOF (Object Management Group 2016). In an implementation of our language, these UML concepts can be replaced by compatible codes. Thus, an execution engine that implements the language specification can be a model-to-model translator or a code generator to support automatic prototyping of generator software from descriptions. It transforms an environment description conforming to the specification to a generator skeleton, following the mapping rules defined by the language.

4 An Example of Code Generation

Figure 1 shows a very simple example. The upper part of this figure shows a piece of an environmental description in our language implemented with textual notations (i.e., its concrete syntax). It describes a phenomenon "Wind" that has a spatial boundary. This "Wind" has one possible *variation* represented by the notation "Dynamics". It corresponds to the type of *variation* that represents the value change in an attributive domain space (i.e., as its *variant*) related to the time (i.e., as its *variable*).

The lower part of Fig. 1 shows the generated *Activity* from the above description in the form of a Java method "computeSpeed". It provides a skeleton to compute the value of the attribute "Speed" at a spatio-temporal location. At the object level of the description, the presence of "extent" results a method to evaluate that if the location is within the influencial area of the "Wind". At the field level, the presence of "Dynamics" results a method to compute the corresponding *variation*. Other parts of the generated code is omitted due to space limitation. The generation result is not optimized but only serves for illustration purpose. Further, the concepts in the language model can serve as variability points and can be extended by application packages in different implementations.

5 Conclusions

This contribution utilizes spatio-temporal conceptualizations in the language-driven software development and extends them with expressions of spatio-temporal variations. It provides linguistic support for prototyping simulated environment genera-

```
SpatialElement Wind
              extent Extent
              AttributiveField Speed{
Dynamics Increase} //Variation (A,T,field)
```

```
package windExample;

import lib.Point;

public class Computation {
    Boolean overlap;
    public Wind w;

    public float computeSpeed(Wind w, Point location){
        Boolean inside = overlap(location, w.extent);
        if(inside)
            {
        w.speed = increase(w.speed);
        return w.speed.value;
            }
        else return 0;
    }

    public Boolean overlap(Point p, Extent e){
        Boolean overlap = null;
        //methods to implement
        return overlap;
        }

    Speed increase(Speed speed){
        //method to implement
        return speed;
        }
}
```

Fig. 1 Generation of the code skeleton from a description

tors from descriptions in the requirement view. The models are expressed in UML terminology. Model-driven development technologies such as code generation tools can be easily adapted for the language implementation.

In addition, our work also identifies domain-specific commonality and variability points to guide the model-level code reuse for computing the state of spatio-temporal entities. This provides interfaces for software developers to implement and modify generators with compatible output formats, or to invoke code libraries following same spatial conceptualization at these variability points.

Acknowledgements The presented work is supported by the graduated school Safe Automation of Maritime Systems (SAMS), funded by the Ministry of Science and Culture of Lower Saxony, Germany.

References

Couclelis H (1992) People manipulate objects (but cultivate fields): beyond the raster-vector debate in GIS. In: Frank AU, Campari I, Formentini U (eds) Theories and methods of spatio-temporal reasoning in geographic space, vol 639. Springer, Berlin, pp 65–77

Galton A (2004) Fields and objects in space, time, and space-time. Spat Cogn Comput 4(1):39–68

Guttag J, Horning J (1978) The algebraic specification of abstract data types. Acta Informatica 10:27–52

Klügl F, Fehler M, Herrler R (2005) About the role of the environment in multi-agent simulations. In: Weyns D, van Dyke Parunak H, Michel F (eds) Environments for Multi-Agent Systems. E4MAS 2004. LNCS, vol 3374, pp 127–149. Springer, Heidelberg

Object Management Group: OMG Meta Object Facility (MOF) Core Specification (2016)

Sierra J (2014) Language-driven software development. In: SLATE 2014. pp 3–12. Bragança, Portugal

Worboys MF (1995) GIS: a computing perspective. Taylor & Francis, London

Yuan M (1997) Knowledge acquisition for building wildfire representation in geographic information systems. Int J Geogr Informat Syst 11(8):723–745

Part II
Rethinking Wayfinding Support Systems

Rethinking Wayfinding Support Systems—Introduction

**Jakub Krukar, Angela Schwering, Heinrich Löwen,
Marcelo De Lima Galvao and Vanessa Joy Anacta**

Personal GPS-based navigation devices have firmly substituted the use of paper maps, public signage, and occasional advice from local residents. The time and effort saved through this shift is unquestionable. However, computerised wayfinding support yielded problems uncommon before. Users fail to remember a route followed repeatedly. Navigators face complete disorientation when the device suddenly malfunctions. Tourists do not recognise scenes from the routes they have travelled. And yet, users trust this new technology even when they are being led into life-threatening situations amid common-sense knowledge suggesting otherwise.

Computerised wayfinding support relies on offloading cognitive activity onto an external aid and delivering the minimum of required information at the right place and time. Self-localisation and spatial updating are skills intrinsically involved in solving wayfinding tasks but are not required by the GPS-based devices. As a result, support provided by computerised wayfinding assistance is incompatible with the natural ways in which humans explore, learn, and interact with new spaces.

This workshop aims at exploring the possibilities for embedding wayfinding support systems in human everyday experiences. In order to achieve that, functional

J. Krukar (✉) · A. Schwering · H. Löwen · M. De Lima Galvao · V.J. Anacta
Spatial Intelligence Lab, University of Münster, Münster, Germany
e-mail: krukar@uni-muenster.de

A. Schwering
e-mail: schwering@uni-muenster.de

H. Löwen
e-mail: loewen.heinrich@uni-muenster.de

M. De Lima Galvao
e-mail: galvao.marcelo@uni-muenster.de

V.J. Anacta
e-mail: v.anacta@uni-muenster.de

© Springer International Publishing AG 2018
P. Fogliaroni et al. (eds.), *Proceedings of Workshops and Posters at the 13th
International Conference on Spatial Information Theory (COSIT 2017)*, Lecture Notes
in Geoinformation and Cartography, https://doi.org/10.1007/978-3-319-63946-8_29

features of wayfinding support need broadening: guiding the user to efficiently and successfully navigate their body is barely the first necessary requirement. The main variables distinguishing between more and less successful systems will be their compatibility with spontaneous cognitive strategies, and integration with context-dependent tasks tied to wayfinding.

This also requires rethinking the performance metrics used to evaluate such systems. Users' efficiency, speed, and number of mistakes are important, but not without considering what they learn, how they incorporate new information into their spatial knowledge of varying certainty, how flexibly they are able to use this knowledge in alternative contexts, and how independent of the wayfinding aid they become as a result. Of crucial importance is the task-related context in which navigation is embedded, since rarely (if ever) navigation is performed for navigation's sole sake.

We hope that the following papers accepted to the workshop proceedings are indicative of the growing appreciation of the aforementioned issues. They represent different methods with which researchers explore in-the-wild wayfinding behaviour and spatial cognition of the users, as well as the focus shifting onto variables indicating learning that happens with support of the wayfinding aids.

Four papers were accepted. Credé and Fabrikant's contribution explores the benefit of visualising global landmarks in a wayfinding aid, focusing on the working memory of the navigator. Bauer and colleagues studied the use of orientation information in sketchmaps that were drawn after an indoor navigation task. Ooms and Van de Weghe describe a crowdsourced set of indoor route instructions and an approach with which they aim to automatise the generation of human-like path descriptions. Closing this book section, a contribution by Wang and colleagues focuses on wayfinding aids that already exist in the environment but remain omitted by the visualisations and instructions produced by personalised technology. We hope the reader will find this selection informative and inspiring.

Supporting Orientation During Indoor and Outdoor Navigation

Christina Bauer, Manuel Müller, Bernd Ludwig and Chen Zhang

Abstract Supporting orientation is one of the major challenges when designing pedestrian navigation systems. Especially for indoor areas, it is still an open question, which landmarks can be used to support the cognitive process of orientation. Furthermore, it has to be analyzed whether the referenced landmark type has to be adapted to different instruction types like verbal and visual displays. In order to address these questions we conducted a study with 132 participants who had to navigate different routes and give verbal instructions supplemented by drawing sketch maps. Our results show that orientation information is important in indoor areas, too. Moreover, the instruction type significantly influences the chosen orientation information. We therefore argue that it is important to incorporate orientation information in future indoor navigation systems and adapt the information based on the instruction type.

Keywords Pedestrian navigation · Orientation · Sketch maps

1 Motivation

Several indoor navigation prototypes were developed in the past years. Nevertheless, navigation interfaces for indoor environments are still less explored compared to outdoor environments (Huang and Gartner 2010). Common navigation prototypes rely on different depiction ranging from augmented or virtual reality interfaces to plain

C. Bauer (✉) · M. Müller · B. Ludwig · C. Zhang
Chair of Information Science, Universität Regensburg Universitätsstraße 31,
93053 Regensburg, Germany
e-mail: christina2.bauer@ur.de

M. Müller
e-mail: manuel-tonio.mueller@ur.de

B. Ludwig
e-mail: bernd.ludwig@ur.de

C. Zhang
e-mail: chen.zhang@stud.uni-regensburg.de

© Springer International Publishing AG 2018
P. Fogliaroni et al. (eds.), *Proceedings of Workshops and Posters at the 13th International Conference on Spatial Information Theory (COSIT 2017)*, Lecture Notes in Geoinformation and Cartography, https://doi.org/10.1007/978-3-319-63946-8_30

map depictions with different levels of complexity (for an overview and discussion of possible depictions see e.g. Butz et al. (2001)). All in all, map depiction in combination with text instructions are the most popular form of communicating route instruction (Huang and Gartner 2010). Moreover, the majority of these systems uses turn-by-turn instructions to guide the wayfinder.

Nevertheless, landmarks, and especially landmarks that can be used to maintain orientation during a complex navigation task are missing in most prototypes (Anacta et al. 2016). Orientation during navigation implies that human beings are able to identify their own position in relation to the objects in the environment and their potential destination in order to reach a specific point (Golledge 1999). Landmarks can help to support this cognitive process. At decision points, landmarks are required to maintain orientation (Michon and Denis 2001). On the other hand, these salient objects can also be used to support (global) orientation at non decision points (Anacta et al. 2016).

The study presented in this paper addresses the research question which landmarks are chosen by wayfinders to support orientation at non decision points while navigating a complex indoor/outdoor area. Therefore, we conducted a study in order to gain a deeper understanding how human beings would describe a route through this environment and which landmarks are used to explain these routes. The participants had very good spatial knowledge of the test area and had to give verbal route instructions while navigating. Moreover, they had to draw a sketch map after the task in order to collect visual route instructions. We analyzed the gained data with two research questions in mind. First, do participants use different landmark types in verbal route instructions compared to their sketch maps? Second, are different landmark types used for indoor areas compared to outdoor areas? Consequently, we are expanding the work and ideas of Anacta et al. (2016) for indoor environments.

The remainder of this paper is structured as follows. First of all, we report on the related work concerning landmark classification, which was used to annotate our data. In the following section our study and the annotation process is described in detail and the results are presented. Finally, the implications of our findings and ongoing work is discussed.

2 Landmark Classification

Basically, the rough subdivision into global and local landmarks is a frequently used classification (Winter et al. 2008). Local landmarks can be located at decision points or along the route, whereas global landmarks are located off the route and are therefore not necessarily immediately visible. Moreover, decision points can be subdivided in potential choice points and route choice points where a turn of direction has to be made (Lovelace et al. 1999). A more detailed framework for landmark classification is proposed in Schwering et al. (2013) and extended in Anacta et al. (2016). The authors additionally distinguish point-like and regional landmarks. They introduce orientation landmarks and subdivide the categorization depending on whether the instruction includes a turning movement or not. The authors found out that more than half of the instructions people give support orientation. Only about a quarter of the instructions are related to a turn information.

We based our annotation on the landmark categories described in Schwering et al. (2013) and Anacta et al. (2016) but simplified them to the following.

- OGL: orientation using global landmark (off the route)
- OLL: orientation using local landmark (at the route)
- NTLL: non turning movment using local landmark (at potential decision point)
- TLL: turning movement using local landmark (at decision point)

Consequently, our annotation focuses on orientation landmarks, which consist of OGL, OLL, and NTLL, as they do not imply a turning action (Anacta et al. 2016). Moreover, we also counted the amount of TLL, as this type is the most prominent landmark category in turn-by-turn instructions.

3 Study

In the next subsections the experimental set-up, the annotation process and the results are described.

3.1 Test Environment

The study took place at the University of Regensburg, which is a large-scale campus covering an area of approximately $0.5 \, \text{km}^2$. It consists of several partly connected buildings which can be reached using either indoor paths or traversing the outdoor campus area (see Fig. 1, left). The buildings consist of several levels with office rooms, lecture halls and public places like cafeterias and shops. The whole campus is already modeled in order to compute routes through the different buildings. An example of the complexity of a building model is shown in Fig. 1 (right).

3.2 Participants

Students of an undergraduate course were instructed to recruit test users for the study described in this paper. The recruitment of the participants was a course requirement. This resulted in a test sample of 132 participants. 81 male and 51 female persons participated in the study. All of the participants were students at the University with different courses of study and have been studying for at least one semester (mean = 5.9) . Their mean age was 23.1 years with a standard deviation of 3.0 (range:19–32). All in all, participants had expert knowledge of the area and visit the place on a regular basis.

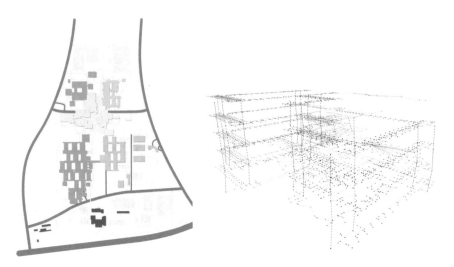

Fig. 1 Overview of the test area (*left*) and graph model of an exemplary building (*right*)

3.3 Procedure

The study consisted of three parts. Before the experiment, participants could freely chose a route on the campus. One restriction was that it had to go through at least two areas of the university (including the buildings and the outdoor area) to ensure sufficient complexity of the collected data. Moreover, participants had to confirm that they are very familiar with the route. The participants were informed beforehand that the route choice would be restricted.

After filling in a form concerning their demographic data, test persons navigated the chosen route accompanied by two test supervisors. The latter meanwhile noted the exact course of the route in architectural floor plans that were not visible to the participants. The data collected during the first part of the experiment was used to analyze whether participants deviate from the shortest route, if they are free to choose their own preferred route. Moreover, we analyzed which factors like the amount of indoor and outdoor path or the number of stairs that have to be taken influence this decision (see Müller et al. 2017).

Afterwards, participant were asked to walk back the same route, this time giving verbal route instructions as if they would explain the route to a stranger. This "thinking aloud" method was already proposed by Sefelin et al. (2005) in order to gain a set of landmarks that can be used to guide a person. The verbal instructions were noted and the position of the participant and the referenced objects were marked in the architectural plans.

Choosing a route and generating route instructions can be cognitively demanding tasks. We divided the experiment into two parts to separate these spatial reasoning tasks in order to unburden the participants during the experiment.

As a last step, the participants were asked to draw a sketch map of the route they have chosen on a A4 sized paper. They were free to add any information they liked. The instruction was to draw a map that could be used by strangers to find the way. No time limit was given.

3.4 Annotation

For the results and considerations presented in this paper, the data of phase two and three of the experiment was used (see Sect. 3.3). First of all, the sketch maps and the verbal route instructions had to be annotated. Figure 2 shows an example of a sketch map. Gray-shaded areas indicate indoor environments. Circles represent global landmarks (OGL). In the exemplary sketch map these are the lecture hall "Audimax" (OGL_1), a pizzeria (OGL_2), a lake (OGL_3), and the computer

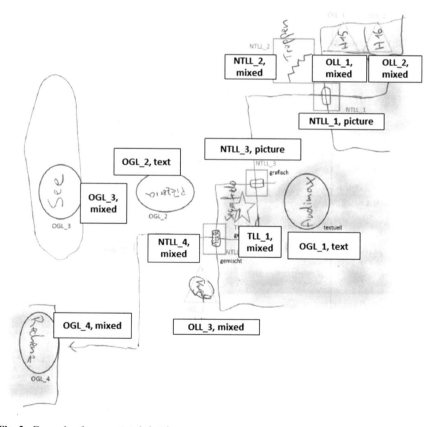

Fig. 2 Example of an annotated sketch map

center (OGL_4). Squares indicate local landmarks without turning movements like doors (NTLL_1, NTLL_3, NTLL_4) or stairs (NTLL_2). Triangles are local landmarks used for orientation (like lecture halls OLL_1 + OLL_2 and artwork OLL_3). Finally, stars represent landmarks where a change of direction is needed (a cafeteria TLL_1). Moreover, we annotated whether the landmark is depicted graphically, using a text label, or whether these techniques are mixed. These different drawing methods are not discussed in this paper, but could be topic of future work. Our classification is only based on the position of the landmark relative to the route. Therefore, the annotation did not differ for indoor and outdoor landmarks. The same annotation process was applied to the transcribed verbal route instruction. The following example is translated from German to English.

- (Outdoor) Go straight ahead up the stairs (NTLL).
- (Outdoor) Enter the building.
- (Indoor) Pass the lecture hall "Audimax" (OGL) and go to the toilets (OLL).
- (Indoor) Go left at the cafeteria (TLL).
- (Outdoor) Pass the pizzeria (OGL), the "Kugel"[1] (OLL) and the library (OGL) are on your left.
- (Outdoor) Go left in front of the lake (TLL).
- (Indoor) Enter the building through the door (NTLL).

3.5   Results

The mean route length of the chosen routes was 350.7 m (SD = 120.6), with a mean indoor part of 206.8 m (SD = 112.8) and a mean outdoor part of 143.9 m (SD = 139.4) in length. We formulated different hypotheses according to our research question if the navigation area (H_1 and H_2) and the instruction type (H_3 and H_4) influence the chosen landmarks.

- H_1: In the sketch maps, the amount of orientation landmarks differs for indoor and outdoor environments.
- H_2: In the verbal instruction, the amount of orientation landmarks differs for indoor and outdoor environments.
- H_3: In indoor areas, the amount of orientation landmarks differs for the verbal route instructions and sketch maps.
- H_4: In outdoor areas, the amount of orientation landmarks differs for the verbal route instructions and sketch maps.

Orientation landmarks are defined as OLL (orientation local landmark), OGL (orientation global landmark), NTLL (non turning local landmark), therefore landmarks at decision points were not taken into account. Each landmark type was analyzed separately. To calculate the differences between the amount of indoor and

---

[1]The "Kugel" is a large and well-known work of art at the university.

Table 1 Mean distribution of landmark type in % (SM = sketch map; VI = verbal instruction)

	Indoor (SM)	Outdoor (SM)	Indoor (VI)	Outdoor (VI)
OLL	35.1	25.2	28.5	30.5
OGL	7.9	19.1	1.2	7.8
NTLL	44.0	36.0	53.5	38.4
TLL	13.0	19.7	16.8	23.3

outdoor landmarks, we calculated the relative amount of landmarks, since the routes contained more indoor parts for the majority of the participants. This means, we calculated the percentage of chosen OLL, OGL, NTLL, and TLL separately for indoor and outdoor areas and different instruction type (see Table 1 for the descriptive statistics).

None of the data was normally distributed, we therefore conducted a Wilcoxon signed-rank test. The results show that concerning H_1, for all three landmark types the nullhypothesis can be rejected ($p < 0.01$). For indoor areas, significantly more OLL and NTLL are drawn, whereas less global landmarks are referenced.

In the verbal route descriptions the same significant results were found for global landmarks and NTLL, therefore H_2 can be assumed for these landmark types.

In outdoor areas, global landmarks are drawn more often in sketch maps than used in route instructions ($p < 0.001$). On the other hand, OLL and OGL differ significantly in indoor environments in both conditions (SM and VI). Local landmarks (OLL) and global Landmarks (OGL) are drawn more often in sketch maps ($p < 0.001$).

The most frequent landmark type for both instruction types is a local landmark at a decision point where no turn of direction is implied. Local landmarks at turning points only make up about 20% of the landmarks. This clearly indicates the need for orientation landmarks for indoor and outdoor navigation.

4 Conclusion and Future Work

To summarize the results, we found out that global landmarks are significantly less relevant in indoor environments for both, sketch maps and verbal instructions. Moreover, sketch maps contain more global landmarks compared to verbal instructions. The second finding is in line with the results for outdoor areas reported in Anacta et al. (2016).

Global landmarks can be used to give an overview of the wayfinding area but have to be well-known and ideally highly visible. These objects are rather rare in indoor environments. When navigating in an indoor area, it seem pointless to refer to a global landmark that is located outdoors. Furthermore, local landmarks at potential decision points are mentioned more often in indoor environments. Our results give

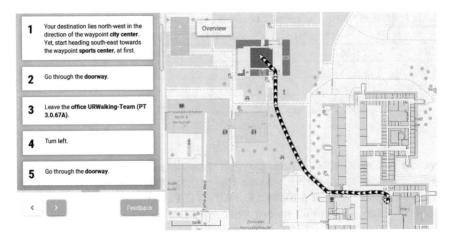

1	Your destination lies north-west in the direction of the waypoint **city center**. Yet, start heading south-east towards the waypoint **sports center**, at first.
2	Go through the **doorway**.
3	Leave the **office URWalking-Team (PT 3.0.67A)**.
4	Turn left.
5	Go through the **doorway**.

Fig. 3 Screenshot of our navigation system incorporating different landmark types

hints that map material should reference global landmarks to maintain orientation, whereas textual instructions should rely on local landmarks for orientation.

In the present study only participants who were very familiar with the test environment were part of the test sample. The related work shows that this factor influences the decision which landmarks and how many objects are chosen (Lovelace et al. 1999). It is part of future work to examine whether the chosen landmarks of our experiment are helpful for unfamiliar wayfinders.

All in all, the results show that orientation landmarks are an important landmark type for both, indoor and outdoor areas. Nevertheless, the type of landmark preferred by the users differs depending on the area and type of instruction. Therefore, our future work will focus on how to integrate the different landmarks for different environments and different instruction types like text and map-based descriptions. An example of our preliminary realization of these concepts is given in Fig. 3. Our next step will be to graphically highlight global landmarks and to analyze at which route points orientation landmarks should be included.

References

Anacta VJA, Schwering A, Li R, Muenzer S (2016) Orientation information in wayfinding instructions: evidences from human verbal and visual instructions. Geo J, pp 1–17. doi:10.1007/s10708-016-9703-5

Butz A, Baus J, Krüger A, Lohse M (2001) A hybrid indoor navigation system. In: Proceedings of the 6th international conference on intelligent user interfaces, IUI '01, pp 25–32. ACM, New York, NY, USA. doi:10.1145/359784.359832

Golledge RG (1999) Human wayfinding and cognitive maps. In: Wayfinding behavior: cognitive mapping and other spatial processes, pp 5–45

Huang H, Gartner G (2010) A survey of mobile indoor navigation systems, Springer, Berlin, Heidelberg, pp 305–319

Lovelace KL, Hegarty M, Montello DR (1999) Elements of good route directions in familiar and unfamiliar environments, Springer, Berlin, Heidelberg, pp 65–82

Michon PE, Denis M(2001) When and why are visual landmarks used in giving directions? Springer, Berlin, Heidelberg, pp 292–305

Müller M, Ohm C, Schwappach F, Ludwig B (2017) The path of least resistance. KI - Künstliche Intelligenz 31(2):125–134. doi:10.1007/s13218-016-0472-6

Schwering A., Li R, Anacta VJA (2013) Orientation information in different forms of route instructions. In: Short paper proceedings of the 16th AGILE conference on geographic information science, Leuven, Belgium

Sefelin R, Bechinie M, Müller R, Seibert-Giller V, Messner P, Tscheligi M (2005) Landmarks: Yes; but which?: Five methods to select optimal landmarks for a landmark- and speech-based guiding system. In: Proceedings of the 7th international conference on human computer interaction with mobile devices &Amp; Services, MobileHCI '05, pp 287–290. ACM, New York, NY, USA. doi:10.1145/1085777.1085834

Winter S, Tomko M, Elias B, Sester M (2008) Landmark hierarchies in context. Environ Plann B Plann Des 35(3):381–398 doi:10.1068/b33106

Let's Put the Skyscrapers on the Display—Decoupling Spatial Learning from Working Memory

Sascha Credé and Sara Irina Fabrikant

Abstract Driven by the increasing evidence of negative effects on spatial learning when using automated navigation systems, the academic and private sectors have become interested in strategies to improve users spatial awareness when relying on such devices during navigation. Global landmarks have been found to support orientation when using navigation devices, particularly for individuals with low spatial abilities. However, there is no empirical evidence on how global landmarks differ from local landmarks when processed and stored in working memory during navigation, and how cognitive demands from concurrent tasks might interfere with learning. In the proposed study, we aim to contribute to the understanding of these processes by investigating how local and global landmarks are mentally processed and integrated into a coherent cognitive map. We observe participants solving assisted navigation tasks with different levels of working memory load in a virtual urban environment. Insights into the efficiency of learning local and global landmarks under high concurrent task demands will be beneficial to develop future design guidelines that improve spatial learning in general, and more particularly in stressful navigation situations.

Keywords Spatial cognition · Navigation · Working memory · Spatial knowledge · Global and local landmarks

1 Introduction

Technological advances in mobile information technology and humancomputer interaction has significantly increased the multi-tasking during navigation. During assisted navigation users typically need to perform a variety of concurrent tasks.

S. Credé (✉) · S.I. Fabrikant
Department of Geography, Geographic Information Visualization and Analysis,
University of Zurich, Zurich, Switzerland
e-mail: sascha.crede@geo.uzh.ch

S.I. Fabrikant
e-mail: sara.fabrikant@geo.uzh.ch

© Springer International Publishing AG 2018
P. Fogliaroni et al. (eds.), *Proceedings of Workshops and Posters at the 13th International Conference on Spatial Information Theory (COSIT 2017)*, Lecture Notes in Geoinformation and Cartography, https://doi.org/10.1007/978-3-319-63946-8_31

163

Beside reading and understanding navigation instructions of an employed system, mobile users also require to match this information with their perceptions of the physical surroundings, while having to monitor proximal pedestrians or physical objects to avoid collisions, just to name a couple of relevant concurrent tasks involved in assisted navigation and wayfinding. Given the complexity of the parallel cognitive processing involved during assisted navigation, it can be considered as an intrinsically demanding task, and this task often imposes elevated workloads on users working memory. Importantly, prior findings indicate that under high working memory load, spatial memory performance is impaired when learning from map representations (Coluccia et al. 2007) and when learning from direct experiences during a navigation task (Garden et al. 2002). Based on prior work in cognitive psychology we argue why global landmarks should be learned more efficiently, compared to local landmarks, specifically in situations of high working memory load. Building on this theoretical line of thought, we propose an ongoing empirical navigation study in a virtual urban environment that aims to investigate our contention.

1.1 Working Memory and Spatial Learning

Working memory is conceptualized as the cognitive structures responsible for maintaining, updating, and manipulating information over time (Repovš and Baddeley 2006). It has been appraised as a central factor in learning about the spatial configuration of environmental spaces from direct experience (Hegarty et al. 2006). Unfortunately, the limited resources of working memory need to be shared out dynamically between the different task demands. Cognitive performance in a primary task can be drastically impaired by a concurrent task which occupies attention. Prior research in psychology could show that with increasing temporal intervals between storage and recall cognitive performance decreased (Towse and Hitch 1995). In the so called time-based resource sharing model of working memory (TBRS), researchers argue that the refreshing of information in working memory over time requires attentional resources. Any concurrent activity that occupies attention for protracted and frequent periods of time prevents the refreshing of decaying memory (Barrouillet and Camos 2012).

We argue that temporal constraints and related spatial updating processes in working memory are also important aspects during the mental formation of large-scale spaces. Large-scale spaces cannot be perceived from a single viewpoint, but naturally require users to change their location and relate multiple views experienced during locomotion (Montello 1993). In this case, some type of temporal mental record is required, which has to maintain information over the course of several seconds or minutes. Likewise, when navigating with digital navigation assistants, temporal storage is important, because navigators need to match the represented (typically exocentric) spatial information displayed on the mobile map device with the enviornmental information directly perceived from an egocentric perspective (Lobben 2007).

Given that enough processing resources are available for refreshing memories, differently acquired pieces of spatial information (e.g., vista spaces, information extracted from a map) can be integrated sequentially into a more coherent mental representation in working memory. This integrated type of spatial knowledge is often referred to as survey knowledge and mentally represents the different environmental aspects within a single reference frame. It allows users to master more complicated spatial tasks such as e.g., shortcutting, pointing or distance estimations (Meilinger et al. 2008). Most likely it is survey knowledge which lacks after digital navigation system use, as contemporary navigation aids typically rely on turn-by-turn instructions, which reinforce route perspective.

How can survey knowledge be successfully formed when attentional resources are split due to activities concurrent to spatial learning during navigation? TBRS assumes that activities that frequently allocate attention for refreshing memory traces should have little impact (Barrouillet and Camos 2012). Indeed TBRS predicts that memory performance in tasks that involve both storage and processing depend on the balance between periods during which memory traces decay, because attention is occupied by concurrent processing, and those periods during which attention is available for refreshing memory traces (Barrouillet and Camos 2012).

Given negative effects of concurrent processing tasks on refreshing temporal information in working memory, we aim to investigate how split attentional resources might affect the learning of local and global landmark configurations, during navigation in a virtual urban environment.

1.2 Global Landmarks

Landmarks are important environmental components that are used to assist navigation and play a critical role in spatial knowledge acquisition. Prior research in the context of mobile navigation assistance focused mostly on landmarks for the purpose of supporting the processes of route choice and local reorientation (Beeharee and Steed 2006; Millonig and Schechtne 2007). Only recently, the highlighting of global landmarks on mobile map displays was recognized and investigated concerning its influence on spatial knowledge acquisition (Li et al. 2016, 2014). In their empirical study, (Li and colleagues 2014) took participants to different locations in an urban district where they gained spatial knowledge using a mobile navigation device. While one experimental group had mobile devices visualizing distant off-screen landmarks, the display of the control group visualized only those landmarks that fell into the mapped area. In a subsequent spatial memory task, the authors observed that individuals who reported high spatial abilities did not benefit much when learning with the visible global landmarks, compared to when only local landmarks were visible. By contrast, individuals who reported low spatial abilities had significant better memory when global landmarks were highlighted (Li et al. 2014). This evidence is especially interesting for the present study idea, because it indicates that global landmarks might be unequally beneficial, depending on the users cognitive processing capabilities.

2   Empirical Study Proposal

We propose a navigation study in a virtual urban environment, in which we aim
to observe the spatial knowledge acquisition performance of participants learning
local and global landmarks during an assisted navigation task with different levels
of working memory load requirements. Relying on evidence from prior research
that found visual spatial working memory as particularly crucial for spatial updating
performance during navigation (Garden et al. 2002) and for the mental representation
of space in a survey manner (Aginsky et al. 1997), we introduce a spatial concurrent
task.

2.1   Experiment Design

We developed a 2×2 mixed factorial design for 48 participants. Two groups of par-
ticipants (high|low workload) perform assisted navigation tasks in a virtual urban
environment. Both groups are instructed to learn the configurations of landmarks
which are highlighted in the environment (within-participant factor). Figure 1 shows
examples of both types of highlighted local (1a) and global (1b) landmarks.

Each trial will have four differently colored highlighted landmarks of the same
type (local or global). Using a virtual reality setup, we can ensure that perused local
and global landmark configurations are identical, comparable between trials, and
equally distributed within trials. While the landmark type will be the within participant factor, the introduction of a concurrent task will be realized as a between subject factor. Participants in the high workload group are asked to solve a concurrent
attention-demanding task during assisted navigation. In the spatial concurrent task,
participants need to indicate from which direction a sound is coming (e.g., left, right,
front, or back) by pressing the a key (Meilinger et al. 2008). By contrast, the control
group will do the same navigation trials without the interference of a concurrent task.
Still, to ensure equally timed learning periods, both groups need to solve the navi-

(a)   **(b)**

Fig. 1 Participants are asked to learn either **a** highlighted local landmarks along a route, or **b** highlighted global landmarks visible in the distance during navigation in a virtual urban environment

gation task within an identical, moderate time constraint. Subsequent to the learning phase, we propose to test navigators spatial knowledge with a landmark-placement task (Frankenstein 2015). Participants get a rectangular panel (30×30 cm) made out of metal and will be asked to place labeled magnets on it, representing the learned landmarks, the starting point, and the destination. Participants pick the labeled magnets one after another in a random order from an opaque bag. During the placement procedure, they are allowed to rearrange the magnet configurations until they are satisfied with their solution. The scoring of participants landmark configurations is based on their accuracy of distances and angles. By metrically evaluating landmark placement, we can compare participants landmark placement to the actual placement in the virtual environment. Metric accuracy measures shall give insights into the quality of the mental representations, and how well different spatial information pieces are integrated on a single, global reference frame.

2.2 Hypotheses

We expect local landmark learning to rely more heavily on working memory update, as these landmarks are not always seen at the same time. Instead, mental integration of local landmarks is restricted, due to limited visibility, and thus must be mentally processed sequentially, over the course of multiple navigation instances and views. Consequently, mental integration of local landmarks strongly relies on updating in working memory, which maintains the spatial information active until integration can take place. When concurrent processing activities occur during critical updating phases, we expect spatial learning performance to decline.

By contrast, global landmarks are often visible simultaneously, and can be seen from several parts of the route. We thus expect global landmarks to rely less on the update function of working memory to be successfully represented in spatial memory. Furthermore, as global landmarks are visible from many vantage points on the route, they can also be more flexibly learned during the entire course of a navigation phase. Should attention be required by concurrent processing demands for a given period during navigation, global landmark encoding can be postponed until attentional resources from working memory are freed up again. Figure 2 schematizes how the concurrent task might interfere with local landmark learning, compared to global landmarks.

In summary: we hypothesize that the constant availability of attentional resources is more essential for the learning performance of local landmarks compared to learning global landmarks. Consequently, high levels of mental workload (e.g., presence of concurrent tasks) have a pronounced negative impact on mentally representing local landmark configurations. By contrast, we expect spatial learning performance of global landmarks to remain unaffected by high levels of mental workload.

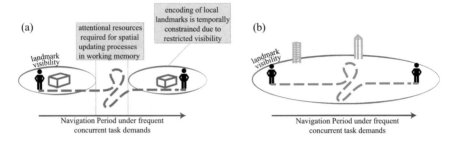

Fig. 2 Attentional resources are required to different extents when learning local and global landmark configurations. **a** In order to integrate local landmarks in a survey representation, attentional resources are particularly required during the temporal intervals without landmark visibility. The spatiotemporal intervals between single landmark exposures require users to update the spatial relationships between their own body and relevant surroundings **b** In order to represent global landmarks in a survey representation, learning is not complicated by long spatiotemporal gaps. In case of concurrent cognitive processing demands learning can be flexibly postponed until attentional resources from working memory are freed up again

3   Discussion

Prior research provides ample evidence of negative effects on pedestrian or vehicle drivers spatial knowledge acquisition when assisted by digital navigation devices during navigation and wayfinding (Parush et al. 2007; Burnett and Lee 2005). While the detrimental effects of assisted navigation is recognized and agreed upon amongst researchers from different fields (Coluccia et al. 2007; Montello 2009), there is little agreement on the underlying mechanisms which lead to the observed memory impairments. We propose to empirically investigate two possible sources that might explain spatial learning decline when using digital navigation assistance.

The first source concerns the particularities of mentally acquiring large scale spaces, particularly urban environment. Due to visibility restrictions the learning of urban spaces requires navigators to integrate information from different places over time. Consequently, their successful learning relies on the availability and allocation of attentional resources. However, different aspects environmental features might rely differently on the constant availability of attentional resources from working memory. Therefore, the present paper proposes to investigate to which extent local and global landmark learning is vulnerable to temporal declines in the availability of attentional resources.

The second source concerns the nature of assisted navigation, which is an intrinsically demanding task that often happens in a multi-tasking context, and thus additionally might result in conditions of high working memory loads. In this context, learning is not prioritized for the user, but users are mainly interested in getting the appropriate route directions to reach a desired destination efficiently and effectively. Following this line of thought, the resulting challenge for navigation assistance researchers is to find strategies how spatial learning can still be maintained with only scarce cognitive resources remaining.

We approach these two challenges by first examining whether global landmarks that play a key role in the formation of accurate mental representations require only minimal effort to be mentally processed in a high workload navigation context. The proposed experimental design investigates whether the learning of global landmarks compared to local landmarks during navigation relies less on the refreshing function of working memory. If this is so, the proper integration of global landmarks on mobile navigation displays could particularly useful to support spatial learning when users are coping with limited cognitive resources (e.g., in situations of stress).

When running the navigation experiment in a virtual urban environment we will emphasize visual information processing for spatial learning. Bodily self-movement cues (proprioceptive and vestibular information) during the learning phase and spatial updating processes will thus be limited (Waller and Greenauer 2007). This might result in small disadvantages to the learning performance of local landmarks, compared to global landmarks. Local landmarks should benefit more from body-based sensory information, as participants effectively navigate to each landmark location. However, spatial learning is a visual process for the most part, and proprioceptive and vestibular information seem to provide only very small advantages relative to visual only spatial learning (Ruddle et al. 2011).

Insights of the proposed empirical study will deepen our understanding of the mental processing of local and global landmarks and of the benefits of these landmarks for spatial learning in different navigation situations. Empirical results shall serve to inform future display designs for specific navigation contexts. For example, context and user responsive map designs on navigation systems could react to a users stress level during navigation in real time, and, for example, adaptively highlight global landmarks or other relevant information during navigation to facilitate spatial knowledge acquisition even with heightened demands on cognitive resources.

Acknowledgements This research is funded by Swiss National Science Foundation (SNF number: 156072).

References

Aginsky V, Harris C, Rensink R, Beusmans J (1997) Two strategies for learning a route in a driving simulator. J Environ Psychol 17(4):317–331

Barrouillet P, Camos V (2012) As time goes by: temporal constraints in working memory. Curr Dir Psychol Sci 21(6):413–419

Beeharee AK, Steed A (2006) A natural way finding exploiting photos in pedestrian navigation systems. In: Proceedings of the 8th conference on human-computer interaction with mobile devices and services pp 81–88

Burnett GE, Lee K (2005) The effect of vehicle navigation systems on the formation of cognitive maps. In: International conference of traffic and transport psychology

Caquard S (2015) Cartography iii: a post-representational perspective on cognitive cartography. Prog Hum Geogr 39(2):225–235

Coluccia E, Bosco A, Brandimonte MA (2007) The role of visuo-spatial working memory in map learning: new findings from a map drawing paradigm. Psychol res 71(3):359–372

Frankenstein J (2015) Human spatial representations and spatial memory retrieval. Doctoral dissertation

Garden S, Cornoldi C, Logie RH (2002) Visuo-spatial working memory in navigation. Appl cognit psychol 16(1):35–50

Hegarty M, Montello DR, Richardson AE, Ishikawa T, Lovelace K (2006) Spatial abilities at different scales: Individual differences in aptitude-test performance and spatial-layout learning. Intelligence 34(2):151–176

Li H, Corey RR, Giudice U, Giudice NA (2016) Assessment of visualization interfaces for assisting the development of multi-level cognitive maps. In International conference on augmented cognition pp 308–321

Li R, Korda A, Radtke M, Schwering A (2014) Visualising distant off- screen landmarks on mobile devices to support spatial orientation. J Locat Based Serv 8(3):166–178

Lobben AK (2007) Navigational map reading: predicting performance and identifying relative influence of map-related abilities. Ann Assoc Am geogr 97(1):64–85

Meilinger T, Knauff M, Bülthoff HH (2008) Working memory in wayfinding a dual task experiment in a virtual city. Cognit Sci 32(4):755–770

Millonig A, Schechtner K (2007) Developing landmark-based pedestrian-navigation systems. IEEE Trans Intell Transp Syst 8(1):43–49

Montello DR (1993) Scale and multiple psychologies of space. In: Spatial information theory a theoretical basis for GIS, pp 312–321

Montello DR (2009) A conceptual model of the cognitive processing of environmental distance information. In: International conference on spatial information theory pp 1–17

Parush A, Ahuvia S, Erev I (2007) Degradation in spatial knowledge acquisition when using automatic navigation systems. Spat inf theory 238–254

Repovš G, Baddeley A (2006) The multi-component model of working memory: explorations in experimental cognitive psychology. Neuroscience 139(1):5–21

Ruddle RA, Volkova E, Bülthoff HH (2011) Walking improves your cognitive map in environments that are large-scale and large in extent. ACM Trans Comput-Hum Interact (TOCHI) 18(2):10

Towse J, Hitch G (1995) Is there a relationship between task demand and storage space in tests of working memory capacity? Q J Exp Psychol 48(1):108124

Waller D, Greenauer N (2007) The role of body-based sensory information in the acquisition of enduring spatial representations. Psychol Res 71(3):322332

Finding the Right Match: Human Cognition via Indoor Route Descriptions Versus Existing Indoor Networks and Algorithms to Support Navigation

Kristien Ooms and Nico Van de Weghe

Abstract This working paper aims to compare existing approaches in indoor navigation. In this we focus on networks and algorithms to respectively model the indoor space and to calculate routes. This is compared with crowdsourced and text-based route instructions. As such the goal is to develop and evaluate an indoor solution that can generate indoor networks and route descriptions which are in line with human intuition, which is a consequence of cognition.

Keywords Indoor navigation · Cognition · Crowdsourced

1 Introduction

Humans move constantly and thus need to make decisions on how to go from one point to the next. Therefore, navigation will remain one of the fundamental problems in human cognition, wayfinding and geospatial research (Montello 2005). The term wayfinding is sometimes used as a synonym of navigation, but it is actually the goal-directed part of navigation based on decision making and planning while moving from one place to another, whereas navigation is a more broad term which also includes path planning before the actual wayfinding task starts (Montello 2005). Users' cognitive processes play a crucial role in this. These processes are diverse and depend on various influencing factors: personal characteristics (e.g. familiarity with the environment, cognitive capabilities, sense of direction, gender), purpose of the trip (e.g. commuter traffic, recreational) (Golledge 1999), environmental characteristics (e.g. indoor, outdoor), mode of locomotion (e.g. car, pedestrian, bike, boat) and manner of orientation (e.g. guided with a map, verbal directions). Information about the structure of the environment is stored in what is

K. Ooms (✉) · N. Van de Weghe
Department of Geography, Ghent University, Krijgslaan 281 (S8), 9000 Ghent, Belgium
e-mail: kristien.ooms@ugent.be

N. Van de Weghe
e-mail: nico.vandeweghe@ugent.be

© Springer International Publishing AG 2018
P. Fogliaroni et al. (eds.), *Proceedings of Workshops and Posters at the 13th International Conference on Spatial Information Theory (COSIT 2017)*, Lecture Notes in Geoinformation and Cartography, https://doi.org/10.1007/978-3-319-63946-8_32

called a cognitive map (Downs and Stea 1977; Golledge 1999). When navigation (the network, the path to follow, etc.) can be linked to this cognitive map, it can be perceived as more intuitive to the user.

Several authors (Li 2008; Giudice et al. 2010; Karimi 2011) have identified that indoor environments raise new challenges when developing navigation solutions compared to existing outdoor systems because of several specific differences between indoor and outdoor spaces. Differences in visual access (e.g. visible information in the line of sight), the degree of architectural differentiation, the availability of signs, and general spatial configuration have proven to be important factors influencing wayfinding (Hölscher et al. 2006, 2011; Li and Klippel 2012). Due to not only physical structural differences, but also due to differences in constraints and usage between indoor and outdoor environments, adaptations of outdoor conceptualizations to the indoor environment are necessary (Vanclooster et al. 2014). To resolve the difficulties raised by the characteristics of indoor environments, three interacting components that define navigation, namely localization, path planning and guidance along the path (Nagel et al. 2010), must be dealt with. In this working paper we only focus on the latter two. Nevertheless, to enable path planning appropriate space models need to be defined as well, which are capable of capturing the special characteristics of indoor environments as described above.

1.1 Path Planning

Path planning is a key element of navigation guidance solutions as it aims at computing an optimal route between an origin and a destination (Montello 2005). Recent literature reviews on indoor path planning (Kwan and Lee 2005; Thill et al. 2011) have demonstrated that existing indoor navigation solutions often only provide users with shortest (Dijkstra 1959) or fastest route alternatives (Vanclooster et al. 2014, 2016; Huang and Gartner 2010). The results of those algorithms often exhibit non-realistic paths (e.g. using complex intersections, avoiding main walking areas), which could easily lead to the failure of the navigation tasks. Furthermore, previous studies give evidence that humans do not exclusively take shortest or fastest paths (Golledge 1995; Hochmair 2005), but that humans value equally as much the form and complexity of the routes, such as definition of angles (Winter 2002), routes with least instruction complexity (Kulik 2003; Richter and Duckham 2008), simplest path (Mark 1986), reliable routes minimizing the number of complex intersections with turn ambiguities (Haque et al. 2007), routes with fewest turns (Jiang and Liu 2011), hierarchical paths (Fu et al. 2006), least risk path algorithm (Grum 2005), and routes avoiding 'uncomfortable' areas (Huang et al. 2014). The aforementioned paths can thus be considered as more intuitive paths for the users. For outdoor navigation, these different route planning algorithms have been proposed in the literature to compute 'optimal' routes other than shortest or fastest ones, but due to the perceived difference between outdoor and indoor environments, their appropriateness has to be evaluated in indoor situations as well. Their matching route descriptions can also vary in, among others, perspective, amount of information, included feature types, descriptor types (Fallah et al. 2013; Gartner 2004).

1.2   Space Models and Networks

An essential part of path planning applications is the availability of well-defined space models as representation of the user's environment. Many authors agree on the need for a routing graph, or network, as underlying space concept to support navigation guidance. The most important ones that can be found in literature are: topological models (Lee 2004; Stoffel et al. 2007, 2008; Sato et al. 2009), corridor derivation networks (Lee 2004), cell-decomposed networks (Lorenz et al. 2006) and visibility-based models (Zheng et al. 2009). See also (Franz et al. 2005) for a comprehensive overview of graph-based models in architecture and cognitive science. Recently, the Open Geospatial Consortium (OGC) approved a new standard IndoorGML[1] for the representation and exchange of geoinformation for indoor applications (Gröger et al. 2008), with the Geometric Network Model as underlying network (Lee 2004). Although this is a promising evolution, this standard is not developed to specifically support indoor navigation. Recent efforts have shown possibilities of automatically assigning nodes to each room object and connecting them when they are connected in reality (Stoffel et al. 2008; Anagnostopoulos et al. 2005; Meijers et al. 2005). However, the development of a comprehensive methodology for automatic network creation requires a thorough foundation and agreement on the appropriate and optimal (i.e. user friendly) network structure of indoor environments, which supports the user in his navigation task (Becker et al. 2009). Network structures that are in line with the user's cognitive map can be perceived as more intuitive compared to purely geometric structures, among others. Up to this point and as far as we know, this is still missing in indoor navigation research.

1.3   Crowdsourced Route Descriptions—SoleWay

In order to evaluate how well algorithmically generated routes correspond to how humans structure spatial information cognitively, the latter must be materialized. This corresponds most closely with route instructions (verbal or text-based) that you receive when you ask someone for directions. This is based on how the person who give the instructions has structured the environment (cognitive or mental map);

---

[1]GML = Geography Markup Language.

how he constructed the 'best' or 'most intuitive' route; and how this is formulated (description). This 'best' route can thus be the shortest, or easiest or 'most pleasant' route. This principle of describing directions in natural language is the starting point of the indoor navigation system SoleWay (see https://soleway.ugent.be/). This system collects text-based route descriptions from the crowd; anyone who wants to enter a route description is welcome to do so. When another user requests a route description from A to B, the description that was entered by a previous user is displayed. Consequently a large collection of route descriptions is constructed over time.

2   Research Goals and Approach

This working paper presents a methodology on how to *develop an approach to automatically generate indoor networks and route descriptions based on crowd-sourced and text based route instructions as an approximation of how humans structure spatial information cognitively.* As a consequence, more intuitive route descriptions can be generated, which should be easier to follow and are thus linked to a lower cognitive load. It is our intention to discuss this idea during the workshop and to implement the methodology in the near future. The main goal is translated into the following research questions:

RQ1—Which existing network (NE) best fits human route descriptions in indoor navigation?

Based on the literature, several existing networks N_E can be selected. It can be evaluated how well these networks match with the structure of our cognitive maps, and are thus more intuitive for humans. The structure of these cognitive maps is included in the crowdsourced human route descriptions available in SoleWay. These are thus in turn translated into a network (N_D): constituted out of the nodes and edges that are mentioned in the route descriptions. As a next step, network N_D can be compared with the networks N_E. Based on these outcomes, the network that best matches the cognitive map (N_C) can be selected.

RQ2—Which existing routing algorithm (AE) best fits human route descriptions in indoor navigation?

Based on the literature, several algorithms A_E can be selected. As these algorithms are inherently linked to an underlying network, only network N_C (selected in RQ1) is used. As a first step, the actual routes for the available human route descriptions (available in SoleWay) have to be determined (also based on the structure of network N_c), resulting in the routes R_D. Second, for the same origin-destination combinations of the routes R_D, the routes R_A need to be calculated using the algorithms A_E. This is repeated for each of the selected algorithms A_E. Third, the routes R_D have to be compared with the corresponding routes calculated with each of the selected algorithms A_E resulting in RI_D versus RI_{A1}; RI_D versus $R1_{A2}$; ... (for origin-destination pair 1, to be repeated for all available OD pairs). The correlation measures between the routes R_D and routes R_A will be based on the one hand on (1) benchmark parameters (such as total path length; number of turns; number of spatial units passed); and on the other hand on (2) edge based comparisons (how many and which of the edges match between the different routes). See (Vanclooster et al. 2014) for a detailed overview on these types of analysis. Based on these outcomes, the algorithm that best matches a path that is intuitive to follow by humans (A_C) can be selected.

RQ3—How to automatically populate the selected network Nc from RQ1 solely based on available human route descriptions in indoor navigation?

In this step, a (limited) set of route descriptions is translated into a network structure (N_c), which in turn can be used to calculate new routes (using A_C) A. First, each human route description has to be automatically converted to a series of nodes and edges on the selected network, resulting in routes on the network. Second, overlaps between all these human route descriptions have to be identified. These overlaps allow 'stitching' the routes on the network to the network N_C that potentially covers the whole building.

RQ4—How to automatically generate human route descriptions?

In this step, routes R_A will be (automatically) calculated using network N_C (see RQ1) and algorithm A_C (see RQ2), also considering the cases where (1) the complete network N_C of the building is available, or (2) network N_C still has to be derived based on the available human route descriptions (see RQ3). The algorithmically generated routes R_A have to be converted into text-based route descriptions. Based on the available network (complete or partial), the selected algorithm that produces more intuitive paths for humans has to be able to generate routes (and descriptions) that are not registered yet in the crowd-based system. The latter will be evaluated in the indoor environment with the actual end users.

3 Conclusion

The proposed research implements a stepwise approach with the ultimate goal to have a fully working system that can automatically generate, besides a network N_C, also routes R_A and the corresponding route descriptions D_A as well. These building blocks are selected as such that they have a close match with human intuition. The latter is a consequence of human cognition and thus how humans structure and process spatial information. The required information is derived from available crowdsourced and text-based route descriptions in SoleWay. As such, missing routes can be completed automatically while creating a network structure for the building.

References

Anagnostopoulos C, Tsetsos V, Kikiras P Hadjiefthymiades SP (2005). OntoNav: a semantic indoor navigation system. In: 1st workshop on semantics in mobile environments. Ayia Napa, Cyprus

Becker T, Nagel C, Kolbe T (2009) A multilayered space-event model for navigation in indoor spaces. In: Lee J, Zlatanova S (eds) 3D geo-information sciences. Springer, Berlin

Dijkstra EW (1959) A note on two problems in connexion with graphs. Numer Math 1:269–271

Downs RM, Stea D (1977) Maps in minds: reflection on cognitive mapping. Harper & Row, New York

Duckham M, Kulik, L (2003) "Simplest" paths: automated route selection for navigation. In: Kuhn W, Worboys M, Timpf S (eds.) Spatial information theory. Foundations of geographic information science, vol 2825. Springer, Heidelberg, pp 169–185

Fallah N, Apostolopoulos I, Bekris K, Folmer E (2013) Indoor human navigation systems: a survey. Interact Comput 25(1):21–33

Franz G, Mallot H, Wiener J (2005). Graph-based models of space in architecture and cognitive science–a comparative analysis. In: Proceedings of the 17th international conference on systems research, informatics and cybernetics. Baden-Baden, Germany

Fu L, Sun D, Rilett LR (2006) Heuristic shortest path algorithms for transportation applications: state of the art. Comput Oper Res 33:3324–3343

Gartner G (2004) Location-based mobile pedestrian navigation services—the role of multimedia cartography. Paper presented at joint workshop on ubiquitous, pervasive and Internet mapping, Tokyo (Japan)

Giudice NA, Walton LA, Worboys M (2010) The informatics of indoor and outdoor space: a research agenda. Paper presented at the second ACM SIGSPATIAL international workshop on indoor spatial awareness, San Jose, CA

Golledge RG (1995) Path selection and route preference in human navigation: a progress report. In: Frank A, Kuhn W (eds) Spatial information theory a theoretical basis for GIS, vol 988. Springer, Heidelberg, pp 207–222

Golledge RG (1999) Wayfinding behavior: cognitive mapping and other spatial processes. The Johns Hopkins University Press, Baltimore

Gröger G, Kolbe T, Czerwinski A, Nagel C (2008) OpenGIS city geography markup language (CityGML) encoding standard, Vol OGC 08-007r1. Open Geospatial Consortium, p 234

Grum E (2005). Danger of getting lost: optimize a path to minimize risk. In: 10th international conference on information & communication technologies (ICT) in urban planning and spatial development and impacts of ICT on physical space. Vienna, Austria: CORP 2005

Haque S, Kulik L, Klippel A (2007) Algorithms for reliable navigation and wayfinding. In: Barkowsky T, Knauff M, Ligozat G, Montello D (eds) Spatial cognition versus reasoning, action, interaction, vol 4387. Springer, Berlin, pp 308–326

Hochmair H (2005) Towards a classification of route selection criteria for route planning tools. In: Developments in spatial data handling. Springer, Heidelberg, pp 481–492

Hölscher C, Meilinger T, Vrachliotis G, Brösamle M, Knauff M (2006) Up the down staircase: wayfinding strategies in multi-level buildings. J Environ Psychol 26(4):284–299

Hölscher C, Tenbrink T, Wiener JM (2011) Would you follow your own route description? cognitive strategies in urban route planning. Cognition 121(2):228–247

Huang H, Gartner G (2010) A survey of mobile indoor navigation systems. In: Gartner G, Ortag F (eds) Cartography in central and eastern Europe. Springer, Berlin, pp 305–319

Huang H, Klettner S, Schmidt M, Gartner G, Leitinger S, Wagner A, Steinmann R (2014) AffectRoute—considering people's affective responses to environments for enhancing route-planning services. Int J Geogr Inf Sci 28(12):2456–2473

Jiang B, Liu X (2011) Computing the fewest-turn map directions based on the connectivity of natural roads. Int J Geogr Inf Sci 25(7):1069–1082

Karimi HA (2011) Universal navigation on smartphones. Springer Science & Business Media

Kwan M-P, Lee J (2005) Emergency response after 9/11: the potential of real-time 3D GIS for quick emergency response in micro-spatial environments. Comput Environ Urban Syst 29(2):93–113

Lee J (2004) A Spatial access-oriented implementation of a 3-D GIS topological data model for urban entities. Geoinformatica 8(3):237–264

Li K-J (2008) Indoor space: a new notion of space. In: Bertolotto M, Ray C, Li X (eds) Web and wireless geographical information systems, vol 5373. Springer, Berlin, pp 1–3

Li R, Klippel A (2012) Explorations of wayfinding problems in libraries: a multi-disciplinary approach. J Map Geogr Libr 8(1):21–38

Lorenz B, Ohlbach H, Stoffel E-P (2006) A hybrid spatial model for representing indoor environments. In: Carswell J, Tezuka T (eds) Web and wireless geographical information systems, vol 4295. Springer, Heidelberg, pp 102–112

Mark DM (1986) Automated route selection for navigation. IEEE Aerosp Electron Syst Mag 1: 2–55

Meijers M, Zlatanova S, Pfeifer N (2005). 3D geo-information indoors: structuring for evacuation. In: The 1st international ISPRS/EuroSDR/DGPF-workshop on next generation 3D city models. Bonn, Germany, p 6

Montello DR (2005) Navigation. In: Shah P, Miyake A (eds) The Cambridge handbook of visuospatial thinking. Cambridge University Press, New York, pp 257–294

Nagel C, Becker T, Kaden R, Li K, Lee J, Kolbe TH (2010) Requirements and space-event modeling for indoor navigation: OGC discussion paper. Open Geospatial Consortium

Richter K-F, Duckham M (2008) Simplest instructions: finding easy-to-describe routes for navigation. In: Cova T, Miller H, Beard K, Frank A, Goodchild M (eds) Geographic information science, vol 5266. Springer, Heidelberg, pp 274–289

Sato A, Ishimaur N, Tao G, Tanizaki M (2009). OGC OWS-6 outdoor and indoor 3D routing services engineering report

Stoffel EP, Lorenz B, Ohlbach HJ (2007) Towards a semantic spatial model for pedestrian indoor navigation. In: Advances in conceptual modeling—foundations and applications, vol 4802/2007. Springer, Berlin, pp 328–337

Stoffel EP, Schoder K, Ohlbach HJ (2008) Applying hierarchical graphs to pedestrian indoor navigation. In: Proceedings of the 16th ACM SIGSPATIAL international conference on advances in geographic information systems. ACM, Irvine (CA)

Thill J-C, Dao THD, Zhou Y (2011) Traveling in the three-dimensional city: applications in route planning, accessibility assessment, location analysis and beyond. J Transp Geogr 19(3): 405–421

Vanclooster A, Ooms K, Viaene P, Fack V, Van de Weghe N, De Maeyer P (2014) Evaluating suitability of the least risk path algorithm to support cognitive wayfinding in indoor spaces: an empirical study. Appl Geogr 53:128–140

Vanclooster A, van de Weghe N, De Maeyer P (2016) Integrating indoor and outdoor spaces for pedestrian navigation guidance: a review. Trans GIS. DOI:10.1111/tgis.12178

Winter S (2002) Modeling costs of turns in route planning. Geoinformatica 6(4):345–361

Zheng J, Winstanley A, Pan Z, Coveney S (2009). Spatial characteristics of walking areas for pedestrian navigation. Paper presented at the 3th international conference on multimedia and ubiquitous engineering, Qingdao, China

Considering Existing Indoor Navigational Aids in Navigation Services

Wangshu Wang, Haosheng Huang and Georg Gartner

Abstract Existing indoor navigational aids such as signs and floor plans are originally designed to assist navigation and to support spatial learning. However, they are often neglected in current navigation services. Integrating such information adequately into indoor navigation services requires a better understanding of their usages. Thus, we conducted an empirical study in two buildings with 28 participants who had to think aloud while performing wayfinding tasks. By analysing the participants' verbal protocols, we distinguished two decision making scenarios and suggested categorizations of their indoor wayfinding tactics. Our results confirmed people's reliance on existing indoor navigational aids and indicated that signs were the most commonly used aids. In addition, the characteristics of targets influenced the choice of aids. Therefore, we recommended that indoor navigation services include signs and present route information adaptively based on distinct destinations.

Keywords Indoor navigation · Navigational aid · Wayfinding tactic · Decision making

1 Introduction

Different indoor navigation systems have been designed to support wayfinding (Kargl et al. 2007; Li et al. 2017; Rehman and Cao 2017). In these systems, floor plans and verbal instructions are the most commonly applied route communication techniques

W. Wang (✉) · G. Gartner
Department of Geodesy and Geoinformation, Vienna University of Technology, Vienna, Austria
e-mail: wangshu.wang@tuwien.ac.at

G. Gartner
e-mail: georg.gartner@tuwien.ac.at

H. Huang
Department of Geography, GIScience Center, University of Zurich, Zurich, Switzerland
e-mail: haosheng.huang@geo.uzh.ch

© Springer International Publishing AG 2018
P. Fogliaroni et al. (eds.), *Proceedings of Workshops and Posters at the 13th International Conference on Spatial Information Theory (COSIT 2017)*, Lecture Notes in Geoinformation and Cartography, https://doi.org/10.1007/978-3-319-63946-8_33

(Huang and Gartner 2010). However, distance based and turn-by-turn instructions draw too much of users' attention and may raise their cognitive load. Navigating with this kind of indoor navigation systems, users are devoid of interactions with the environment, thus fail to remember routes travelled and become disoriented easily when the system is not available (Reilly et al. 2008).

Current indoor navigation systems ignore existing navigational aids in buildings (e.g., wall-mounted floor plans, signs, and kiosk maps), which are intended for facilitating indoor wayfinding and supporting spatial learning (Reilly et al. 2008). Navigational aids represent explicit information in the overall configuration and structure of the environment (Conroy 2001; Vilar et al. 2012). Reilly et al. (2008) demonstrated that using kiosk maps alongside a mobile wayfinding application can help users gain spatial knowledge. Passini (1984) and O'Neill (1991a, b) pointed out that signs, floor plans and architectural features are the key factors affecting human wayfinding. These factors were studied by environmental psychologists in virtual reality (Conroy 2001; Vilar et al. 2012, 2014). They showed that people rely on signs rather than on architectural features as long as signs are available and that signs are preferred to floor plans in certain situations. However, compared with physical movement, walking virtually lacks important bodily cues (e.g., vestibular information especially during turns), which may lead to inconsistent findings with reality (Meilinger et al. 2008; Klatzky et al. 1998). Therefore, an in situ study is needed to confirm the reliance on existing indoor navigational aids by wayfinders.

To address the above issues, a research project is being carried out, which aims at investigating the possibility of providing indoor navigation services by considering existing navigational aids. During the first part of this research, an experiment was conducted in two buildings with distinct navigational aids to confirm the reliance on existing indoor navigational aids. This paper reports on the methodology and preliminary results of the experiment, which will allow us to answer the following questions:

1. Do people rely on existing indoor navigational aids while navigating? What do they refer to when making indoor wayfinding decisions? What is the most frequently used indoor navigational aid?
2. What environmental features do people employ in indoor route depictions and descriptions?

A follow-up study will then aim to conceptualize indoor navigation services, and integrate existing indoor navigational aids into route planning and communication.

2 Methodology

Unlike in the virtual environment, it is difficult to control and compare existing indoor navigational aids in experiments conducted in the real environment. Consequently, we decided to conduct an exploratory experiment to find out the relevance of existing indoor navigational aids to human wayfinding. Verbal protocols have often been used to obtain a deep insight on human wayfinding behavior. Lynch (1960) explored the mental representation of city-scale spaces in verbal protocols and sketch maps. Passini (1984) investigated detailed decision making process of wayfinding by verbal comments of participants. Best (1970) noted that people are more likely to get lost at decision points. Therefore, we employ the thinking aloud method (Ericsson and Simon 1993) to understand participants' indoor wayfinding decision making in terms of what information they refer to at decision points. Following similar definitions from O'Neill (1991a) and Butler et al. (1993), we define a decision point as a location where people have at least two outgoing options. To represent indoor wayfinding decision making and reasoning process at a decision point, we introduce the term *indoor wayfinding tactic*. This is the process of careful planning of the following action during an indoor wayfinding task, performed at a decision point. Compared with an indoor wayfinding strategy, which is a long-term and overall planning of the whole wayfinding task (Hölscher et al. 2006), it is short-term and specific to the current decision point. We therefore name it as "tactic" (Bates 1979).

2.1 Experiment Setup

2.1.1 Selection of Buildings

In this study, two buildings, the Albertina and the main building of Vienna University of Technology (TU), were selected as our experiment venues. The buildings differ in floor plan configuration complexity, function and signage system used. Built in the 18th century, the Albertina was originally a palace in the Neoclassical style (Albertina 2017). It has been renovated and transferred into a modern museum. The signage system is well designed (Fig. 1a), with floor plans mounted on the wall (Fig. 1b), as well as printed floor plans freely available at the ticket office.

The main building of Vienna University of Technology was built in the 19th century (Wien 2017). Later, several separate buildings nearby were joined by staircases into a whole complex. Due to its configuration, different sections in this building differ in floor heights and signage systems. The staircases are numbered and utilized as the major wayfinding guidance by local students. Students report getting lost easily inside the building owing to its complexity, a lack of floor plans, and multiple, inconsistent signage systems.

2.1.2 Participants

28 participants (16 female, 12 male, age: M = 29, SD = 6.77) were recruited through our website, social media and flyers posted on other universities' campuses. They hold different professions such as mathematicians, English teachers, musicians, painters, etc. All participants confirmed to have no visiting experience to either build-

(a)   **(b)**

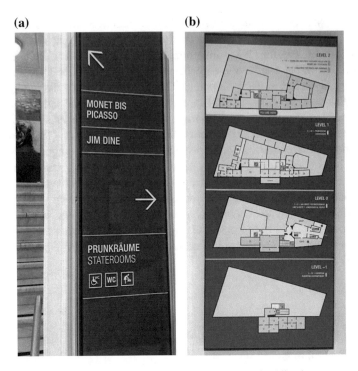

Fig. 1 **a** A sign in the Albertina. **b** A wall-mounted floor plan in the Albertina

ing. They were all fluent in English and at least understood basic German. All participants signed the informed consent before participating in the experiment. Each participant received a free entry to the Albertina and a small gift.

2.1.3 Procedure

An experimenter met a participant in front of one of the test buildings. After a short opening conversation, the participant was lead to the starting point of that building. The experimenter read an introduction informing the participant to find three targets in each building. The participant was then given the first target and asked to describe their strategy to find it. The participant was also told to stop at every decision point on their way to the target. At each decision point, the participant was encouraged to describe their wayfinding options, hints revealing the destinations of these options, to make their decision, and to justify it (Fig. 2). After reaching the first target, the participant was asked to describe the route from the starting point to target one as if someone unfamiliar to the building was asking them for help. The participant was then asked to draw a sketch map. Then, the second target was given. Each target was the starting location of the following task for the verbal description and sketch

Fig. 2   A participant was justifying her decision with a sign at a decision point in the TU

map. The wayfinding process ended when all the targets were found. The participant was enquired afterwards about their general impression of the existing wayfinding aids in both buildings. Except asking the experimenter, the participant was encouraged to think aloud during the process and to use whatever information available to help wayfinding. The experimenter followed the participant and recorded the whole process using a head-mounted GoPro Hero 4 Silver[1] and a voice recorder.

To avoid learning effects, half of the participants started from the Albertina and the other half from the TU. All participants received the targets in the same order. In the Albertina, the three targets were "The entrance of exhibition *Contemporary Art*", "Room 8 of exhibition *Monet bis Picasso*", and "The toilet close to exhibition *Wege des Pointillismus*". The targets in the TU were a ceremony hall "Festsaal", and two lecture rooms "Hörsaal 15" and "Hörsaal 17".

2.2   Data Processing and Analysing

The videos and the audios were synchronized and merged using a video editor.[2] Afterwards, the merged videos were transcribed into verbal protocols and both were analysed qualitatively. In order to analyse indoor wayfinding tactics and the relevance of existing indoor navigational aids to human wayfinding, we focused on participants' justifications of their decisions. In this study, we only considered the verbal justifications expressed at decision points.

---

[1]https://gopro.com/.

[2]http://www.videosoftdev.com/.

We applied the structuring method of qualitative content analysis to code the verbal protocols. This is a content-analysis method that starts with a defined category system from existing theory, and keeps revising these categories while coding (Mayring 2014). Passini (1984) categorization of wayfinding strategies and tactics for decision making served as our initial category system.

3 Categorization of Indoor Wayfinding Tactics

Prior to categorizing indoor wayfinding tactics, we distinguish two scenarios at a decision point encountered by a wayfinder: (1) when relevant wayfinding support information is explicitly accessible and perceived, and (2) when *no* relevant wayfinding support information is perceived.

Based on Passini (1984) wayfinding decision making theory and our analysis, we propose the following categorization of indoor wayfinding tactics, which will be iteratively evaluated and eventually adopted.

1. Tactics used when relevant wayfinding support information is explicitly accessible and perceived:

   - Sign: Wayfinders make a decision based on identification signs, directional signs, and floor directories without a graphical floor plan, e.g., *"There is a sign straight ahead for HS 17, so I'll follow that."*
   - Floor plan: Wayfinders make a decision based on floor plans, e.g., *"That will be on level 2 according to this plan."*
   - External help: Wayfinders make a decision based on external help from other people or electronic devices, e.g., *"He said that I need to go this way to find the 4th."*

2. Tactics used when no relevant wayfinding support information is perceived:

   - Memory: Wayfinders make a decision based on certain information they memorized about the setting or similar settings, e.g., *"Now I'm going out of the bridge. And with my memory then I should go right."*
   - Inference: Wayfinders make a decision based on their mental manipulation and reasoning without relevant information available, e.g., *"I'm going to decide to go up another floor because 14 and 15 are on this floor and a higher number should be on the floor above."*
   - Searching: Wayfinders make a decision because they wish to find signs, floor plans, someone to ask or any relevant information on the option they choose, e.g., *"I'm going to come down to see if there is some signage out here."*
   - Preference
     - Architectural feature: Wayfinders make a decision based on corridor width, room size, brightness, color, and their preferred direction, e.g., *"I took this side because I think this room is bigger. That attracts me to go inside."*

  – Non-spatial related preference: Wayfinders make a decision based on cultural and personal preferences, e.g., *"Don't want to go through this crowd."*
- Random: Wayfinders make a decision randomly, without specifying any reasons, e.g., *"This is just a sort of random decision to get my bearings."*

4   Preliminary Results and Discussion

4.1   Indoor Wayfinding Tactics in the Albertina and the TU

The frequency and proportion of different indoor wayfinding tactics used in the Albertina and the TU are shown in Table 1. In both buildings, referring to signs was the most frequently used tactic, despite the inadequate signage system in the TU. Comparing the two scenarios, we discovered that in the Albertina, 62% of the decisions were made in Scenario 1, while in the TU, the proportion dropped to 55%, due to the limited availability of existing navigational aids in the building.

Considering the first scenario, in the Albertina, although both wall-mounted and paper-printed floor plans were available, the usage of sign tactic (63%) was twice as much as floor plan tactic (30%). This is probably due to a higher cognitive load of using floor plans comparing to using signs (Passini 1984). Since the TU did not offer floor plans to its visitors, no floor plan tactic was employed in the TU. As the sole information source other than asking other people or checking mobile phones, the sign tactic dominated in the TU with a percentage of 85%. The tactic of seeking external help was not popular in either building.

Table 1   Indoor wayfinding tactics and their frequency and proportion in the Albertina and the TU

Scenarios	Indoor wayfinding tactics	Frequency (n)		Proportion (%)	
		Albertina	TU	Albertina	TU
Scenario 1	Sign	121	189	39	47
	Floor plan	58	0	19	0
	External help	14	34	5	8
Scenario 2	Memory	4	8	1	2
	Inference	80	80	26	20
	Searching	26	80	8	20
	Architectural feature	1	0	0	0
	Non-spatial related preference	3	1	1	0
	Random	3	10	1	2

In the second scenario, compared to other tactics, the inference and the searching tactics were more regularly employed by our participants in both buildings. However, the distribution of the inference tactic and the searching tactic differed dramatically between the two buildings. In the Albertina, 68% of the decisions were made by inference, while 22% of the decisions were made because participants wished to search for information in the chosen route. This can be explained by the simple structure of the Albertina and its well-designed guiding system that wayfinding support information is available at almost all possible decision points. A different situation was found in the TU. The usage of the inference tactic and the searching tactic were similar. Participants searched for wayfinding support information more frequently owing to the absence of floor plans and the sparsely placed signs in the TU. The complexity of the building and a lack of overview that could be provided by floor plans made it very hard to build a mental map of the building on first exposure. Hence, the decisions made by inference sometimes led participants to disorientation, which in turn contributed to the high rate of the searching tactic.

4.2   The Choice of Indoor Wayfinding Tactics Among Tasks

The choice of tactics is influenced not only by the architecture and the existing navigational aids of the buildings but also by the tasks. Distinct targets evoke the distinct sense of the corresponding tasks, thus affect the selection of the tactics.

In the Albertina, the tactics used in different scenarios have almost identical distributions among different tasks, according to Table 2. However, the ratio of using signs to floor plans varies substantially among tasks. In the Task 1, participants were required to find the entrance of an exhibition. There was no wall-mounted floor plan

Table 2 Indoor wayfinding tactics and their frequency and proportion among tasks in the Albertina

Indoor wayfinding tactics	Frequency (n)			Proportion (%)		
	Task 1	Task 2	Task 3	Task 1	Task 2	Task 3
Sign	39	46	36	52	40	30
Floor plan	4	23	31	5	20	26
External help	8	1	5	11	1	4
Memory	0	2	2	0	2	2
Inference	21	32	27	28	28	22
Searching	3	9	14	4	8	12
Architectural feature	0	1	0	0	1	0
Non-spatial related preference	0	1	2	0	1	2
Random	0	0	3	0	0	2

Table 3 Indoor wayfinding tactics and their frequency and proportion among tasks in the TU

Indoor wayfinding tactics	Frequency (n)			Proportion (%)		
	Task 1	Task 2	Task 3	Task 1	Task 2	Task 3
Sign	56	95	38	75	49	28
Floor plan	0	0	0	0	0	0
External help	0	20	14	0	10	10
Memory	0	4	4	0	2	3
Inference	17	28	35	23	15	26
Searching	2	42	36	3	22	27
Architectural feature	0	0	0	0	0	0
Non-spatial related preference	0	0	1	0	0	1
Random	0	3	7	0	2	5

before finishing this task. The sign tactic was therefore applied extensively. While the targets became more specific and wall-mounted floor plans started to appear from the Task 2 on, the ratio of using signs to floor plans decreased. There were two decision points with signs and floor plans, which could be passed by the participants during both Task 2 and Task 3. By analysing the tactics used at these decision points, we discovered that the only situation more floor plan tactics were employed was at the landing of Level 2 during the Task 3. In that situation, the participants just left the Exhibition *Monet bis Picasso* and were to find the toilet close to Exhibition *Wege des Pointillismus*. Since the description of the Target 3 contained spatial relation, the participants preferred floor plans to signs to get an overview of the spatial structure. Once the spatial information was acquired, participants would tend to refer to signs for the immediate directions at later decision points.

In the TU, the tactics used in different scenarios were greatly influenced by the tasks (Table 3). The use of signs dropped considerably from the Task 1 to the Task 3, while the proportion of applying the searching tactics grew, as a result of the insufficient existing navigational aids in the building. The Target 3 was spatially close to the Target 2, but there was no direct sign from the Target 2 to the Target 3. Hence, there was a high possibility that the participants would not be able to find the Target 3 without enough inference and searching.

5 Conclusion and Outlook

In summary, after preliminary interpretation of the experimental data, we reflect on our first research question and draw an initial conclusion that people rely on existing indoor navigational aids, in spite of their quality. Although the preference of existing indoor navigational aids varies individually, signs are the most commonly used.

Different tasks influence the choice of indoor wayfinding tactics. The requirements of survey knowledge provided by floor plans increase while the targets become more specific, especially when spatial relations are contained in the targets. In designing navigation systems, this is worth considering for presenting route information adaptively, according to distinct destinations.

We are currently investigating on the impact of different types of decision points on the choice of indoor wayfinding tactics. The relevance of individual sense of direction to the tactics is another direction we are working towards. To address the second research question, the verbal descriptions and the sketch maps are also being analysed.

Based on the findings from the experiment, we are going to include signs in indoor route planning and to generate route instructions accordingly. One example is, rather than providing turn-by-turn instructions, navigating users to a certain sign which displays the destination and instructing them to follow the succeeding signs. An appropriate visualization of the indoor environment with relevant signs highlighted is a further direction for research.

References

Albertina (2017) The history of Albertina. Available via http://www.albertina.at/en/palace/history. Accessed 10 May 2017

Bates MJ (1979) Information search tactics. J Am Soc Inf Sci 30:205–329

Best G (1970) Direction finding in large buildings. In: Canter DV (ed) Architectural psychology-Proceedings of the conference at Dalandhui. RIBA, London, pp 72–75

Butler D, Acquino AL, Hissong AA, Scott PA (1993) Wayfinding by newcomers in a complex building. Hum Fact 25(1):159–173

Conroy R (2001) Spatial navigation in immersive virtual environments. Unpublished doctoral dissertation, University of London

Ericsson KA, Simon HA (1993) Protocol analysis: verbal reports as data. MIT Press, Cambridge, MA

Huang H, Gartner G (2010) A survey of mobile indoor navigation systems. In: Gartner G, Ortag F (eds) Cartography in central and eastern Europe. Lecture notes in geoinformation and cartography, pp 305–319. Springer, Berlin/ Heidelberg

Hölscher C, Meilinger T, Vrachliotis G, Brösamle M, Knauff M (2006) Up the down staircase: wayfinding strategies and multi-level buildings. J Env Psychol 26(4):284–299

Kargl F, Geßler S, Flerlage F (2007) The iNAV indoor navigation system. In: Ubiquitous computing systems, pp 110–117. Springer, Berlin/ Heidelberg

Klatzky R, Loomis J, Beall A, Chance S, Golledge R (1998) Spatial updating of self-position and orientation during real, imagined, and virtual locomotion. Psychol Sci 9(4):293–298

Li L, Xu Q, Chandrasekhar V et al (2017) A wearable virtual usher for vision-based cognitive indoor navigation. IEEE Trans Cybern 47(4):841–854

Lynch K (1960) The image of the city. MIT Press, Cambridge, MA

Mayring P (2014) Qualitative content analysis: theoretical foundation. Basic procedures and software solution. Beltz, Klagenfurt

Meilinger T, Knauff M, Bülthoff H (2008) Working memory in wayfinding—a dual task experiment in a virtual city. Cogn Sci 32(4):755–70

O'Neill MJ (1991a) Evaluation of a conceptual model of architectural legibility. Env Beh 23(3):259–284

O'Neill MJ (1991b) Effects of signage and floorplan configuration on wayfinding accuracy. Env Beh 23(5):553–574

Passini R (1984) Wayfinding in architecture. Van Nostrand Reinhold Company, New York

Rehman U, Cao S (2017) Augmented-reality-based indoor navigation: a comparative analysis of handheld devices versus google glass. IEEE Trans Hum Mach Syst 47(1):140–151

Reilly D, Mackay B, Inkpen K (2008) How mobile maps cooperate with existing navigational infrastructure. In: Map-based mobile services, pp 267–292. Springer, Berlin/ Heidelberg

TU Wien (2017) About TU Wien. Available via https://www.tuwien.ac.at/en/about_us. Accessed 10 May 2017

Vilar E, Rebelo F, Noriega P (2012) Indoor human wayfinding performance using vertical and horizontal signage in virtual reality. Hum Fact Ergon Manuf Serv Ind 24(6):601–615

Vilar E, Rebelo F, Noriega P, Duarte E, Mayhorn CB (2014) Effects of competing environmental variables and signage on route-choices in simulated everyday and emergency wayfinding situations. Ergonomics 57(4):511–524

Part III
Speaking of Location: Future Directions in Geospatial Natural Language Research

Speaking of Location: Future Directions in Geospatial Natural Language Research—Introduction

Kristin Stock, Chris B. Jones and Maria Vasardani

Research into the description of location using human (natural) language has been approached from linguistics, geospatial and computer science perspectives and has addressed a wide range of topics. The papers presented in this Section reflect this diversity. Understanding the ways in which people use spatial language is an important focus in the presented papers, and is addressed from several different angles. One important theme explores the frames of reference (FoR) that apply in different languages and settings, with particular attention given to the way in which changes from one FoR to another (e.g. between ego-centric and geo-centric) depend on whether a setting is urban or rural (Palmer et al.). There is also emphasis on exploitation of local landmarks as reference objects, again in (Palmer et al.) and also in (Supriyono and Scheider), who cast Balinese descriptions in a FoR that is encoded as a semantic web ontology, with a view to transforming expressions into geo-spatial coordinates.

A second broad theme addresses spatial mental models, with a study exploring the malleability of mental representations adopting survey and route perspectives presented by Meneghetti and Muffato, who investigate whether practice with different strategies can enable increased mental flexibility in perspective. Davies and Tenbrink demonstrate cognitive grouping of locations, combining evidence from latent semantic analysis (LSA) of text descriptions of places and, after dimensionality reduction, mapping them to geographic space, in combination with locations obtained from

K. Stock (✉)
Institute of Natural and Mathematical Sciences, Massey University,
Auckland, New Zealand
e-mail: k.stock@massey.ac.nz

C.B. Jones
School of Computer Science and Informatics, Cardiff University, Cardiff, Wales
e-mail: jonescb2@cardiff.ac.uk

M. Vasardani
Department of Infrastructure Engineering, University of Melbourne,
Victoria, Australia
e-mail: maria.vasardani@unimelb.edu.au

© Springer International Publishing AG 2018
P. Fogliaroni et al. (eds.), *Proceedings of Workshops and Posters at the 13th International Conference on Spatial Information Theory (COSIT 2017)*, Lecture Notes in Geoinformation and Cartography, https://doi.org/10.1007/978-3-319-63946-8_34

people's perceptions of the location of the same places. Egorova and Purves explore perceptions of so-called fictive motion, revealing differences between types of fictive motion descriptions of paths provided by alpine hikers, in which static objects are portrayed as moving or changing in size and visibility, the study being an initial stage in developing an annotation scheme that distinguishes different types of frame relative path descriptions.

A third theme addresses spatial relations and context, with Stock and Hall presenting a typology of contextual factors of spatial relations based on an analysis of verbal explanations of the use of a set of spatial prepositions. In an experiment in an indoor environment, Doore et al. explore the variation in the preferred use of prepositions, such as on, against and near to describe contact, meronymy and disjointness relations between pairs of static and mobile objects. The importance of context in the interpretation of language is also recognised in other papers, including the previously mentioned Supriyono and Scheider, and Palmer et al.

These papers sit at the nexus of linguistics, cognitive science and computer science, with longer term aims on a spectrum ranging from understanding human language and cognition to the implementation of automated systems that can use spatial language to communicate with humans, and each offering interesting and informative contributions to this fascinating field of study.

Socioculturally Mediated Responses to Environment Shaping Universals and Diversity in Spatial Language

Bill Palmer, Alice Gaby, Jonathon Lum and Jonathan Schlossberg

Abstract This paper reports on an empirical experiment-based study testing the extent to which systems of linguistic spatial reference correlate with aspects of the physical environment in which a language community lives. We investigated linguistic spatial behaviour in two unrelated languages in both similar and contrasting locations, ranging from atoll islands to urban environments, using standardised tests whose results were subject to quantitative analysis. Our findings reveal significant variation in spatial referential strategy preference in the two languages. Some preferences correlated with environment (e.g. island vs. urban). However, others correlated with degree and nature of interaction with environment, and others with linguistic resources available to speakers. The findings demonstrate that spatial behaviour reflects a complex interplay of responses to environment; sociocultural interaction with environment; and speakers' linguistic repertoire.

Keywords Spatial language · Spatial frames of reference · Topography · Cross-cultural research · Dhivehi · Marshallese

B. Palmer (✉) · J. Schlossberg
Faculty of Education and Arts, School of Humanities and Social Science, University of Newcastle (Australia), Callaghan, NSW 2308, Australia
e-mail: bill.palmer@newcastle.edu.au

J. Schlossberg
e-mail: schlossberg.jonathan@gmail.com

A. Gaby · J. Lum
Faculty of Arts, School of Languages, Literatures, Cultures and Linguistics, Monash University, Clayton, VIC 3800, Australia
e-mail: alice.gaby@monash.edu

J. Lum
e-mail: lum.jonathon@gmail.com

© Springer International Publishing AG 2018
P. Fogliaroni et al. (eds.), *Proceedings of Workshops and Posters at the 13th International Conference on Spatial Information Theory (COSIT 2017)*, Lecture Notes in Geoinformation and Cartography, https://doi.org/10.1007/978-3-319-63946-8_35

1   Introduction

Considerable diversity in spatial reference across languages is well attested (Levinson 2003; Levinson and Wilkins 2006; Pederson et al. 1998). Nonetheless, universal tendencies can be detected within this diversity, and salient landscape and other external-world features seem to play a role in the detail of systems involving absolute Frame of Reference (FoR) (Palmer 2002, 2015), and even in FoR choice (Bohnemeyer et al. 2014; Dasen and Mishra 2010; Majid 2004). However, those aspects of the environment that are perceived as salient vary across cultures, and the nature of the interaction between humans and their environment plays a crucial role, as seen in demographic variation within individual languages in tendencies in FoR choice (Pederson 1993) and in geocentric versus egocentric strategies more generally (Palmer et al. 2016).

2   The Topographic Correspondence Hypothesis

Spatial relations of any type can be expressed in any language. However, in perhaps all languages some spatial concepts are lexicalised or expressed in a grammaticized way, while others are relegated to periphrastic expression. These lexicalized and grammaticized expressions are key to understanding the extent to which spatial reference displays universal tendencies, and the extent to which variation is systematic.

Geocentric spatial reference, including the use of absolute FoR as well as environmental landmarks, invokes aspects of the external world, suggesting that linguistic systems are responsive to the environment in which a language is spoken (Palmer 2002). This in turn predicts that aspects of systems of spatial reference will correlate with salient aspects of the physical environment. Palmer (2015) formulates this as the Topographic Correspondence Hypothesis (TCH), a tool to test the extent to which linguistic spatial systems correlate with environment in ways that can account for aspects of spatial reference that are universal or vary in systematic ways. To test TCH, Palmer (2015) proposes the Environment Variable Method (EVM), an approach that treats environment as a controlled variable. TCH makes predictions along two parameters: (A) that a single language spoken in diverse environments will display commensurate diversity in spatial reference; and (B) that diverse languages spoken in a single environment will display commensurate similarities in spatial reference. EVM tests (A) by holding the language constant and varying the environment. Prediction (B) is harder to test, because while the environment is to be held constant and the language varied, the environment cannot be held constant to the extent of investigating diverse languages in a single location, as it would be impossible to rule out similarities between languages arising from contact. Instead, language loci that are as similar as possible are to be used.

3 Spatial Behaviour in Atoll-Based Communities

To test TCH and cast light on the relationship between spatial language, non-linguistic spatial behaviour and the environment, we investigated spatial reference in languages spoken in the topographic environment of the atoll, in a field-based three year project funded by Australian Research Council Discovery Project Grant DP120102701. Atolls are an unusual environment for human habitation, comprising narrow strips of land around a central lagoon. A preliminary study of spatial reference in four atoll-based languages (Palmer 2007) found similarities in their spatial systems that are anchored in aspects of the physical environment in which the languages are spoken, including aspects tailored specifically to the topography of the atoll, principally an atoll-specific lagoonside-oceanside axis.

4 Methodology

Our research tested TCH in atoll-based languages by investigating spatial reference in Marshallese (Austronesian, Marshall Islands) and Dhivehi (Indo-Aryan, Maldives). Following EVM, we identified a baseline language-environment pairing of Marshallese spoken on an atoll (Jaluit). This was compared: along parameter (A) with Marshallese spoken (a) on a non-atoll island in the Marshall Islands, and (b) in urban Springdale, Arkansas, USA. Along parameter (B) the baseline Marshallese pairing was compared with Dhivehi spoken on Laamu, an atoll selected for its topographic similarity to the primary Marshallese site. Within each field site, experimental task-based data was elicited from participants across a range of demographic variables including age, gender, education, and occupation, and in a range of locations of varying dominant subsistence modes and degrees of population density. To complete the coverage, comparative Dhivehi data was collected from urban Addu, and densely urban Malé, the Maldivian capital.

Identical task-based elicitation techniques were used in all locations to ensure maximal comparability of data. Once gathered, the data was subject to statistical analysis. In addition to established elicitation techniques such as the Man and Tree task (Senghas version: (Terrill and Burenhult 2008)) and Route Description task (Wilkins 1993) we developed and deployed several new experimental tasks, including an Object Placement task (Schlossberg et al. 2016), a Virtual Atoll Task (Lum and Schlossberg 2015), and a new verbal Animals-in-a-Row task.

We report here on results of the Man and Tree task. Two participants are separated by a screen. Each has an identical set of 16 photographs of a toy man and tree in various configurations. A 'director' selects images to describe to a 'matcher', who identifies the corresponding image from their own set. The matcher may ask questions during the task. Once all images have been identified, the participants exchange roles and repeat the task. Participant pairs were matched in gender and approximate age, and 59 Dhivehi and 48 Marshallese-speaking pairs participated.

Table 1 Man and Tree metadata

	Dhivehi				Marshallese			
Location	Laamu (fishing)	Laamu (non-fishing)	Addu	Malé	Jaluit (Jabor island)	Jaluit (Jaluit island)	Kili	Springdale
Type	atoll	atoll	urban atoll	island city	atoll	atoll	single island	inland city
Pop. density	medium	medium	medium-high	very high	medium	low	low-medium	medium
Pairs/tasks	28	22	5	4	16	11	12	9
Location descriptions	1269	1181	293	270	640	490	452	351
Orientation descriptions	880	775	176	170	541	340	336	223
Total descriptions	2149	1956	469	440	1181	830	788	584

The distinction between fishing and non-fishing villages on Laamu in Table 1 refers to dominant subsistence mode. In fishing villages the dominant occupation is fishing. In non-fishing villages the dominant occupations involve indoor work.

5 Findings

The study's findings provide some support for TCH. In terms of linguistic resources available to speakers, both languages employ a landward-seaward axis correlating to the boundary between land and sea. However, in Marshallese this is only used at sea, while in Dhivehi it is used on land, with only one term also used at sea. Further, the distinction between an island's lagoonside and oceanside is lexicalised in both languages. However, in Dhivehi these terms cannot participate in grammaticized constructions, while in Marshallese they frequently do. Some of our quantitative findings also support TCH. For example, analysis of our Man and Tree data reveals a strong preference for egocentric referential strategies in urban settings, but a preference for geocentric strategies in most less urban island locations, supporting previous findings of an urban preference for egocentric reference (Dasen and Mishra 2010; Majid 2004; Pederson 1993; Pederson et al. 1998). Our results are discussed in more detail below.

While our findings provide some support for TCH, quantitative analysis reveals a more nuanced picture than TCH alone allows. While Marshallese and Dhivehi provide speakers with a similar range of strategies for spatial reference, strategy preference varies significantly between the languages, and within each language on the basis of demographic and locational variables. A range of linguistic resources are available to speakers of both languages, including absolute, intrinsic and relative FoR; reference to topographic landmarks; reference to ad hoc landmarks (features of the room the task was run in, buildings elsewhere in the village, etc.); Speech Act Participant (SAP) as landmark (away from me, towards you, etc.); and reference anchored in the toy tree in orientation descriptions (facing away from the tree, etc.). Variation took the form of tendencies in strategy preference, rather than exclusive choice of one or more strategy over others. Our findings introduce a crucial caveat to TCH: social and cultural factors mediate between language and environment, such that a simple predictable relationship between the two does not exist. Lexicalized and grammaticized systems of spatial reference may correlate to aspects of the environment, but the extent to which they do, and which aspects of the environment are invoked, varies on the basis of affordance, and degree and nature of cultural interaction with environment.

The Man and Tree data was subjected to quantitative analysis on the basis of the nature of the referential anchor. The results are summarised below. In some references, the anchor is internal to the figure/ground array, in other it is external. Strategies involving an array-internal anchor include intrinsic FoR, and references anchored in the tree in orientation descriptions. Strategies involving an array-external anchor include relative and absolute FoR, and topographic, ad hoc, and SAP land-

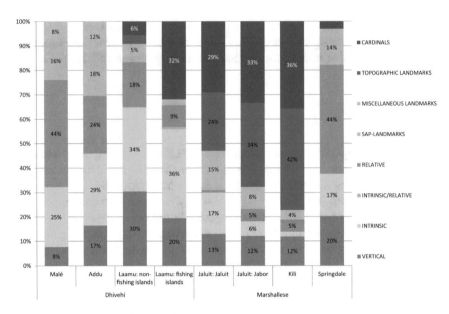

Fig. 1 Man and Tree location descriptions

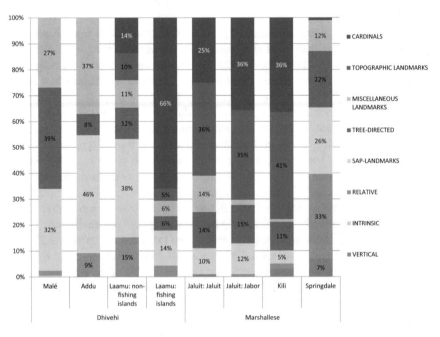

Fig. 2 Man and Tree orientation descriptions

marks. References involving an array-external anchor were further analysed on the basis of whether they were egocentric (relative FoR, SAP landmarks), or geocentric (absolute FoR, topographic and ad hoc landmarks). The results showed some variation on the basis of language, and other variation that crosscut language on the basis of demographic variables in which degree and nature of interaction with the environment appears to play a role. Figures 1 and 2 show proportion of references in each location employing each available strategy, separated into location references (e.g. the man is standing to the right of the tree) and orientation references (e.g. the man is facing away from the tree).

In location references, array-internal references involve intrinsic FoR. As Fig. 1 shows, intrinsic FoR accounts for a significantly higher proportion of references in Dhivehi than in Marshallese, regardless of location. Preference for this array-internal strategy therefore correlates with language, not environment. The same is not true of the orientation-specific array-internal strategy of anchoring the reference in the tree. In Fig. 2 the array-internal tree-directed strategy is significantly more highly represented in the principally urban environments of Dhivehi-speaking Malé and Marshallese Springdale. Here the variation correlates with environment.

Once array-internal strategies are removed, patterns of preference for egocentric versus geocentric in array-external strategies emerge. Figures 3 and 4 show array-external references only, with egocentric and geocentric strategies generalized. For Marshallese, Fig. 3 shows an overwhelming preference for geocentric strategies in atoll and island locations in the Marshall Islands, contrasting with an equally strong preference for egocentric strategies in urban Springdale. The pattern for Dhivehi is less crisp. However, a significant preference for egocentric strategies correlates with the dense urban environment of Malé, while Laamu fishing villages more resemble Marshallese atoll/island locations in a significant preference for geocentric strategies. Preferences in relatively urban Addu lie between Malé and Laamu. The variation between fishing and non-fishing communities on the same atoll, Laamu, is striking. Here the variation correlates with dominant subsistence mode, not environment: communities dominated by fishing occupations, involving intensive interaction with the environment, display a preference for geocentric strategies, while communities dominated by white-collar work on the same atoll display a preference for egocentric strategies. Figure 4, showing egocentric versus geocentric preference in array-external orientation descriptions, gives a clearer instance of the same pattern. Dhivehi displays a moderate preference for egocentric strategies overall, with the exception of Laamu fishing communities, which share geocentric preference with atoll/island-based Marshallese. Again, urban Springdale displays a significant preference for egocentric strategies.

Similar patterns emerge when the demographic variables are age or gender. Men, who traditionally worked as fishermen or sailors, display higher use of geocentric strategies than women. Older speakers, who are more likely to have spent at least part of their lives in outdoor occupations, display higher use of geocentric strategies than younger speakers, who are more likely to have only experienced indoor work. The overall picture is one in which environment plays a role, but nature and degree of engagement with environment is also significant in strategy preference.

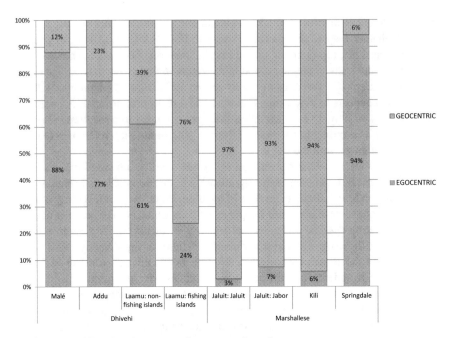

Fig. 3 Man and Tree location egocentric v geocentric preference

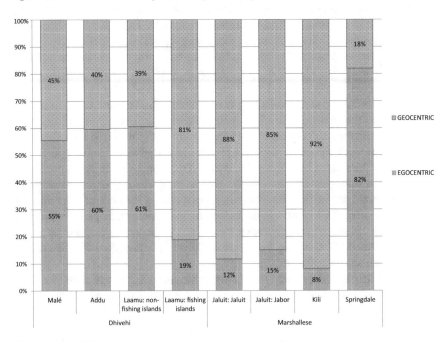

Fig. 4 Man and Tree orientation egocentric v geocentric preference

Other aspects of the variation in Figs. 1 and 2 do not correlate with environment or demographic factors, but to linguistic resources of the language itself. While the range of strategies available to speakers of Dhivehi and Marshallese are similar, they differ in which spatial concepts are lexicalised or grammaticized, as well as which terms participate in specialised spatial constructions. In Marshallese, terms for the lagoonside and oceanside of an atoll island, wilderness areas, and land and sea, as well as cardinal terms, and front, back, left, right (FBLR) terms, are grammaticized as local nouns and as directional enclitics (Palmer 2007). These participate in two specialized constructions: as local nouns in a syntactically reduced nominal spatial construction, and as directional enclitics tightly bound to the verb in high frequency path expressions. In Dhivehi, only one clearly specialised spatial construction exists, a locative dative construction that occurs alongside a general dative construction. Participation in this specialized construction is confined to cardinals, FBLR terms, and terms for beach and inland, all of which occur more frequently in non-specialized constructions as non-grammaticized landmarks. These grammatical differences between the two languages correlate to differences in geocentric strategy preferences in Figs. 1 and 2. While Laamu fishing villages and Marshallese atoll/island locations display a similar preference for geocentric strategies, the proportion of geocentric references involving topographic landmarks is much higher in Marshallese than in Dhivehi, where more use is made of cardinals. However, the figures mask the extent to which grammaticized terms and specialized constructions are involved. A majority of Marshallese topographic landmark references involve the grammaticized lagoonside/oceanside terms in their specialized constructions. Fewer topographic terms are grammaticized in Dhivehi, and these occur with lower frequency in fewer specialized constructions.

6 Conclusions

Our findings reveal significant variation in spatial referential strategy preference in two languages which afford their speakers a similar range of available strategies. Some preferences correlate with environment, such as a preference for egocentric strategies in urban Malé and Springdale, in contrast with a geocentric preference in atoll/island Marshallese. Others correlate with degree and nature of interaction with environment, seen in variation between Dhivehi fishing and non-fishing villages. Still others correlate with linguistic resources, as seen in the high proportion of topographic landmarks in geocentric references in Marshallese in the form of grammaticized terms, contrasting with a higher proportion of cardinals in Dhivehi. Some preferences, such as the higher proportion of intrinsic references in Dhivehi compared with Marshallese, do not correspond to environment, demographic variables, or linguistic resources, but simply to language community. Our findings demonstrate that human spatial behaviour cannot be understood by appeal solely to arbitrary cultural choice, or linguistic structure, or environment alone. Instead, spatial behaviour reflects a complex interplay of responses to salient features of the natural and built environment; its affordances; sociocultural interaction with the environment including uses, associations and meanings attached to it; and the linguistic repertoire available to speakers.

Acknowledgements We are grateful for the comments of three anonymous reviewers. Any errors remain ours.

References

Bohnemeyer J, Donelson K, Tucker R, Benedicto E, Capistrán Garza A, Eggleston A, Hernández Green N, Hernández Gómez MS, Herrera Castro S, O'Meara CK, Palancar E, Pérez Báez G, Polian G, Romero Méndez R (2014) The cultural transmission of spatial cognition: evidence from a large-scale study. In: CogSci 2014 Proceedings, pp 212–217

Dasen PR, Mishra RC (2010) Development of geocentric spatial language and cognition: an ecocultural perspective. Cambridge University Press, Cambridge

Levinson SC (2003) Space in language and cognition: explorations in cognitive diversity. Cambridge University Press, Cambridge

Levinson SC, Wilkins D (eds) (2006) Grammars of space: explorations in cognitive diversity. Language culture and cognition. Cambridge University Press, Cambridge. bibtex: levinson_wilkins_2006 bibtex[place=Cambridge;collection=Language Culture and Cognition]

Lum J, Schlossberg J (2015) The virtual atoll task: a spatial language elicitation tool. In: Harvey M, Antonia A (eds) The 45th Australian Linguistics Society conference proceedings. University of Newcastle, Newcastle, pp 82–103

Majid A, Bowerman M, Kita S, Haun DBM, Levinson SC (2004) Can language restructure cognition? the case of space. Trends Cogn Sci 8(3):108–114. http://repository.ubn.ru.nl/handle/2066/57358

Palmer B (2002) Absolute spatial reference and the grammaticalisation of perceptually salient phenomena. In: Bennardo G (ed) Representing space in Oceania: culture in language and mind. Pacific Linguistics, Canberra, pp 107–157

Palmer B (2007) Pointing at the lagoon: directional terms in Oceanic atoll-based languages. In: Siegel J, Lynch J, Eades D (eds) Language description, history and development: linguistic indulgence in memory of Terry Crowley. John Benjamins Publishing Company, Amsterdam, pp 101–117. doi:10.1075/cll.30.14pal, https://benjamins.com/catalog/cll.30.14pal

Palmer B (2015) Topography in language: absolute frame of reference and the Topographic Correspondence Hypothesis. In: De Busser R, LaPolla RJ (eds) Language structure and environment: social, cultural, and natural factors. John Benjamins Publishing Company, Amsterdam, pp 177–226. doi:10.1075/clscc.6.08pal, https://benjamins.com/catalog/clscc.6.08pal

Palmer B, Gaby A, Lum J, Schlossberg J (2016) Topography and frame of reference in the threatened ecological niche of the atoll. In: Geographic grounding. Place, direction and landscape in the grammars of the world, Copenhagen

Pederson E (1993) Geographic and manipulable space in two Tamil linguistic systems. In: Frank AU, Campari I (eds) Spatial information theory a theoretical basis for GIS: European conference, COSIT'93 Marciana Marina, Elba Island, Italy. Springer, Heidelberg, pp 294–311. doi:10.1007/3-540-57207-4$_2$0. bibtex: Pederson1993

Pederson E, Danziger E, Wilkins D, Levinson S, Kita S, Senft G (1998) Semantic typology and spatial conceptualization. Language 74(3):557–589. doi:10.2307/417793, http://www.jstor.org/stable/417793

Schlossberg J, Lum J, Poulton T (2016) Interpreting front, back, left, right: evidence from Marshallese, Dhivehi and English. Talk at University of Bergen. http://www.academia.edu/25871685/Interpreting_front_back_left_right_Evidence_from_Marshallese_Dhivehi_and_English

Terrill A, Burenhult N (2008) Orientation as a strategy of spatial reference. Stud Lang 32(1):93–136. doi:10.1075/sl.32.1.05ter, http://www.jbe-platform.com/content/journals/10.1075/sl.32.1.05ter

Wilkins D (1993) Route description elicitation. In: Levinson SC (ed) Cognition and space kit (version 1.0). Max Planck Institute for Psycholinguistics, Nijmegen, pp 15–28

Translating Verbally Communicated Local Geographic Knowledge Using Semantic Technologies: A Balinese Example

Pandu Supriyono and Simon Scheider

Abstract Using cognitive linguistic strategies, people can verbally encode and convey their spatial realities with little effort (i.e. *"my house is right across the street from the grocery store"*). However, to date there are a limited number of ways to transform such spatial information into forms that are useful for computational analysis in a geographic information system (GIS), and for sharing across research communities. This paper uses a case study in the Balinese language to investigate the spatial and linguistic information necessary to compute such transformations. That is, to transform verbally communicated spatial scenes into GIS-suitable data. We propose an ontology which captures reference frames used in certain Balinese locative expressions together with the parameters (ground, direction and template) required for transformation. The approach allows for the sharing of translation methods and the reuse of contextual information on the Web. Based on this model, we identify open research questions on the way to supporting approximate transformations of locative expressions.

Keywords Semantic translation · verbally communicated geographic knowledge · Linked data · Semantic web

1 Introduction

In daily communication, people encode their perceptions of spatial properties and arrangements in natural languages using locative expressions such as *"my house is right across the street from the grocery store"*. Qualitative interviews or natural language narratives provide flexible means of collecting such spatial information since people are free to make use of whichever cognitive and linguistic strategies they

P. Supriyono (✉) · S. Scheider
Department of Human Geography and Planning, Utrecht University,
Utrecht, The Netherlands
e-mail: pk.supriyono@gmail.com

S. Scheider
e-mail: s.scheider@uu.nl

© Springer International Publishing AG 2018
P. Fogliaroni et al. (eds.), *Proceedings of Workshops and Posters at the 13th International Conference on Spatial Information Theory (COSIT 2017)*, Lecture Notes in Geoinformation and Cartography, https://doi.org/10.1007/978-3-319-63946-8_36

prefer to employ in order to verbally describe a scene they are familiar with. The challenge of making use of such information lies in the limited number of ways in which it can be transformed into data models suitable for computational analysis (i.e. through a geographic information system) and sharing across research communities. Research about transformation methods is therefore necessary to exploit the collected spatial information, even though such methods currently have a very limited quality due to the fact that locative concepts are still poorly understood. The present paper investigates the requirements for computing an approximate transformation with a GIS and shows the role that semantic technologies can play in sharing transformation methods. We discuss this based on a use case to transform a simple locative expression in the Balinese language into a spatial representation that is "GIS-suitable" and shareable. We specify the information necessary to compute a translation—which we call the parameters—in an ontology, allowing for a Web based repository that makes translation methods and parameter values shareable across research communities. Finally, we discuss a simplistic transformation script for our use case and identify a list of open problems that any transformation approach has to deal with in order to actually do justice to the chosen locative expressions.

2 Related Work

Since the early days of knowledge representation in artificial intelligence, there have been attempts to capture "naive" spatial concepts with corresponding formalisms (Hayes 1978), and naive geography in particular (Egenhofer and Mark 1995). This has led to formalizations of qualitative spatial concepts as expressed in spatial language (Mark 1991; Frank 1996; Freksa 1991; Frank and Mark 1991). Furthermore, it was recognized that there are various ways to understand concepts across cultures and languages with respect to—for instance—geographic landmarks (Smith and Mark 2003), and that this understanding is dynamic and contextually sensitive (Stock 2010).

The varieties in which different languages encode spatial arrangements—through linguistic frames of reference—has been subject of numerous accounts in the field of linguistics (Levelt 1996). Frames of reference have commonly been distinguished between (1) relative, (2) intrinsic and (3) absolute frames, with each frame having an egocentric and allocentric variation (see Levinson 2003). However, few attempts have been made to "map" verbally, textually or visually communicated narratives in cartesian spaces, with the exception of some natural language processing approaches which map spatial references from (Web) texts (Vasardani et al. 2013; Khan et al. 2013). Computational models of linguistic frames of reference have indeed been suggested (Frank 1998; Tenbrink 2011; Tenbrink and Kuhn 2011; Scheider 2016; Kiefer et al. 2015). Furthermore, the field of robotics has made attempts at disambiguating frames of reference from locative expressions (Moratz and Tenbrink 2006; Spranger and Steels 2012). Nonetheless, at present there is an absence of methods that could specify the linguistic frame of reference employed in a spatial utterance and identify the information elements necessary to translate between linguistic (local) and coordinate reference frames. The latter is needed to bring local geographic knowledge out of its linguistic context into a GIS.

3 The Challenge of Transforming Balinese Expressions into GIS

To illustrate the transformation challenge, this paper makes use of locative expressions in the Balinese language. Balinese spatial concepts are grounded in cosmological symbolism based on Balinese-Hindu religion, although this aspect falls beyond the scope of this paper.[1] In Balinese, geographic objects in the natural environment are used as references to orient oneself or another through language. Consider the following locative statements:

> Pakir motor medajo uli deriki. Tia lakar ke pasar ... wenten pasar nika ring kelod.

In English, these can be paraphrased as:

> From here, my motorbike is parked in the direction towards the mountain. I am going to the shop ... there are shops in the direction of the coast.

Simply put, *medajo* (conjugated from *kaja*) refers to the general direction towards a particular sacred mountain. Usually this is the closest or most visible mountain from a community. In the second sentence, *kelod* refers to the general direction towards the coast. In both *kaja* and *kelod*, the geographic objects that underly the semantics of the terms are contextually dependent as they are agreed upon by local convention. As such, the meanings vary across communities: when uttering "*kaja*", one community might refer to a certain mountain whilst another community refers to a different one. For the sake of example, in this paper we imagine this interview to have taken place in the village of Sidemen, where *kaja* points to Mount Agung.

The idea we follow in this paper is illustrated in Fig. 1. The situation is a fictional qualitative interview in the village of Sidemen. A Balinese respondent utters the locative expressions exemplified earlier in this section. Based on knowing the language culture (and asking the respondent themselves) it is possible to identify a corresponding reference frame for this expression from a repository, where each frame is described in terms of an ontology (see Sect. 3.1). In this case, the utterance of "*medajo*" calls for a reference frame that is based on *kaja*. However, since the interview was taken in Sidemen, the local *variation* of the *kaja* reference frame should be used. The encoding of the frame necessitates the instantiation of two empty parameter slots, namely the egoground (since the utterance mentions "from here") and direction slots. The specification of these parameters are explained in the next

---

[1]See Wassmann and Dasen (1998) for a comprehensive overview of Balinese cosmology and its relationship with spatial orientation.

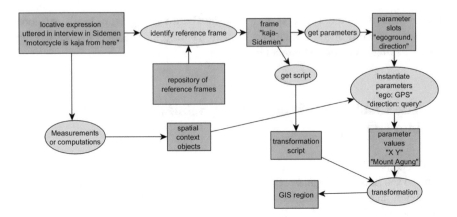

Fig. 1 General approach we follow in this paper. *Yellow circles* are computational steps and *boxes* are their inputs and output

subsection. Once the parameter values are available, a script associated with the corresponding frame can be retrieved and executed to transform the locative expression to an approximate region that can be used in a GIS.

3.1 Specifying and Instantiating Transformation Parameters

In this section, we discuss the basic information elements necessary for transforming the discussed Balinese reference frames to a spatial coordinate system by way of an ontology pattern. The challenge is to identify "anchors" of the frame which can serve as the *parameters of the transformation*. That is, instead of specifying geometric conditions of spatial references upfront, we identify their transformation parameters (Scheider 2016) and operationalize their semantics. In general, we can distinguish the following kinds of transformation parameters, based on the survey of cognitive reference frames described in Levinson 2003; Frank 1998; Scheider 2016:

1. The *ground object*, which corresponds to the origin of the frame. In egocentric frames, the ground object is the speaker or the observer of a scene. In allocentric frames, it is another known object.
2. The *direction* of the frame, which corresponds to its primary axis. The direction can be given by the face of the ground object (intrinsic frame) or in terms of other directional objects.
3. The *spatial template* which circumscribes the relative location. The template does not have a crisp boundary and is scaled with respect to the ground object and the directional parameter. Furthermore, it may also depend on other objects which constrain the form of the template.

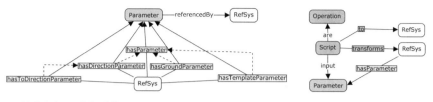

(a) Ontology of the different parameters of a reference frame.   (b) Transformation scripts.

Fig. 2 Ontology design pattern used in this article. *Arrows* denote RDF properties, and *rounded rectangles* denote classes. *Dotted arrows* denote RDF(S) subproperty relations and "are" *arrows* denotes subclass relations

We represent these parameters in terms of an ontology depicted in Fig. 2.[2] We reuse here some of the concepts from the *AnalysisData* ontology[3] (Scheider and Tomko 2016). Reference systems (or reference frames) are denoted by the class *RefSys* and are linked to their parameters by the arrow *hasParameter*, where different parameter types can be distinguished by subproperties (indicated here by dotted arrows). For example, *hasToDirectionParameter* links reference frames to an object that indicates the "to" direction of the frame's primary axis.

In order to make use of these parameters, a tiny pattern (Fig. 2b) can be used to link transformation scripts to reference frames. To compute a particular transformation, parameters need to be instantiated. The difficulty is that in some cases, parameters cannot be determined upfront, but need to be filled in based on local context. Also, note that a Balinese spatial term such as "*kaja*" may refer to different reference frames in different communities.

In the case of the two Balinese directional systems *kaja* (Fig. 3) and *kelod*, the ground object (*hasGroundParameter*) can either be the speaker or an object that anchors an allocentric scene. In our example, this parameter needs to be measured on location (of the utterance) in terms of, for example, a GPS point or an address that can eventually be geocoded to a GPS point *WGS84*[4] since the speaker's own location anchors the scene through the utterance "from here".

Both reference frames make use of a second geographic object as a direction parameter (*hasToDirectionParameter*) and, at the same time, use it as a way to constrain the extent of the spatial template (*hasTemplateParameter*). They thus fall under Levinson's absolute systems (Levinson 2003). In case of *kaja*, this parameter sometimes corresponds to the nearest mountain in the collection of inner mountains of Bali e.g. Mount Agung, which can be identified by a DBpedia[5] Uniform Resource Identifier (URI) http://dbpedia.org/page/Mount_Agung. In other cases, however, this direction parameter can also be specified in terms of a *definite description* like "the nearest inner mountain in Bali", leaving open which mountain exactly is meant until

---

[2] Also available online, see http://geographicknowledge.de/vocab/FrameOfReference.rdf.

[3] http://geographicknowledge.de/vocab/AnalysisData.rdf.

[4] https://en.wikipedia.org/wiki/World_Geodetic_System.

[5] This is a Web resource of linked data based on Wikipedia.

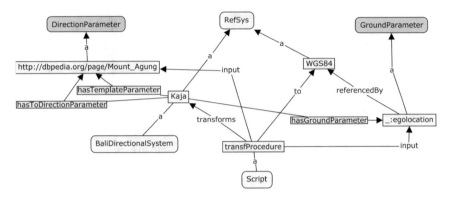

Fig. 3 Linked data model of Balinese cosmological directional system *kaja* as understood in Sidemen

a script evaluates this description when running the transformation. For this purpose, a subset of the inner mountains in Bali can be retrieved by a SPARQL query on DBpedia, along with corresponding information with regards to its coordinate locations[6]:

```
PREFIX dbpedia: <http://dbpedia.org/resource/>
PREFIX dbpedia-owl: <http://dbpedia.org/ontology/>
PREFIX geo: <http://www.w3.org/2003/01/geo/wgs84_pos#>

SELECT DISTINCT ?m ?lat ?long ?geometry
WHERE {
    {{?m a dbpedia-owl:Volcano} UNION {?m a dbpedia-owl:Mountain}}
        ?m dbpedia-owl:locatedInArea dbpedia:Bali.
        ?m  geo:lat ?lat ; geo:long ?long ; geo:geometry ?geometry.
}
```

Listing 1 SPARQL query retrieving all inner mountains in Bali

In the case of *kelod*, this directional parameter is the Bali Sea, which can be obtained from an existing source, such as the VLIZ Maritime Boundaries Geodatabase (Claus et al. 2017) by the International Hydrographic Organization.

3.2 Computing Transformations

Once parameter values for a frame are determined, it is necessary to have a procedure to compute the transformation. Here, we are confronted with a range of difficulties that go largely beyond the scope of this paper (compare next section). However, a simplistic approximation of our scenario can be computed by way of

---

[6]Try it out here: http://yasgui.org/short/rJPCmwxjg.

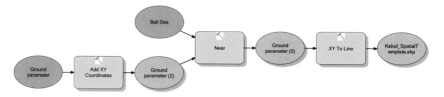

Fig. 4 Geoprocessing script which generates the spatial template *kelod* from the ground parameter and the nearest point to the coast

simple geoprocessing scripts. The scripts were made with ModelBuilder, which is a visual programming language for chaining geoprocessing operations available with ArcGIS.[7]

To compute the spatial location denoted by *kaja*, it is necessary to georeference the two parameters of the transformations (the egocentric "ground" and the direction parameters). Suppose the location of the egocentric ground (the place of the interview) be positioned at latitude −8,46991 and longitude 115,444092. The direction parameter is based on an inner Balinese mountain collection as retrieved from the SPARQL query presented in Sect. 3.1, from which the "nearest" mountain (from the ground location) or a particular name needs to be selected. From there, it becomes possible to generate a line denoting the spatial template *kaja* between the ground parameter and the nearest mountain, which in this case is Mount Agung. The script to approximate the spatial location denoted by *kelod* is depicted in Fig. 4. The ground parameter remains the same. Instead, the *Near* operation[8] is used to pin a point along the coast closest to the ground parameter and record it in the attribute table of the ground parameter. Finally, a line denoting the spatial template *kelod* was generated between the new direction parameter and ground parameter using the *XY to Line* operation.[9]

The map in Fig. 5a shows the results of running corresponding scripts on a map in which the spatial templates *kaja* and *kelod* are represented in a simplified way as lines.

3.3 Open Problems

Our investigation into information requirements and simplistic GIS transformations exposed a number of open problems:

1. *Distinguishing appropriate frames of reference for relevant locative expressions in Balinese.* The difficulty lies in the fact that a term like "*kaja*" refers to different

---

[7]http://desktop.arcgis.com/en/arcmap/10.3/analyze/modelbuilder/what-is-modelbuilder.htm.

[8]http://desktop.arcgis.com/en/arcmap/10.3/tools/coverage-toolbox/near.htm.

[9]http://pro.arcgis.com/en/pro-app/tool-reference/data-management/xy-to-line.htm.

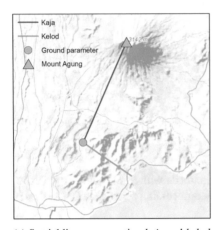

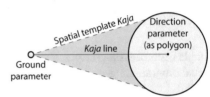

(a) Spatial lines representing *kaja* and *kelod*. The speaker's motorbike is located roughly along the pink line, and the shops along the green line.

(b) Spatial template *kaja* which takes into account the width of a polygon representing the mountain (which is the direction parameter).

Fig. 5 Representations of templates in a GIS

frames throughout local communities, and so Web identifiers need to be generated for variants of a term. In general, locative expressions often have ambiguous semantics (in addition to being spatially vague), which requires empirical investigation.

2. *Identifying necessary parameter slots, extending ontologies accordingly and filling an online repository.* For this purpose, we would need more in-depth investigations of the different kinds of transformations underlying locative expressions in order to add nuance to our model, extending the approach as exemplified in this paper.

3. *Adding procedures to instantiate a given parameter slot of a frame with a value depending on context of utterance.* For example, egocentric positions need to be determined on the fly, and direction parameters may need to be selected from a set of objects (Mount Agung), as we have shown in Listing 1. A problem here is incomplete information on the Web (hardly all mountains have a DBpedia entry, and there is no information available as far as their religious—and thus linguistic—significances are concerned).

4. *Approximating the vague location that is meant by a locative expression.* This also needs empirical and experimental research for evaluation. The methodological difficulty is manyfold (Scheider 2016): How can a cognitive template be vague and which inherent forms should be used? How can the transformation take into account extended and vague instead of point-like parameters (i.e. extended ground or direction objects)? How can the template be scaled with respect to the size of parameter objects? How can the transformation take into account non-Euclidean directions (e.g. along a road or a sea shore)? For example, a general-

ized mountain polygon would have allowed for a cone-shaped spatial template of *kaja*, the base width of which could be determined by the width of the mountain (which is, of course, vague in itself) (see Fig. 5b).

4   Conclusion

In this paper, we have investigated the information necessary to transform locative expressions in Balinese into a GIS suitable form. We have proposed a workflow that captures the transformation process, starting from a local interview situation and ending with a GIS representation. Our investigation was based on two examples (*kaja* and *kelod* in the village Sidemen), for which we worked out in detail how transformation parameters can be encoded into an ontology to build up a repository of local frames supporting transformation processes. Also, we discussed a simplistic transformation procedure in GIS. Based on this example we discovered a range of open scientific challenges that need to be solved in order to actually realize a practically useful transformation procedure that does justice to the concepts underlying spatial utterances. While ontologies can serve as a way to document and share reference frames, their parameters and transformation procedures, a difficulty lies in incomplete knowledge about frame variants and their parameters. Another challenge is that parameters can be context dependent, and thus depend on the situation of the utterance. For this purpose, we suggested open parameter slots and instantiation procedures which allow for flexible and locally sensitive adaption of the transformation methods. One of the biggest challenges, however, is that current GIS technology offers only very limited transformation procedures, missing aspects such as vagueness, extended parameter objects, object dependent scalings, and non-linear directions. Future work should therefore concentrate on these challenges and empirically test the suitability of such transformations.

References

Claus S, de Hauwere N, Vanhoorne B, souza Dias F, Garcia PO, Hernandez F, Mees J (2017) Marineregions.org. http://www.marineregions.org

Egenhofer MJ, Mark DM (1995) Naive geography. In: International conference on spatial information theory. Springer, pp 1–15

Frank AU (1996) Qualitative spatial reasoning: cardinal directions as an example. Int J Geogr Inf Sci 10(3):269–290

Frank AU (1998) Formal models for cognition taxonomy of spatial location description and frames of reference. In: Freksa C, Habel C, Wender K (eds) Spatial cognition, vol 1404. Lecture notes in computer science. Springer, Heidelberg, pp 293–312

Frank AU, Mark DM (1991) Language issues for geographical information systems. In: Principles and applications, geographic information systems, pp 129–150

Freksa C (1991) Qualitative spatial reasoning. In: Mark DM, Frank AU (eds) Cognitive and linguistic aspects of geographic space, vol 63. Springer, Netherlands, pp 361–372

Hayes PJ, et al (1978) The naive physics manifesto. Institut pour les études sémantiques et cognitives/Université de Genève

Khan A, Vasardani M, Winter S (2013) Extracting spatial information from place descriptions. In: Scheider S, Adams B, Janowicz K, Vasardani M, Winter S (eds) ACM SIGSPATIAL International workshop on computational models of place, COMP 2013, 5 Nov 2013. Orlando, Florida, USA

Kiefer P, Scheider S, Giannopoulos I, Weiser P (2015) A wayfinding grammar based on reference system transformations. In: Fabrikant SI, Raubal M, Bertolotto M, Davies C, Freundschuh SM, Bell S (eds) Spatial information theory—12th international conference, COSIT 2015, Santa Fe, NM, USA, 12–16 Oct 2015, Proceedings, pp 447–467

Levelt WJ (1996) Perspective taking and ellipsis in spatial descriptions. In: Language and space, pp 77–108

Levinson SC (2003) Space in language and cognition: explorations in cognitive diversity, vol 5. Cambridge University Press

Mark DM, Frank AU (1991) Cognitive and linguistic aspects of geographic space, vol 63. Springer

Moratz R, Tenbrink T (2006) Spatial reference in linguistic human-robot interaction: Iterative, empirically supported development of a model of projective relations. Spat Cogn Comput 6(1):63–107

Scheider S (2016) Neural GIS computing with cognitive spatial references. In: 2016 Specialist meeting universals and variation in spatial referencing

Scheider S, Tomko M (2016) Knowing whether spatio-temporal analysis procedures are applicable to datasets. In: 9th international conference on formal ontology in information systems (FOIS 2016). IOS Press

Smith B, Mark DM (2003) Do mountains exist? towards an ontology of landforms. Environ Plan B 30(3):411–428

Spranger M, Steels L (2012) Emergent functional grammar for space. In: Experiments in cultural language evolution. John Benjamins, Amsterdam. doi:10.1075/ais3

Stock K (2010) Describing spatial relations using informal semantics. In: Proceedings of GIS Research UK (GISRUK 2010), pp 14–16

Tenbrink T (2011) Reference frames of space and time in language. J Pragmat 43(3):704–722

Tenbrink T, Kuhn W (2011) A model of spatial reference frames in language. In: Spatial information theory. Springer, Heidelberg, pp 371–390

Vasardani M, Winter S, Richter KF (2013) Locating place names from place descriptions. Int J Geogr Inf Sci 27(12):2509–2532

Wassmann J, Dasen PR (1998) Balinese spatial orientation: some empirical evidence of moderate linguistic relativity. J R. Anthropol Inst, 689–711

Place as Location Categories: Learning from Language

Clare Davies and Thora Tenbrink

Abstract How do people refer to places in their environment, and to what extent do the underlying spatial concepts correspond to officially defined regions? We exemplify some types of evidence that may help to determine local vernacular place concepts. The output of latent semantic analysis (LSA) on a web-scraped text corpus was compared with mapping and linguistic data from a pilot experiment, to see how localities within the same geographic area tended to be clustered, how far the spatial geography is similarly distorted, and how far participants' verbal protocols revealed a tendency to group places together (and how). Finally, we list some challenges for future triangulation of such data sources, in deriving vernacular place data.

Keywords Place · Vernacular geography · Latent semantic analysis (LSA) · Spatial cognition · Spatial language · Regions · Categorical reasoning

1 Introduction

The human tendency to name and store knowledge of places that are regions, i.e., larger than a single landmark or point, may be considered as a cognitively efficient means of categorizing known locations in space. This is true both at the scale of single point locations experienced through wayfinding, and at higher hierarchical levels such as grouping localities within cities, and thence into larger geographic regions. Categorization allows assumptions, reasoning and linguistic references to be applied across a range of category members all at once, saving effort in processing and communication (Hahn and Ramscar 2001).

C. Davies (✉)
Department of Psychology, University of Winchester, Winchester SO22 4NR, UK
e-mail: clare.davies@winchester.ac.uk

T. Tenbrink
School of Linguistics and English Language, Bangor University, Bangor LL57 2DG, UK
e-mail: t.tenbrink@bangor.ac.uk

© Springer International Publishing AG 2018
P. Fogliaroni et al. (eds.), *Proceedings of Workshops and Posters at the 13th International Conference on Spatial Information Theory (COSIT 2017)*, Lecture Notes in Geoinformation and Cartography, https://doi.org/10.1007/978-3-319-63946-8_37

Officially (i.e., administratively) defined places tend to have rigid, non-overlapping boundaries, with a degree of hierarchy. However, the usage of place names and place-based reasoning by their human inhabitants, even within relatively modern and planned New World cities, is much messier. The key challenge for so-called 'vernacular geography' is to capture people's fuzzy, context-changeable and possibly spatially distorted understanding of such places, so that the typically intentional extent of a given toponym can be modeled, and so that locations or localities which are seen by local people to 'belong' together (or not) can be identified even where their groupings lack a recognized toponym.

In this paper, we discuss ways of addressing this challenge through analyzing natural language data mentioning local places. After defining the key aspects of human cognitive categories and applying them to vernacular places, two relevant data sets will be compared: data from a pilot study in which participants were recorded talking through a mapping task for localities in their home area, and the outputs of latent semantic analysis (LSA) performed on a web-scraped text corpus mentioning the same set of places. Ideally, in language data, place-as-category information may be extractable not only from the co-occurrences of place names (toponyms) within sentences, but also from additional verbal cues implying clustering of localities (villages and suburbs) into groupings which may, but may not always, have a collective toponym applied to them.

The category information implied by such verbal data can also be triangulated with that from further sources, and synthesized into a composite model of the common cognitive groupings of localities in a given area. In the future, learning 'classifier' algorithms may be employable to 'learn' the local vernacular geography from a variety of sources, as available. First, however, we need to improve our understanding of how and when different phrases can be taken as indicating spatial 'place' categorization—and when they cannot.

2   Places as Categories of Locations

In what sense, can places be viewed as locational categories? Arguably, any place larger than a single point must contain a collection of individual locations—be they navigational landmarks and intersections, or a string of villages along the shore of a large water body. Montello (2003) argued that cognitively, regions reflect the general human tendency to organize knowledge categorically, trying to minimize within-category variation and maximize between-category differences—often to the point of stereotyping or over-generalization. As Montello pointed out, this tendency apparently aids cognitive efficiency and avoids spurious precision in our assumptions and speech.

However, half a century of research into categorical cognition has gone way beyond this general observation. Categories in human cognition have a range of well-established properties, emerging from several decades of research, which in turn have specific implications for our understanding of place. Such properties include fuzziness (Hampton 2007), graded membership (often depending on 'ideal' comparison rather than 'typicality'—see (Barsalou 1985; Davies 2009), and classification using characteristic but not necessarily defining (necessary and sufficient) features. These kinds of properties may help us to identify suitable machine-learning classifiers for building place knowledge based on human-sourced data, since they would in this case need to be cognitively plausible rather than factually accurate.

A few studies have shown evidence that categorical thinking about locations within places can also be influenced in similar ways to experiments on semantic categorization (e.g., Davies 2009). Meanwhile, a body of work mainly focused at what Montello (1993) defined as figural-scale spaces (Newcombe et al. 1999), and figural-scale representations of geographic spaces (Tversky 1981; Friedman 2009) has demonstrated what might be thought of as 'category errors' in spatial memory and thinking. Both in adults and children, locations and shapes of dots, lines and geometric forms tend to be mentally simplified and distorted to more regular or distinctly clustered patterns (Newcombe et al. 1999; Tversky 1981). Similarly, we see similar spatial distortions in mental representations of geographic locations: individual items may be clustered together more distinctively, along straighter lines and with broader separations between clusters, than in physical reality (Hirtle and Jonides 1985; Lloyd and Heivly 1987).

If this is how people remember and reason about locations—such that their grouping into places distorts the space into one akin to the more semantic 'mapping' of categories in non-spatial domains—then it would be useful to know whether we might see the same patterns and tendencies show up in different sources of (vernacular) data about the same set of places. For a given geographic area, we should expect that the locational place category memberships obtained from linguistic data will indicate similar patterns to the results of an experimental mapping task, and to corpus data scraped from the web or social media and reduced to a 'semantic map'. We may later be able to triangulate such information sources together (Gao et al. 2017), to build predictive models of place grouping and toponym referents. Furthermore, if spatial regions can be viewed as a special case of general cognitive categorizing, then we may be able to apply to models of 'place' many of the findings and models from half a century of cognitive science research on, and models of, semantic memory and reasoning.

The next section will compare such example outcomes from two very different data sources: one set re-analyzed from a previous project in the Southampton area of southern England, and another collected as a pilot human-subjects experiment.

3   Comparing Sample Data Sources: Pilot Evidence

3.1   Web-Sourced Text Corpus

The example data shown in this section was previously extracted and analyzed as part of a project presented at COSIT'13 (Davies 2013). The project, as previously reported, was attempting to replicate work by Max Louwerse (Louwerse and Zwaan

2009), demonstrating that the geographic positions of named places could be replicated via latent semantic analysis (LSA Landauer et al. 2007[1]) on a corpus of non-georeferenced online text that mentioned them. The text consisted of web pages 'scraped' from the internet via a corpus builder, and the pages were not selected as 'spatial' descriptions at all. They merely had to mention at least one of the identified locality names together with 'Hampshire' (the enclosing administrative county—to minimize false positives from identical toponyms elsewhere), and within the context of continuous text rather than a list.

Following Louwerse's method as far as he had specified, the corpus was processed using an implementation of LSA, and the resulting matrix of associations among the toponyms was subjected to multidimensional scaling (MDS) to reduce the dimensionality to two. The final 'map' of the toponyms was then geometrically transformed using an affine transformation, and the result compared to the true geographic locations, using Tobler's bidimensional regression technique (Tobler 1994). The whole process was run iteratively, eventually optimizing the final r^2 value in the bidimensional regression to around 0.8. Therefore, the final map derived from the original webpage corpus was as close to overall topographic accuracy as possible.

As discussed in Davies (2013) and illustrated in Fig. 1, this 'optimized' map still showed certain patterns of distortion compared to the localities' true locations:

1. A preservation, but also a broadening, of the central geographic divide that exists within this area due to Southampton Water—an unbridged sea inlet approximately 2–3 km wide (see Fig. 1).
2. Exaggerated clustering of geographically close localities, which in reality are much more evenly distributed through the space. Thus, neighboring localities were moved closer together, and further away from other clusters.
3. Some degree of 'cardinalizing' of spatial directions: Southampton Water seemed rotated to a north-south axis, and localities along its irregular shoreline were arranged in straighter linear configurations than in reality.

The question we now ask about these distortions is whether they might correspond to a genuine cognitive geography, in the minds of local people such as those who (usually) wrote the online texts which were scraped for the corpus. Otherwise, the above could simply reflect limitations and flaws in the computational analysis method, or the lack of genuine spatial locational information inherent in the online texts themselves. Thus, for present purposes, this data serves as a basis for comparison with the more directly human-sourced pilot data presented below.

---

[1]LSA was adopted here because of Louwerse's previous success, and because other methods tend to presume groupings of points exist from the start; we did not.

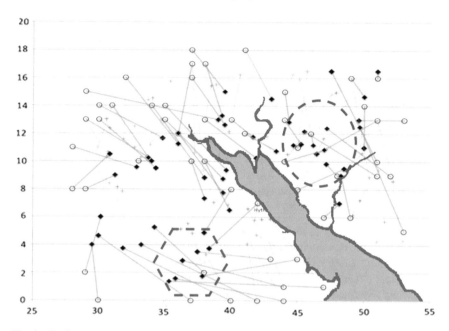

Fig. 1 Configuration of Southampton area localities: true locations (*open circles*), mean placement by pilot study participants (*black diamonds*), and coordinates extracted from LSA on web-sourced text corpus (fainter + signs)

3.2 Pilot Empirical Data: Dot-Placement Task

To begin to 'ground-truth' the above data, by comparing it to individual local human perceptions of the geography of the same area, a pilot empirical study has been performed. This involved a small field experiment in which, after briefing, participants had to arrange and place labeled foam counters onto a blank sheet of white card, cut to match the proportions of the area in question. Each counter bore a number, and had a label attached giving the name of the locality it represented. Participants were encouraged to talk aloud during the task, to provide a verbal account of their decision processes while placing the counters. When they declared themselves finished (without placing any completely unknown names, as guessing was discouraged), the 'map' was photographed.

To compare the relative locational patterns, the pilot data from eight participants was transformed to the same scale as the above LSA data and the 'true' coordinates of the set of localities. Figure 1 combines the true locations of each locality (hollow circles) with the mean coordinates of each locality's placement (black diamonds) in the mapping task, across the eight pilot participants (all long-term residents of, and tested in, Hythe—a village shown near the center of the map), and with the LSA-derived locations ('+' symbols) discussed earlier. For simplicity, cartographic detail is excluded apart from the approximate shoreline of Southampton Water.

The lack of points across the center-left and towards bottom right reflect part of the New Forest National Park and Southampton Water, respectively.

It will again be noted that relative to the original distribution of localities (the circles), both the task- and LSA-based locations are more closely clustered. For discussion purposes, two of the task-based clusters are tentatively outlined with red dashes in Fig. 1, representing:

1. Hexagon: a string of seven settlements (including Hythe) on or near the western shore of Southampton Water, known locally (though not in any map or gazetteer) as 'the Waterside'. (Note that the fainter lines in Fig. 1 also show the same group of places clustering at a slightly different location in the LSA-based map.)
2. Oval: a similarly tight clustering of eleven eastern Southampton suburbs on the other side of Southampton Water.

The second cluster is also close to the placement of the city of Southampton itself, and to its western suburbs of Shirley and Millbrook—in other words, the city's urban geography (less well known to Hythe residents) is effectively shrunk. The clusters effectively move the two shores further apart, emphasizing the role of (bridgeless) Southampton Water in dividing the area.

It appears that while to some extent the clustering is similar to that found in the LSA data, the relative locations and memberships of the clusters have partly shifted in some cases, although this may simply reflect this small and hence unreliable pilot sample. The geography of the area in general is again clearly distorted—with Southampton Water not only widened but also rotated almost 90°—and this reflects the tendency by most participants to leave a vertical north-south space for the inlet, rather than its true northwest-southeast orientation. This presumably reflects the tendency to rotate and simplify axes and coastlines, noted by many previous cognitive studies (Tversky 1981; Lloyd and Heivly 1987). Thus, this data, and possibly the LSA data, does seem to reflect known cognitive biases in spatial representation. More importantly, it clearly begins to reveal the clustering of places at this scale by local inhabitants.

3.3 Pilot Linguistic Data

As mentioned earlier, participants in the pilot study were encouraged to provide simultaneous verbal protocols while laying out the counters. If the clustering and distortions in the layout reflect a categorizing tendency within people's stored mental geography, rather than ignorance or lazy shortcutting in performing the spatial task, then we would expect to see this also reflected in the way participants described the locations in question. Although the transcripts have not been formally (let alone quantitatively) analyzed as yet, it is easy to find hints of categorical thinking about groupings of places, using various words for such groupings as shown in the examples below. Note that often, the groupings are initially linear along major roads or shorelines, but still ending up closer together than their real-life positions (Author's italics.).

P01: "so mostly *stacked up* along the waterfront, south-west of Southampton … right by Fawley, Dibden and Marchwood y'know getting stacked up…"

P02: "Everything's in *groups*. Groups because I know, if I've travelled around that area I'll recognize the names…"

P03: "Marchwood I've heard of and I think of that as being sort of on the way to Totton so I'd probably *put them together*… I'd probably put that with Bucklers Hard in a New Foresty *part of the world*… These are all places that I think of as being *along the A326* [road] although some of them aren't exactly."

P04: "Winsor is somewhere *in amongst this lot* I think… Durley is over with *this little batch*."

P05: "… have to imagine the Southampton water going down there and I'll try and put *all of these uh these foreign east of the water places*…"

P07: "So what I'm doing I'm now looking for *places in the Waterside* as I can use it as *a lateral line* up to Southampton… Just *bunching* um Brockenhurst Lyndhurst and Beaulieu up closer together… So I'll be using *the line of the M27* [highway] to get a lot of the *places along there*…"

P08: "Pooksgreen that can be *part of the Waterside*… Exbury mm just need to go here in a *little clump* this can go there with *the foresty ones*… have to move *the Waterside* over a bit… Bartley that can go over with *the Winsor Cadnam lot*… Calmore can *snuggle in there with* Totton and Testwood."

4 Discussion and Challenges

The above data is being presented mainly to stimulate discussion; clearly there is a long way to go before reliable empirical and linguistic data can be triangulated with web-sourced data to enable derivation of local vernacular geography. Some key questions arise, not least the extent to which the apparent groupings of places in both the task and verbal data were an artifact of the counter-placing task (given the sheer number of counters —55—that needed to be placed somewhere on the card). Even within the small pilot sample, it was also evident that different strategies might be applied—e.g., participant 06 gave no indication of clustering places during the task performance. This corresponds to earlier findings indicating different spatial strategies across individuals (e.g., cluster-based vs. trajectory-based navigation strategies in Tenbrink and Wiener (2009)).

Nevertheless, the tendency to simplify spatial memory via categorical encoding has already been demonstrated at various other spatial scales (e.g., Newcombe et al. 1999; Tversky 1981; Friedman 2009; Hirtle and Jonides 1985; Lloyd and Heivly 1987; Lansdale 1998), so it would be surprising if people did not also tend to group localities in this way, notwithstanding their direct experience of them as navigated regions in themselves (unlike the data points used in most previous studies).

And indeed, the parallels between our different data sets in this respect were striking. Although the actual locations on the map did not necessarily coincide, the cognitive biases were clearly present across the board, in terms of clustering, simplifying and 'shrinking' the available space—even though at least half of the space was highly familiar to participants from an immersed, egocentric perspective.

Data such as this, when collected from wider samples, may help us to model why and how place categories form in familiar spaces (as opposed to those only experienced as haptic representations), and how far those categories match the 'messy' aspects of human categories mentioned at the start of this paper. It appears that toponyms are not essential for clustering to occur—we can cope with spatial categories which are effectively nameless—so under what circumstances do toponyms get applied consistently enough, by enough local residents, to become established as verifiable vernacular geography?

Scale is also a consideration. Here, unusually in spatial cognition studies, we were considering the scale above that of an urban streetscape or university campus but below national level; the places that were apparently being grouped into categories were suburban and village settlements or localities, themselves each already a collection of locations which afford the individual 'vistas' or 'reference frames' discussed by work such as Meilinger et al. (2014). Those in turn, of course, tend to represent a collection of different points (and potential viewpoints) within a single scene or 'vista' space. In this study we were considering a relatively high level of the place hierarchy—an area covering approximately 432 km^2. To what extent do different principles and heuristics apply at different scales, for the formation of place clusters or categories? Will these be reflected in linguistic utterances about them?

This reminds us of still further questions, concerning the role of linguistic data in trying to identify vernacular place categories. In the above pilot study, a wide range of phrases was used to indicate groupings of places—making it difficult to imagine the use of any kind of automatic language parsing to help to identify them. Even so, trawling data for similar 'grouping' hints in people's discussions of local places—even if it could never be exhaustive—might in future help to augment a cruder, more co-occurrence-based approach to toponym groupings.

References

Barsalou LW (1985) Ideals, central tendency, and frequency of instantiation as determinants of graded structure in categories. J Exp Psychol Learn Mem Cogn 11:629–654

Davies C (2009) Are places concepts? Familarity and expertise effects in neighborhood cognition. In: Wrac'h A Spatial information theory: 9th international conference COSIT 2009, vol. 5756, France, 21–25 Sep 2009. LNCS. Springer, Berlin, pp 36–50

Davies C (2013) Reading geography between the lines: extracting local place knowledge from text. In: Spatial information theory: 11th international conference COSIT 2013, vol 8116, Scarborough, UK, 2–6 Sep 2013, Proceedings. LNCS. Springer, Berlin pp 320–337

Friedman A (2009) The role of categories and spatial cuing in global-scale location estimates. J Exp Psychol Learn Mem Cogn 35(1):94–112

Gao S, Janowicz K, Montello DR, Hu Y, Yang J-A, McKenzie G, Yan B (2017) A data-synthesis-driven method for detecting and extracting vague cognitive regions. Int J Geogr Inf Sci 31(6):1245–1271

Hahn U, Ramscar M (2001) Similarity and categorization. Oxford University Press, Oxford

Hampton JA (2007) Typicality, graded membership and vagueness. Cognitive Science 31:355–384

Hirtle SC, Jonides J (1985) Evidence of hierarchies in cognitive maps. Mem Cogn 13:208–217

Landauer TK, McNamara DS, Dennis S, Kintsch W (2007) Handbook of latent semantic analysis. Lawrence Erlbaum, Mahwah, NJ

Lansdale MW (1998) Modeling memory for absolute location. Psychol Rev 105(2):351–378

Lloyd R, Heivly C (1987) Systematic distortions in urban cognitive maps. Ann Assoc Am Geogr 77(2):191–207

Louwerse MM, Zwaan RA (2009) Language encodes geographic space. Cogn Sci 33:51–73

Meilinger T, Riecke BE, Bülthoff HH (2014) Local and global reference frames for environmental spaces. Quart J Exp Psychol 67(3):542–569

Montello D (1993) Scale and multiple psychologies of space. In: Frank AU, Campari I (eds) Spatial information theory: a theoretical basis for GIS, Proceedings of COSIT '93. Springer, Berlin, pp 312–321

Montello DR (2003) Regions in geography: Process and content. In: Duckham M, Goodchild MF, Worboys MF (eds) Foundations of geographic information science. Taylor & Francis, London, pp 173–189

Newcombe N, Huttenlocher J, Sandberg E, Lie E, Johnson S (1999) What do misestimations and asymmetries in spatial judgment indicate about spatial representation? J Exp Psychol Learn Mem Cogn 25(4):986–996

Tenbrink T, Wiener J (2009) The verbalization of multiple strategies in a variant of the traveling salesperson problem. Cogn Process 10(2):143–161

Tobler W (1994) Bidimensional regression. Geogr Anal 26:186–212

Tversky B (1981) Distortions in memory for maps. Cogn Psychol 13:407–433

Frame-Relative Constructions in the Description of Motion

Ekaterina Egorova and Ross S. Purves

Abstract Fictive motion in language is the manifestation of our cognitive bias towards dynamism and a good example of the metaphoric nature of much of our thought and linguistic effort. This paper investigates one of its types, frame-relative constructions ("I sat in the car and watched the scenery rush by") in a corpus of alpine narratives. We report on the types of constructions found and suggest the communicative motivations behind their use, followed by the examination of the variety of their linguistic encodings. Finally, we raise the question how the nuanced concepts represented by such constructions relate to those already embedded in an existing spatial language annotation scheme.

Keywords Fictive motion · Frame-relative paths · Alpine corpus

1 Introduction

The recent growth of data in the form of digitized corpora and online user-generated content offers multiple opportunities for answering questions related to human perception of, and experiences in, diverse places. Working with large corpora, however, requires that we can capture a variety of space-related concepts including references to locations, directions and motion events (e.g. Jones and Purves 2008; Mani et al. 2010; Moncla et al. 2016; Pustejovsky and Yocum 2013). In Geographic Information Science the annotation and extraction of space-related concepts from

E. Egorova (✉) · R.S. Purves
Department of Geography, University of Zurich, Winterthurerstrasse 190,
8057 Zurich, Switzerland
e-mail: ekaterina.egorova@geo.uzh.ch

R.S. Purves
e-mail: ross.purves@geo.uzh.ch

E. Egorova · R.S. Purves
URPP Language and Space, University of Zurich, Freiestrasse 16,
8032 Zurich, Switzerland

© Springer International Publishing AG 2018
P. Fogliaroni et al. (eds.), *Proceedings of Workshops and Posters at the 13th International Conference on Spatial Information Theory (COSIT 2017)*, Lecture Notes in Geoinformation and Cartography, https://doi.org/10.1007/978-3-319-63946-8_38

227

natural language text remains a challenging tasks and is currently mostly grounded in prototypical linguistic structures. Apart from the crucial role that context plays in the adequate interpretation of spatial language (Bateman 2010; Pustejovsky and Moszkowicz 2014), the metaphoric nature of thought (Lakoff and Johnson 1980) and the fact that much of our language production does not describe "reality" directly (Langacker 2005) hinder the progress in the development of spatial annotation schemes.

In particular, motion, as one of our basic experiential domains, is argued to be responsible for a general "cognitive bias towards dynamism" (Talmy 2000). One of the manifestations of the latter is fictive motion, that is to say linguistic representation of some static phenomenon as dynamic. For example, in (1a), the scenery is actually static; the literal meaning of the sentence, however, represents it as moving.

1. a. I sat in the car and watched the scenery rush by (Talmy 2000).

This is an example of a so-called frame-relative path, one of the types of fictive motion constructions proposed by Talmy (2000). Essentially, such constructions refer to the situation of a moving observer; however, instead of representing the observer as moving in relation to stationary surroundings, they adopt an observer-centered perspective, representing the surroundings as moving. Although fictive motion has gained significant scientific attention, research has largely focused on coextension paths (e.g. "The road goes along the ridge"), using elicitation experiments (Rojo and Valenzuela 2016) or samples of general corpora (Taremaa 2013). Examination of fictive motion in discourse remains a largely neglected dimension (Matlock and Bergmann 2014). Further, although the existence of fictive motion has been acknowledged by the introduction of the motion_sense tag in ISO-space (Pustejovsky and Yocum 2013), a closer investigation of its linguistic encoding and the concepts behind has not yet found its way into this line of work.

To address this gap, we set out to examine the nature of frame-relative constructions in a very specific natural discourse—alpine narratives, as part of a larger project investigating space conceptualization in the context of mountaineering using a corpus-based approach (Egorova et al. 2015, 2016). The context is characterized by a close interaction through actors (mountaineers) with space and the resulting conceptual importance of motion (typically in describing ascents). Examining patterns of frame-relative constructions use in such a focused, specialized corpus could provide clues to their nature and uncover specifics of space and motion conceptualization in this particular context. In this exploratory study, we aim to answer the following questions:

1. What is the nature and communicative motivation of frame-relative paths in our corpus—i.e., which spatial concepts do they represent?

2. How are these constructions linguistically encoded—i.e., would they be easily operationalizable for automatic annotation?

Note that though fictive motion and metaphoric use of language in general have seen much attention in some fields, the implications and applications which explicitly take account of these effects are still important challenges in current research attempting to extract and analyse spatial language from a geographic perspective.

2 Data and Methodology

We compiled a small corpus describing ascents from the webpage of the American Alpine Club[1] (sections "Featured Articles" and "Climbs and Expeditions"). The corpus consists of 33 articles, with 55 665 tokens in total.

The preliminary annotation described here was performed by the first author. In line with the principles of content and discourse analysis (Krippendorff 2004), several iterations and modifications were conducted before the scope of relevant constructions was identified. As a starting point, constructions depicting static surroundings of the observer as moving were sought. These were identified by verbs of motion (e.g. *come*, *appear*) associated with landscape terms (e.g. *glacier*, *mountain*) as subjects. However, it was soon noted that landscape terms used with verbs of motion can represent not only (distant) surroundings, but also the immediate *path* of the observer. Furthermore, cases were encountered, where the dynamicity was encoded not by a verb of motion, but by a verb (or verbal phrase) representing the *change* of some space property (e.g. *steepen*, *widen*). Those were included as boundary cases, based on the reasoning that the dynamicity component, though more subtle, is still present. Finally, the constructions found were examined for the presence of further markers that could reveal their communicative goal and give potential insights into spatial thinking.

3 Results

In total, 45 constructions were identified as frame-relative paths. They are represented by four types, based on the combination of the two parameters: proximity to the observer's path and the saliency of dynamicity (motion versus change). As mentioned above, prototypical structures encode the motion of the observer through the depiction of (distant) surroundings as moving (we further refer to them as vista constructions). The three boundary cases we encounter are the depiction of the fictive change of certain properties of the surroundings and the path, and finally the fictive motion related to the path. In what follows, we describe each of these types in detail.

Frame-relative vista motion makes up roughly one quarter of all cases and depicts large-scale landscape features moving towards or away from the observer (2a) or (dis)appearing in their vista (2b). Such constructions appear to either primarily encode the progress of the observer (2a) or introduce a new object into the description of a vista (2b). The range of types of verbs and verbal phrases in these constructions is rather limited: *come into view*, *get closer*, *appear*, *disappear*.

2. a. With the glacier getting closer after every rappel, I begin to unwind and relax.
   b. To the northwest in particular, exciting unclimbed objectives were coming into view.

---

[1] http://publications.americanalpineclub.org/search.

Frame-relative path motion refers to (dis)appearance of small-scale landscape features or types of terrain on the path itself (3a–b). Although implicitly describing the motion of the observer (in 3a, edges are static, it is the progress of the observer that makes them appear), they seem to primarily encode those aspects of space that are relevant for locomotion—note "luckily" and "thankfully". The only two types of verbs we encountered in such constructions are: *appear*, *disappear*.

3. a. Luckily, a few edges appeared.
    b. Thankfully, the ice soon disappeared.

Frame-relative vista change depicts the changing properties of distant objects—note "at the base of the face" and "far below" in (4a–b)—as a function of increasing distance. As in case of frame-relative vista motion, such constructions reveal the progress of the observer, highlighting the increasing distance between her current location and the starting point of the motion event. (4b) represents an interesting case, demonstrating that the interpretation of motion and increasing distance is not limited to visual perception.

4. a. The trees at the base of the face were starting to get small.
    b. The Chilliwack River became nothing more than a faint whisper far below.

Frame-relative path change describes the change of the path as the observer moves along it, which can be further reinforced by locative phrases—note "between my feet" and "I was in" in (5c–d). One type of such constructions encodes the spatial termination of a landscape feature or type of terrain (5a), essentially communicating the accomplishment of a route segment. Linguistically it is represented by the verbs *run out*, *diminish*, *peter out*, *end*. On a still finer level of granularity, it is some locomotion-relevant property of the path that can be represented as changing: steepness (5b), dimension (5c–d) or type of terrain (5e). Linguistically, frame-relative change is represented by a diverse set of verbs and verbal phrases, which pertains, on the one hand, to the variety of space properties out there in the "real world", and on the other hand, to several options for their verbalization. Thus, change in steepness can be referred to through semantically-related verbs such as *steepen* used with nouns representing landscape features (as in 5b); but can also be referred to by nouns such as *angle* and verbs such as *ease*, *back off*, *decline* (e.g. "The angle declined"). Encountered encodings of the change of dimension include *widen*, *close in*, *open up*, *narrow down*; of the type of terrain—*change, thin out, become loser, become solid, begin to be covered by*.

5. a. The rock ends, and we follow a short snow ridge.
    b. The pillar now steepened.
    c. The crack I was in petered out.
    d. Between my feet the dihedral diminished.
    e. Then it [ice] thinned out.

Frame-relative path change appears to be the least dynamic type of all the encountered constructions, which clearly puts it at the boundary of the phenomenon. On the one hand, they reflect the local, observer-based perspective, essentially adopting "a moving proximal perspective point with local scope of attention" (Talmy 2000). On the other hand, they are highly conventionalized in language which does not offer many further options for the eloquent verbalization of this type of scene.

4 Discussion

Several interesting observations made during this exploratory study of frame-relative paths suggest the following discussion and future work.

Subjectification, involved in the production of fictive motion, inevitably makes the principal motion event component—dynamicity—lose its saliency, resulting in the elusive character of sense of motion in certain cases (Langacker 2005). The identification of boundary cases (in particular, frame-relative change, where the sense of motion of the observer is very subtle) can thus be dependent on the researcher's mental representation, evoked by the structure. Thus, the inter-coder agreement measurements are an important step in future work going beyond this preliminary investigation. Even more important, though, is to link this type of analysis to linguistically-grounded research, discussing similar phenomena from different perspectives. For example, in (Gawron 2009), the semantics of *widen* (which we identified as representing frame-relative change) is investigated, and its ability to denote properties that are spatially dynamic, but static in time, is interpreted through the prism of contextually available spatial axis. Engaging with this line of research in the future could provide a more linguistically-grounded framework for the definition and identification of frame-relative paths.

Our preliminary findings, based on a very specific corpus, suggest that frame-relative constructions can vary in relation to the immediate path of the observer. Vista constructions signal progress of the observer by referring to large-scale distant landscape features coming into the view and moving towards them. Path constructions describe the dynamically changing properties of the path itself (e.g. terrain, slope). The latter can be specific to the corpus—the role of locomotion in the mountaineering context and the relevance of spatial information of a fine level of granularity has been noted before (Egorova et al. 2015). Correspondingly, the range of possible linguistic encodings of frame-relative change, which attains to the variety of space properties it can refer to, might be corpus-specific. With the slow pace of locomotion described, it is not surprising that we have not encountered verbs such as *rush*, as in the example "I sat in the car and watched the scenery rush by" (Talmy 2000). Thus, the examination of a corpora describing another mode of motion (e.g. travel in some form of vehicle) might uncover different patterns (e.g. absence of frame-relative path change) and a different set of verbs (e.g. presence of verbs semantically related to speed) and would be an interesting step for the future work.

Which conceptual inventory can frame-relative constructions be translated into? To take ISO-space, its latest annotation scheme suggests the `motion_sense` attribute for the MOTION tag, which can have the following values: LITERAL (e.g. "The balloon rose above the building"), INTRINSIC_CHANGE (e.g. "The river rose above the levy"), FICTIVE (e.g. "The mountain rises above the valley") (Pustejovsky and Yocum 2013). None of these senses captures the essence of frame-relative constructions—presence of the moving observer. At the same time, frame-relative constructions do not encode directly any of the concepts captured through the MOVELINK tag attributes, such as source, goal, mover, ground, goal_reached. As an example, "The ice disappeared", "The crack I was in petered out" can be interpreted as the end of the current route segment. However, this is different from the concept behind the goal_reached, since the focus is on the path (the goal is not even mentioned). It can also be interpreted as a property of the ground, but then the latter has to be allowed to have attributes. The major question is if the inclusion of nuanced concepts represented by such constructions into spatial language annotation schemes has an added value. To answer it, we have to follow the strategy of corpus-based development of annotation specifications—namely, examine frame-relative paths in diverse set of small corpora and see the pervasiveness of the phenomena, its conceptual and linguistic range (Pustejovsky and Moszkowicz 2014). Such efforts appear to be necessary, if we want to deal with the complexity of spatial language in the context of the growing demand of spatially annotated data.

References

Bateman JA (2010) Language and space: a two-level semantic approach based on principles of ontological engineering. Int J Speech Technol 13:29–48

Egorova E, Tenbrink T, Purves RS (2015) Where snow is a landmark: route direction elements in alpine contexts. In: Fabrikant SI, Raubal M, Bertolotto M et al (eds) COSIT 2015. Springer International

Egorova E, Boo G, Purves RS (2016) "The ridge went north": did the observer go as well? Corpus-driven investigation of fictive motion. In: International Conference on GIScience Short Paper Proceedings, vol 1, pp 96–100

Gawron JM (2009) The lexical semantics of extent verbs. San Diego State University

Jones CB, Purves RS (2008) Geographical information retrieval. Int J Geogr Inf Sci 22:219–228

Krippendorff K (2004) Content analysis: an introduction to its methodology, 2d edn. Sage Publications, London

Langacker RW (2005) Dynamicity, fictivity, and scanning: the imaginative basis of logic and linguistic meaning. In: Pecher D, Zwaan RA (eds) Grounding cognition: the role of perception and action in memory, language, and thinking. Cambridge University Press, Cambridge

Lakoff G, Johnson M (1980) Metaphors we live by. University of Chicago press, Chicago

Mani I, Doran C, Harris D (2010) SpatialML: annotation scheme, resources, and evaluation. Lang Res Eval 44:263–280

Matlock T, Bergmann T (2014) Fictive motion. In: Divjak D (ed) Dbrowska E. Handbook of cognitive linguistics. DeGruyter Mouton, Berlin

Moncla L, Gaio M, Nogueras-Iso J, Mustière S (2016) Reconstruction of itineraries from annotated text with an informed spanning tree algorithm. Int J Geogr Inf Sci 30:1137–1160

Pustejovsky J, Moszkowicz J (2014) The role of model testing in standards development: the case of iso-space. In: Proceedings of LREC12, pp 3060–3063

Pustejovsky J, Yocum Z (2013) Capturing motion in iso-spacebank. In: Proceedings of the workshop on interoperable semantic annotation, pp 25–33

Rojo A, Valenzuela J (2016) Fictive motion in English and Spanish. Int J Engl Stud 3:123–150

Talmy L (2000) Toward a cognitive semantics, 1st, paper edn. MIT Press, Cambridge, Massachusetts

Taremaa P (2013) Fictive and actual motion in Estonian: encoding space. SKY J Linguist 26: 151–183

When Environmental Information Is Conveyed Using Descriptions: The Role of Perspectives and Strategies

Chiara Meneghetti and Veronica Muffato

Abstract Verbally-conveyed spatial information can typically adopt a survey or a route perspective (i.e. a map view or a personal point of view, respectively). This paper examines the role of spontaneous strategy use in learning from survey and route descriptions (Study 1), and the effect of practicing with route and survey strategies on route description learning (Study 2). In Study 1, participants listened to route or survey spatial descriptions. In Study 2, three groups listened to route description before and after practicing with the use of a survey strategy, a route strategy, or without practicing with any strategy (Survey practice, Route practice, Control groups respectively). In both studies, after listening to each description, participants reported on their strategy use and answered true/false survey and route questions. The results of Study 1 showed a greater accuracy for descriptions conveyed from the same perspective as the one learnt, and survey descriptions elicited a greater use of survey strategies. The results of Study 2 showed that the group which had practiced with a survey strategy were more accurate for survey descriptions, and reported a greater use of the survey strategy than other strategies, or than the other groups. The results are discussed to expand on the spatial cognition framework and its implications.

Keywords Spatial descriptions · Route perspectives · Survey perspectives · Imagery strategy self-reported · Strategy instructions

1 Introduction

One way of acquiring the spatial information we need every day is to use language, by reading or hearing the description of a path, or the location of landmarks in an

C. Meneghetti (✉) · V. Muffato
Department of General Psychology, University of Padova, via Venezia 8, Padova, Italy
e-mail: chiara.meneghetti@unipd.it

V. Muffato
e-mail: veronica.muffato@gmail.com

© Springer International Publishing AG 2018
P. Fogliaroni et al. (eds.), *Proceedings of Workshops and Posters at the 13th International Conference on Spatial Information Theory (COSIT 2017)*, Lecture Notes in Geoinformation and Cartography, https://doi.org/10.1007/978-3-319-63946-8_39

environment. Spatial language is of interest in several disciplines, to prepare devices capable of handling spatial language in order to transfer knowledge of a route and instructions from a user to a robotic system, for instance (Kim et al. 2015). When people process a verbally-conveyed spatial description, they form a mental model, an abstraction that resembles the structure of the corresponding state of affairs in the outside world (Johnson-Laird 1983), in which spatial relations between objects (landmarks) are represented in their mind's eye. The information conveyed by spatial language can typically take two perspectives, i.e. a route (that presents landmarks and their relative positions from a person's point of view, using egocentric terms) or a survey perspective (that presents landmarks and their relations from a bird's-eye view and use canonical terms) (Taylor and Tversky 1992). Some studies found that mental models incorporated the view adopted in such descriptions, and that people responded more accurately to spatial questions expressed from the same perspective (Meneghetti et al. 2011a; Perrig and Kintsch 1985). Other researchers found instead that mental models incorporated multiple views (Taylor and Tversky 1992; Brunye and Taylor 2008), and their accuracy was uninfluenced by the perspective of the descriptions prompting them.

Among the several aspects contributing to differences in the features of people's mental models (such as to be perspective dependent), there are also individual visuo-spatial factors (Meneghetti et al. 2014b). One such individual factor concerns the strategies an individual reports using to deal with spatial recall demands, such as mentally visualizing a path (route strategy) or forming a mental map (survey strategy), or mentally repeating the words of a description, which is a verbal strategy (Gyselinck et al. 2007). The few studies comparing survey and route descriptions found that survey description learning was associated with the use of survey strategies, while route description learning was associated with the use of both survey and route strategies (Meneghetti et al. 2016).

In other studies, participants were specifically instructed to use strategies, and this affected their accuracy in recalling visually-acquired information, which resembles the representation of verbally-acquired spatial information in some aspects (Peruch et al. 2006). Indeed, studies using maps (Taylor et al. 1999) or navigation (Magliano et al. 1995; Kraemer et al. 2017) showed that instructions adopting a survey perspective improved performance in allocentrically-based tasks, while instructions from a route perspective improved performance in egocentrically-based tasks (Taylor et al. 1999). In addition, instructing (Gyselinck et al. 2007) or training (Gyselinck et al. 2009; Meneghetti et al. 2013) participants to use mental images during description learning led to their better spatial recall than if they used a repetition (verbal) strategy.

Given that spontaneous representation can be influenced by the perspective presented during the learning phase (Perrig and Kintsch 1985; Meneghetti et al. 2011a), and that visuo-spatial strategy usage influences recall accuracy (Gyselinck et al. 2007; Meneghetti et al. 2014b), one research issue that remains to be investigated is how practice with using strategies can enable the formation of mental representations uninfluenced by the perspective of the spatial description presented during the learning phase. Therefore, one way to enable individuals to form more malleable mental representations from descriptions could be to instruct them specifically to use a different strategy from the one spontaneously prompted by the spatial description from which they have to use.

Study 1 ascertained the spontaneous use of a survey or route strategy when learning survey or route descriptions. Our aim was to confirm the relationship between self-reported strategy use and accuracy (Meneghetti et al. 2011a), with some differences in the strategies used being a function of the type of description learnt. Study 2 newly examined the effect on the features of mental representations deriving from participants practicing with the use of specific route or survey strategies when the description to learn adopted a route perspective. Our hypothesis was that this would enable participants to form flexible representations that could handle both a route view (induced by the perspective adopted in the description) and a survey view (as specifically instructed).

2 Study 1

2.1 Method

Participants. Study 1 involved 116 undergraduates (58 females; M age 20.84 years, SD 1.48), divided equally into a Route description group, and a Survey description group.

Materials and procedure. Participants were tested individually at a single session lasting an hour and a half. They were asked to listen twice to spatial descriptions from a survey or route perspective (lasting 6 min in all). Then participants scored their self-reported strategy use and completed a verification test.

Spatial descriptions. Four descriptions of two fictitious outdoor environments were recorded (adapted from (Meneghetti et al. 2011b)), each presented from a route perspective and from a survey perspective. All descriptions included the position of 14 landmarks, were of similar length (around 300 words each), and were equally well recalled.

Strategy use scale. Three strategies were considered (as in (Meneghetti et al. 2014b)): survey (I form a mental map); route (I imagine the path to cover); and verbal (I mentally repeat the information). Participants judged their strategy use on a Likert scale from 1 (not at all) to 5 (very much).

Verification test. For each description, 20 true/false sentences (half of them true, half false) (Meneghetti et al. 2011b) described the spatial relations between the landmarks and the protagonist (10 route sentences), or between landmarks in the environment (10 survey sentences).

Table 1 Mean scores and standard deviations for survey and route sentence accuracy, and for self-reported survey, route or verbal strategy use by Route and Survey description groups

	Route description group		Survey description group	
	M	SD	M	SD
True/false route sentences	6.78	1.09	5.99	1.18
True/false survey sentences	6.07	1.11	7.18	1.21
Survey strategy	3.61	1.33	4.33	0.94
Route strategy	3.94	1.28	2.66	1.29
Verbal strategy	3.43	1.11	3.18	1.13

2.2 Results

Survey and route sentence accuracy by Route and Survey description groups. The 2 (Group: survey vs. route description) $\times$ 2 (Type of sentence: survey vs. route) analysis of variance revealed the main effect of the type of sentence, $F(1, 114) = 4.20$, $p = 0.04$, $\eta2p = 0.04$; this was better qualified by Group $\times$ Type of sentence interaction, $F(1, 114) = 63.95$, $p \leq 0.001$, $\eta2p = 0.36$ (with a large effect size). Post hoc comparisons (see Table 1), considering those with $ps \leq 0.001$ as significant in Studies 1 and 2, showed that the Route description group performed better for route sentences than for survey sentences ($p \leq 0.001$), while the Survey description group performed better for survey sentences than for route sentences ($p \leq 0.001$).

Survey and route strategy use ratings by Route and Survey description groups. The 2 (Group: survey vs. route description) $\times$ 3 (Type of strategy: survey vs. route vs. verbal) analysis of variance showed a main effect of the type of strategy, $F(2, 228) = 16.55$, $p \leq 0.001$, $\eta2p = 0.13$; this was better qualified by the significant Group $\times$ Type of strategy interaction, $F(2, 228) = 28.12$, $p \leq 0.001$, $\eta2p = 0.20$ (with a large effect size). Post-hoc comparisons (see Table 1) showed that the Route description group did not differ in participants' rated use of strategies, apart from a tendency for a route strategy being reportedly used more than a verbal strategy ($p = 0.02$); on the other hand, the Survey description group reported using a survey strategy more than either of the other two strategies ($ps \leq 0.001$).

2.3 Discussion

This study showed that participants handled route sentences better after hearing spatial descriptions from a route perspective, and were more accurate with survey sentences if they had heard spatial descriptions from a survey perspective, thus confirming that mental representations are influenced by the perspective learnt (Meneghetti et al. 2011b). In addition, survey descriptions elicited a greater use of survey strategies than route or verbal strategies, whereas route descriptions did not seem to prompt the use of a specific strategy, confirming the elective use of a survey strategy in association with a survey description (Meneghetti et al. 2014b).

Given that knowledge gained from a route perspective can potentially be adapted to a survey view, and that route descriptions can produce more flexible representations (Brunye and Taylor 2008), the aim of our Study 2 was to explore the features of mental representations derived from route descriptions after practicing with the use of route or survey strategies. We predicted that the group practicing with a survey strategy would would develop mental models capable of incorporating multiple views, i.e. able to handle both route sentences (information directly learned from descriptions) and survey sentences (induced by the survey strategy practice).

3 Study 2

3.1 Method

Participants. The study involved 180 undergraduates (119 females; M age 21.59, SD 1.46), equally divided between three groups: 60 participants practiced with using a Route strategy, 60 with a Survey strategy, and 60 served as controls (no practice with strategy use). To ensure that the three groups were comparable in terms of visuospatial preferences, all participants were administered the Sense of Direction and Spatial Representation questionnaire (SDSR) (Pazzaglia et al. 2000), and the three groups had similar scores in all five factors $F \leq 1$ to $F = 1.94$, $ps = 0.14$; see Table 2.

Table 2 Mean scores and standard deviations for the three groups (route and survey strategy practice, and controls) in the five factors of the Sense of SDSR

	Survey strategy practice group		Route strategy practice group		Control group	
	M	SD	M	SD	M	SD
Sense of direction	13.40	3.90	12.84	2.81	12.72	3.57
Knowledge and use of cardinal points	6.55	3.15	6.18	2.76	6.20	2.56
Preference for survey mode	5.53	2.47	5.30	1.48	5.15	1.89
Preference for route mode	6.58	1.64	6.67	1.36	6.85	1.29
Preference for landmark-based mode	7.33	1.52	7.83	1.18	7.55	1.46

Procedure

Practice phase. The Survey and Route strategy practice groups gained experience of using the strategies, and the control group was involved in alternative activities, for approximately one hour during a single individual session. Practice with the survey and route strategies involved: (i) proposing the use of mental imagery with words and sentences; (ii) practicing with the use of a mental map-based strategy (Survey group) or a mental tour-based strategy (Route group) while listening to two descriptions in route view. The control group was involved for the same amount of time in watching a video-lesson on the psychology of personality, and making activities related to it.

Pre- and post-practice measures. Participants listened twice to two route descriptions (the same as in Study 1), one before and the other after practicing with strategy use. After hearing each description, they scored their self-reported strategy use (as in Study 1, but on a Likert scale from 1 to 7), and completed the verification test (as in Study 1, except that 8 survey and 8 route sentences were used).

3.2 Results

Survey and route sentence accuracy by Survey and Route strategy practice and control groups. The 2 (Type of sentence: survey vs. route) × 3 (Group: survey strategy practice vs. route strategy practice vs. controls) × Session (pre- vs. post-test) showed a main effect of Type of sentence, $F(1, 177) = 121.04$, $p \leq 0.001$, $\eta2p = 0.41$, and of Session, $F(1, 177) = 9.30$, $p = 0.03$, $\eta2p = 0.05$, and a 3-way interaction for Session × Type of sentence × Group, $F(2, 177) = 4.89$, $p = 0.009$, $\eta2p = 0.05$ (small effect size). Post hoc comparisons (see Table 3) showed that: the Survey strategy practice group handled survey sentences more accurately at post- than at pre-test ($p \leq 0.001$); and the Survey strategy practice group scored better for route sentences than for survey sentences at pre-test ($p \leq 0.001$), while they scored similarly for route and survey sentences at post-test ($p = 0.33$). In addition, while at pre-test the three

Table 3 Mean scores and standard deviations for survey and route sentence accuracy by group (Survey or Route strategy practice or Controls), before and after practicing with strategies

Type of sentence	Session	Survey strategy practice group		Route strategy practice group		Control group	
		M	SD	M	SD	M	SD
True/false route sentences	Pre	5.23	1.53	5.10	1.18	4.88	1.40
	Post	5.03	1.44	5.55	1.25	5.20	1.53
True/false route sentences	Pre	3.72	1.66	3.83	1.62	3.80	1.55
	Post	4.78	1.38	3.98	1.30	3.93	1.19

groups did not differ when compared between type of sentence accuracy, at post-test the Survey strategy practice group had higher scores in survey sentences than the other two groups ($p = 0.001$).

Survey and route strategy use ratings by group (Survey or Route strategy practice or Control groups). The 3 (Type of strategy: survey vs. route vs. verbal) $\times$ 3 (Group: Survey strategy practice vs. Route strategy practice vs. Controls) $\times$ 2 (Session: pre- vs. post-practice) showed a main effect of Type of strategy, $F(2, 354) = 112.14$, $p \leq 0.001$, $\eta 2p = 0.39$, Session $F(1, 177) = 50.19$, $p \leq 0.001$, $\eta 2p = 0.21$. There was also (other than 2-way interactions) a significant 3-way interaction for Session $\times$ Type of strategy $\times$ Group, $F(4, 354) = 8.44$, $p \leq 0.001$, $\eta 2p = 0.09$ (good effect size). Post hoc comparisons (see Table 4) showed that: the Survey strategy practice group reported having made greater use of survey strategies at post-test than at pre-test ($p \leq' 0.001$); and the Control group tended to do the same ($p = 0.006$; this value identifies a trend towards the cut-off $p \leq 0.001$).

Also, while the Route strategy practice and Control groups both reported using a route strategy the most, at both pre- and post-test, the Survey strategy practice group reported using a survey strategy more than a verbal or route strategy, but only at post-test, not at pre-test ($p \leq 0.001$). Finally, the Survey strategy practice group used a survey strategy more at post-test than the other two groups, while the latter two reported using a route strategy more than the former ($p \leq 0.001$).

3.3 Discussion

Study 2 showed the positive effect of practicing with a survey strategy on route description recall accuracy. In fact, the group that practiced using a survey (mental map-based) strategy to process spatial descriptions in route view had previously shown evidence before practice of the congruency effect (they had handled route

Table 4 Mean scores and standard deviations for reported survey, route and verbal strategy use by group (Survey or Route strategy practice or Controls), at pre- and post-test

Type of strategy	Session	Survey strategy practice group		Route strategy practice group		Control group	
		M	SD	M	SD	M	SD
Verbal	Pre	2.93	1.77	2.82	1.57	3.32	1.73
	Post	3.35	1.71	3.10	1.43	3.82	1.60
Route	Pre	5.02	1.96	5.65	1.72	5.78	1.37
	Post	4.72	1.85	5.88	1.46	5.78	1.43
Survey	Pre	3.40	2.24	3.10	1.90	3.33	2.04
	Post	5.97	1.21	3.65	1.88	4.07	2.11

sentences more accurately than survey sentences, as in Study 1), whereas after the practice session it was equally accurate with both survey and route sentences. In addition after practicing with a survey strategy, this group also performed better survey sentences than the other groups.

Analyzing self-reported strategy use consistently showed that, at post-test, the Survey strategy practice group reported using a survey strategy more than the other two groups, which both reported using a route strategy most, at both pre- and post-test. These results showed that instructing participants to learn route description using a strategy based on a mental map enabled them to form perspective-free representations.

4   Conclusion

Studies 1 and 2 explore the contribution of self-reported individual spontaneous (Study 1) and specifically taught (Study 2) strategy use on mental models formed from spatial descriptions. The results of both studies mainly showed: (i) that participants' mental models maintained the features of the type of description, in terms of the perspective adopted (Perrig and Kintsch 1985; Meneghetti et al. 2011a); and newly (ii) that they were able to incorporate both perspectives when participants were specifically instructed to use a survey strategy to handle a route description. This was supported by participants' self-reported strategy use. Previous studies (Perrig and Kintsch 1985) showed the influence of the spatial perspective on the features of a mental representation (as confirmed in the present study), and that spatial learning is related to visuo-spatial strategy usage (Gyselinck et al. 2007; Meneghetti et al. 2014b). The present study fills a gap in our understanding of the impact of practicing with strategies on the flexibility of mental representations. It was demonstrated that the representations can change perspective if individuals receive specific strategy use instructions and practice.

It is important to acquire the ability to manage survey (allocentric) information because this is associated with a useful approach to the environment, such as short-cuts (an elective measure of environment knowledge), and wayfinding (Montello 2005). Our findings suggest that identifying ways to develop survey knowledge (such as practicing with survey strategies) is important to the theoretical understanding of the malleability of environment learning, as well as having implications in other, related disciplines in which spatial language is of interest. For instance, when programming navigation services or identifying ways to approach the environment, we should bear users' or speakers' individual differences in mind, as well as the value of practicing with a (survey) strategy before using a device or approaching environment learning tasks.

Our studies have some limitations, such as: the fact that we did not include practice with survey descriptions; or our use of fictitious environments; or our having adopted the survey versus route presentation dichotomy, which can only be partially representative of real life. These aspects need to be better managed in further studies.

To conclude, these two studies showed the importance of taking into account the different spatial strategies that individuals use to approach spatial language tasks and their malleability.

References

Brunye TT, Taylor HA (2008) Extended experience benefits spatial mental model development with route but not survey descriptions. Acta Psychol 127:340–354

Gyselinck V, Beni R, Pazzaglia F (2007) Working memory components and imagery instructions in the elaboration of a spatial mental model. Psychol Res 71:373–382

Gyselinck V, Meneghetti C, De Beni R, Pazzaglia F (2009) The role of working memory in spatial text processing: What benefit of imagery strategy and visuospatial abilities? Learn Individ Differ 19:12–20

Johnson-Laird PN (1983) Mental models: towards a cognitive science of language, inference, and consciousness. Harvard University Press

Kim J, Vasardani M, Winter S (2015) From descriptions to depictions: a dynamic sketch map drawing strategy. Spat Cogn Comput 5868:29–53

Kraemer DJM, Schinazi VR, Cawkwell PB (2017) Verbalizing, visualizing, and navigating: the effect of strategies on encoding a large-scale virtual environment. J Exp Psychol Learn Mem Cogn 43:611–621

Magliano JP, Cohen R, Allen GL, Rodrigue JR (1995) The impact of a wayfinder's goal on learning a new environment: different types of spatial knowledge as goals. J Environ Psychol 15:65–75

Meneghetti C, Borella E, Muffato V (2014) Environment learning from spatial descriptions: the role of perspective and spatial abilities in young and older adults. Spatial cognition, vol IX. Springer International Publishing, pp 30–45

Meneghetti C, Ronconi L, Pazzaglia F, De Beni R (2014b) Spatial mental representations derived from spatial descriptions: the predicting and mediating roles of spatial preferences, strategies, and abilities. Br J Psychol 105:295–315

Meneghetti C, De Beni R, Gyselinck V, Pazzaglia F (2013) The joint role of spatial ability and imagery strategy in sustaining the learning of spatial descriptions under spatial interference. Learn Individ Differ 24:32–41

Meneghetti C, Labate E, Pazzaglia F (2016) The role of visual and spatial working memory in forming mental models derived from survey and route descriptions. Br J Psychol

Meneghetti C, De Beni R, Gyselinck V, Pazzaglia F (2011a) Working memory involvement in spatial text processing: what advantages are gained from extended learning and visuo-spatial strategies? Br J Psychol 102:499–518

Meneghetti C, Pazzaglia F, De Beni R (2011b) Spatial mental representations derived from survey and route descriptions: when individuals prefer extrinsic frame of reference. Learn Individ Differ 21:150–157

Montello DR (2005) Navigation. In: Shah P, Miyake A (eds) The cambridge handbook of visuospatial thinking. Cambridge University Press, Cambridge, pp 257–294

Pazzaglia F, Cornoldi C, De Beni R (2000) Differenze individuali nella rappresentazione dello spazio e nell'abilita di orientamento: presentazione di un questionario autovalutativo. G Ital di Psicol 27:627–650

Perrig W, Kintsch W (1985) Propositional and situational representations of text. J Mem Lang 24:503–518

Peruch P, Chabanne V, Nesa M-P (2006) Comparing distances in mental images constructed from visual experience or verbal descriptions: the impact of survey versus route perspective. Q J Exp Psychol 59:1950–1967

Pestun MV, Galaktionov VA (2016) Algorithms for the construction and recognition of navigational route descriptions for cartographic computer systems. Program Comput Softw 42:341–346

Siegel AW, White SH (1975) The development of spatial representations of large-scale environments. pp 9–55

Taylor HA, Naylor SJ, Chechile NA (1999) Goal-specific influences on the representation of spatial perspective. Mem Cognit 27:309–19

Taylor HA, Tversky B (1992) Spatial mental models derived from survey and route descriptions. J Mem Lang 31:261–292

The Role of Context in the Interpretation of Natural Language Location Descriptions

Kristin Stock and Mark Hall

Abstract Research into methods for automated interpretation of human language descriptions of location has recognized the importance of spatial relations (usually represented by prepositions in English). However, the role of context has been widely acknowledged in natural language processing research and in linguistic studies of spatial language. There are a large number of different aspects of context that may be important in automated interpretation of location descriptions. In this paper, we present a summary of the contextual factors that have been discussed in the literature, and also describe and test a methodology for identifying contextual factors that respondents consider important in the use of specific spatial relations. We combine these sources to present a broad typology of contextual factors in the interpretation of geospatial natural language to set the scene for future research.

Keywords Geospatial · Natural language · Context · Location

1 Introduction

The ability to automate the interpretation of human language descriptions of location, effectively mapping a string of text to a coordinate location, is useful for a number of applications. Social media, blogs, scientific reports and logs are all examples of potential sources of geographic data that is currently untapped. Expressions like *there has been an accident outside the post office* and *the specimen was collected on the hillside above Arthurs Pass* contain geographic information, but need to be processed in order to make use of it.

K. Stock (✉)
Massey University, Auckland, New Zealand
e-mail: k.stock@massey.ac.nz

M. Hall
Edge Hill University, Ormskirk, UK
e-mail: Hallmark@edgehill.ac.uk

© Springer International Publishing AG 2018
P. Fogliaroni et al. (eds.), *Proceedings of Workshops and Posters at the 13th International Conference on Spatial Information Theory (COSIT 2017)*, Lecture Notes in Geoinformation and Cartography, https://doi.org/10.1007/978-3-319-63946-8_40

Several research directions are being pursued with the goal of automating the interpretation process, and also performing the reverse conversion, in which text is generated to describe a location. These directions include but are not confined to: so-called spatial role labelling to identify the key elements of a spatial expressions, focused on spatial preposition, locatum and relatum (e.g. (Kordjamshidi et al. 2011)); the development of mathematical models for the interpretation of specific spatial prepositions (e.g. (Hall et al. 2015; Shariff et al. 1998)) and linguistically-based discussions of how spatial prepositions are used (e.g. (Talmy 2000)). A number of researchers have identified specific factors that may have an impact on the meaning of geospatial natural language (and thus interpretation and generation), but there has been no broad view taken across a range of contextual factors relevant for geospatial language at different levels. We aim to address this gap, and to test the use of an empirical methodology to try to identify contextual factors that may not already be well understood. We limit the scope of our work to the use of context in location descriptions in the English language. Variations are likely in the role and importance of different factors in different languages.

2   Geographic Information and Context

In the broader computer science and natural language processing fields, the importance of context is well recognized, and has been explored in some detail. For example, Porzel (2010) identifies four types of contextual information: domain, discourse, interlocutionary and situational context, and identifies some of the contextual challenges posed by spatial information descriptions.

In the geographic information literature, researchers studying semantics and ontologies have addressed the subject, and in one of the most extensive treatments, Cai (2007) defines situation as the circumstances that surround a particular action, and context as the collection of situations that vary in ways that do not impact on the user's behavior. His schema for context includes: user task; location; expected features of the context; domain ontologies; goals, subgoals and events.

Other broad schemes are proposed by Souza et al. (2006), who identify four types of context: user context; data context; association context and procedure context; and Brodaric (2007) who represents context using dimensions, which consist of origins, uses and effects, and involve entities with specific roles. These schemes identify the importance of pragmatics, a fact that is also clearly demonstrated by Herskovitz (1985). Attention has also been given to the role of object affordance (the uses to which an object is put) (Sen 2008).

Other researchers have focused on one or a small subset of factors that might be considered part of the broader picture of context, rather than attempting to define a comprehensive scheme. For example, context has been considered closely related to application domain, combining object types of interest, the relations between them and the functions that the user wishes to perform on them (Rodriguez and Egenhofer 2004). In geospatial natural language research, Lautenshütz et al. (2006)

explores object liquidity or solidity as one aspect of context, while Kray et al. (2013) discusses the influence of indoors, outdoors or transitional space.

There have also been various efforts to define formal and semi-formal structures into which language can be slotted, considering different aspects of context, including Hornsby and Li (2009) and Zwarts (2005), who define representations of paths and Kracht (2002), who defines a structure for locative expressions. In addition, SpatialML (2009), ISO-Space (Pustejovsky et al. 2011) and GUM-Space (Bateman et al. 2010) are more comprehensive structures for the representation of spatial language, incorporating some contextual components.

Given the varying use of the term context, in this paper, we adopt a wide definition, encompassing anything that is not directly stated in the textual expression itself, including the wider situation and environment, as well as the characteristics of the terms used in the expression (e.g. the relatum and locatum), and potentially incorporating both parts of the two-level semantics discussed by Lang and Maienborn (2011).

3 Contextual Factors Discussed in the Literature

The following paragraphs present a set of contextual factors that have been identified by researchers, either by directly specifying their importance in geospatial natural language interpretation, or indirectly by discussing how they vary across expressions. They are presented in no particular order.

Image schema is one of the most commonly studied contextual factors in geospatial language (Lakoff and Johnson 1980; Mark and Frank 1996), and is particularly connected to the preposition. *The house is on Main Street, the house is on the island* and *the house is on the hill* show different interpretations of the preposition *on* with varying image-schemas.

The importance of **perspectival mode** is another factor that is commonly discussed (Bateman et al. 2010; Coventry and Garrod 2004; Hubona et al. 1998; Talmy 2000), with the interpretation varying depending on whether the viewer has a survey (bird's eye) or a route perspective. In *there were some houses in the valley* the valley is viewed from above and seen as a container, while in *there were houses now and then through the valley* it is a path viewed as if moving through the scene, and time is used to indicate spatial location.

Frame of reference is well recognized as an important factor, in which the expression may take an intrinsic (frame of reference is the object itself), absolute (frame of reference is external) or relative (frame of reference is relative to observer or other object) frame of reference. *The kiosk is in front of the building* may refer to the front relative to the building (e.g. the entrance) or front relative to the observer (Landau and Jackendoff 1993; Levinson 2003).

Shape (or **geometry type**) may impact on the way an expression is interpreted (Landau and Jackendoff 1993). For example, some prepositions only make sense with some shapes, whilst others are interpreted differently depending on the shape.

The road goes beside the river implies a degree of alignment while *the church is beside the river* does not.

Some spatial expressions refer to parts of an object that are related to the notion of an **axial structure** of an object (front, back, left, right, top, bottom). The axis may be inherent in the object (the front of a building is normally the main entrance) or may be contextually imposed (the front of a moving flood is the side facing its direction of travel). For example: *the city is in front of the hurricane* versus *the kiosk is in front of the building* (Bateman et al. 2010; Landau and Jackendoff 1993).

Scale has been identified as a factor that may influence the meaning of a spatial expression (Lautenschütz et al. 2006). For example: *the road runs across the park* versus *the canal runs across the country*. Lautenshütz et al. (Lautenschütz et al. 2006) also discuss the importance of whether an object is **liquid or solid** in interpretation of meaning, as in *the river runs to the sea* versus the *road runs to the sea*.

Coventry and Garrod (2004) discuss the importance of **force dynamics** that exist between objects (e.g. how they push against each other) on the way language is used, along with the nature of objects and the purpose of the expression. For example, they assert that the semantics of the *in* preposition includes the notion of location control (*the car is in a traffic jam, the car is in a queue*). GUM-Space classifies spatial modalities in ways that reflect this theory (Bateman et al. 2010).

The **domain** (or more specifically, the geographic feature types) involved in the spatial expression may affect its interpretation (Klippel et al. 2011). However, the characteristics of the objects in the domain (for example, the factors that have been presented above) and the types of expressions that make sense with particular feature types may explain these domain variations.

Spatial expressions may refer to objects that are **bounded or unbounded**, and spatial relations may be conceptually bounded or unbounded. For example: *the plane crossed the lake in 3 min* versus *the plane crossed the water for 3 min* (Talmy 2000; Zwarts 2005).

Spatial expressions may refer to objects that are single items, or that are collections of items (and thus divided in nature). Factors such as **Dividedness, Quantity** and **Plexity** fall into this category (Landau and Jackendoff 1993; Talmy 2000). For example: *there were houses in the valley* versus *there was a house in the valley*.

Pattern of distribution of objects in space can impact on language interpretation (Talmy 2000). For example, *every second shop had flags hung outside*.

4 Empirical Identification of Contextual Factors

While a number of factors have been investigated in these studies, and several broad schemes for representing context have been developed, little attention has been given to the contextual factors that are specific to a given spatial relation. The role of the surrounding environment has been incorporated in defining mathematical models to represent the meaning of relations such as near (Gahegan 1995) and opposite (Bartie et al. 2011), but this has not been extended to other relations, and little work has been done to ensure that we have a clear picture of the range of factors that might determine the meaning of many spatial relations.

In order to address this gap, we developed and tested a methodology to empirically collect information on the factors that might determine whether and why a particular preposition is used by an individual in a given situation. We focus on prepositions as they are an important way of expressing spatial relations, but we acknowledge that spatial relations are often expressed through other parts of speech, and that this approach provides only a partial picture.

We conducted a study in which we asked participants to provide a brief written explanation of the reasons a particular preposition was applicable in a given situation. The data was collected as part of a wider study, in which participants were shown an aerial photographic image of part of a city and asked to judge the level of applicability of a preposition to describe locations relative to a specified reference object (e.g. *photo taken at the Millenium Centre*) by rating a marker on the image on a scale of 1–9. We do not report this first part of the survey here as it has been reported elsewhere (Hall et al. 2015), but instead focus on the following step, in which participants were asked to provide qualitative free text descriptions of how they interpreted the preposition used in the applicability question ("How do you define < spatial preposition >?"). The survey was distributed by email to all staff and students at Cardiff University with an incentive of a 50 lb voucher prize-draw for those who completed it. Over a six week period, 1210 responses were collected. The study investigated four different prepositional phrases (*at, near, next to, between*) and one phrase that combined a preposition and relatum (*at the corner*). The study was conducted in English only. All of the location descriptions were given in English and referred to geographic locations in Cardiff, Wales.

We analyzed a randomly selected sample of 100 responses for each of the given phrases, summarizing and grouping together similar reasons for the selection of a particular applicability measure for a given phrase. The following are some examples of the responses received for *next to*:

- "So that there is little or nothing between the photograph and the object photographed"
- "Very close to the location (no more than 10 metres approx)"
- "Stood outside a specified building, or with nothing but space between you and the building, and either in sight of a street or on an adjacent/very close street."

In analyzing the results, we performed a simple count of the number of times a particular contextual factor (or reason) was identified for each phrase and across all phrases, as we were interested in identifying factors that were more broadly important than for just a single phrase. The reasons/factors were manually identified in a bottom up fashion by one of the authors. We recognize this as a limitation of the current work, and in future work, inter-annotator agreement will be evaluated. Descriptions that used similar text or were synonymous were grouped together, as shown in Table 1. In some cases, similar reasons were grouped together to create a single factor. For example, visibility was identified as important for several phrases,

Table 1 Examples of reasons given for each factor

Factor	Examples of reasons given
Proximity	Within a very short distance, short walking distance, immediate vicinity, within a 30 s walk
Visibility	Object is clearly visible, clear line of sight, see it face to face, clearly visible, in clear sight of, in full view, if you can still see it
Immediacy, similar immediacy, buildingImmediacy	Right beside, no buildings in between, the closest thing to you, a few buildings between, closest notable landmark, no streets in between
Centrality	Near the centre point, equidistant between two points, the nearer it was to the middle the more I agreed
Convergence	Where two road join, where two streets met/crossed, where two points meet
Physical containment	In, inside, within, within its boundaries, from named object looking out, right inside, exactly in the location
projection, visibility of projection	In front of, in front of the entrance, by the doors, at the side of, beside, can see the front, on the same side, side by side, close to the side of an object
Contact	At the precise point referenced, exactly at the point, at the physical corner not 20 m away from it
Collinearity approximateCollinearity travelCollinearity	In or into the space that separates two things, on the line connecting two points, in some space between, within the two boundaries set by the two locations
Immediate non-containment	just outside, literally outside, immediately outside, right outside
Object shape	90 degree angle, shape of corner, definite point on the corner, where two points meet at an angle, right angle
Vertical contact	On top of, right on top of, on, directly located on
Collocation	Directly there, actually there, at, it's right there
Surroundedness, specific surroundedness	The streets and sidewalks around the building, in an area surrounding the venue, in the general area
Termination	At the end, at the ends of both locations, no further and not actually in the streets

but the nature of visibility varied (from the locatum, to the locatum, etc. Similarly, immediacy (whether or not there were intervening objects) had several different types, including whether or not the intervening objects were of the same type, were large, etc.

We ignored clauses within the responses that used the name of the spatial relation that was being defined (e.g. in "next to or nearby with no large object in between", *next to* was ignored). Figure 1 shows the number of times a particular factor was mentioned for each spatial relation for those that were mentioned 5 or more times across all spatial relations. Infrequently mentioned factors were excluded as they were only mentioned by one or two people for a given spatial relation, and were thus thought to be less critical in the determination of meaning of the relation.

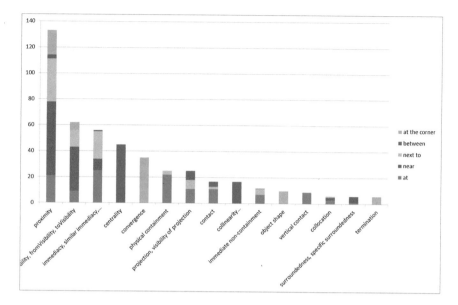

Fig. 1 Counts of factors in reasons given for applicability of spatial phrases

The results indicate that the factors that are important for *next to, near* and *at* are much more similar than those for *between*. *At the corner* also has its own profile in terms of the factors that are considered important.

5 A Typology of Contextual Factors

The factors involved in context have been categorized in various ways, as described in Sect. 2. However, the broadest schemes (e.g. (Porzel 2010)) do not address spatial aspects, and while some of the spatial schemes (e.g. (Brodaric 2007; Cai 2007; Souza et al. 2006)), incorporate spatial aspects, they do not include the level of detail identified in our empirical study. On the other hand, Talmy (Souza et al. 2006) lists factors that relate to specific spatial relations, but does not include the broader view. We therefore propose the typology shown in Fig. 2, which groups together many of the factors identified by previous researchers, and in our empirical study. We identify six broad types, moving from the most general environmental level factors with aspects such as indoor/outdoor, incorporating the observer and his or her goals and tasks as identified in the pragmatics research, down to the more detailed levels of specific spatial relation and object (geographic feature) factors. This is a preliminary typology and requires more detailed development and specification, both within each of the types and to define linkages and relationships between types.

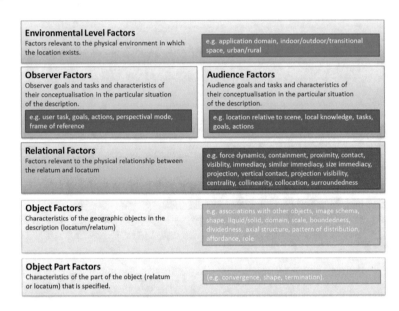

Environmental Level Factors
Factors relevant to the physical environment in which the location exists.
e.g. application domain, indoor/outdoor/transitional space, urban/rural

Observer Factors
Observer goals and tasks and characteristics of their conceptualisation in the particular situation of the description.
e.g. user task, goals, actions, perspectival mode, frame of reference

Audience Factors
Audience goals and tasks and characteristics of their conceptualisation in the particular situation of the description.
e.g. location relative to scene, local knowledge, tasks, goals, actions

Relational Factors
Factors relevant to the physical relationship between the relatum and locatum
e.g. force dynamics, containment, proximity, contact, visiblity, immediacy, similar immediacy, size immediacy, projection, vertical contact, projection visibility, centrality, collinearity, collocation, surroundedness

Object Factors
Characteristics of the geographic objects in the description (locatum/relatum)
e.g. associations with other objects, image schema, shape, liquid/solid, domain, scale, boundedness, dividedness, axial structure, pattern of distribution, affordance, role

Object Part Factors
Characteristics of the part of the object (relatum or locatum) that is specified.
(e.g. convergence, shape, termination)

Fig. 2 A typology of contextual factors in geospatial natural language

6 Conclusions

Considerations of context are essential for accurately determining the meaning of geospatial location expressions, and in this paper, we have defined context very widely, to consider both aspects of the broader environment in which a description occurs, as well as characteristics of the observer, the objects involved and the spatial relations between them. Based on a review of the literature and an empirical study, we identify a range of contextual factors that influence the interpretation of spatial expressions and propose a typology that groups the factors into six types.

In future work we propose to apply the empirical approach presented here more widely to gain a more comprehensive picture of the contextual factors that influence the use of spatial language. Additional work is needed to develop the typology, specifying the factors within each type and the relationships between types.

References

Bateman JA, Hois J, Ross R et al (2010) A linguistic ontology of space for natural language processing. Artif Intell 174:1027–1071

Bartie P, Reitsma F, Clementini E et al (2011) Referring expressions in location based services: the case of the 'opposite' relation. Advances in Conceptual Modeling. Recent Dev New Dir 231–240

Brodaric B (2007) Geo-Pragmatics for the Geospatial Semantic Web. Trans GIS 11:453–477

Cai G (2007) Contextualization of geospatial database semantics for human–GIS interaction. Geoinformatica 11:217–237

Coventry KR, Garrod SC (2004) Saying, seeing and acting: the psychological semantics of spatial prepositions. Psychology Press, East Sussex

Gahegan M (1995) Proximity operators for qualitative spatial reasoning. In: Spatial Information theory: a theoretical basis for GIS, COSIT '95. Lecture notes in computer science 988. Springer, Berlin

Hall MM, Jones CB, Smart P (2015) Spatial Natural language generation for location description in photo captions. In: Spatial information theory: COSIT 2015. Lecture notes in computer science 9368. Springer

Herskovits A (1985) Semantics and pragmatics of locative expressions. Cogn Sci 9:341–378

Hornsby KS, Li N (2009) Conceptual framework for modeling dynamic paths from natural language expressions. Trans GIS 13:27–45

Hubona GS, Everett S, Marsh E et al (1998) Mental representations of spatial language. Int J Human-Comput Stud 48:705–728

Klippel A, Xu S, Li R, Yang J (2011) Spatial event language across domains. In: Proceedings of the 2nd workshop on computational models of spatial language interpretation (CosLI-2), pp 40–47

Kordjamshidi P, Van Otterlo M, Moens MF (2011) Spatial role labeling: Towards extraction of spatial relations from natural language. ACM Trans Speech Lang Process 8:4

Kracht M (2002) On the semantics of locatives. Linguist Philos 25:157–232

Kray C, Fritze H, Fechner T, Schwering A, Li R, Anacta VJ (2013) Transitional spaces: between indoor and outdoor spaces. In: Spatial information theory. Springer, Berlin

Lakoff G, Johnson M (1980) Conceptual metaphor in everyday language. J Philos 77:453–486

Landau B, Jackendoff R (1993) Whence and whither in spatial language and spatial cognition? Behav Brain Sci 16:255–265

Lang E, Maienborn, C (2011) Two-level Semantics. In: Semantics: an international handbook of natural language meaning. Mouton de Gruyter, pp 709–740

Lautenschütz A-K, Davies C, Raubal M et al (2006) The Influence of Scale, Context and Spatial Preposition in Linguistic Topology. In: Proceedings of the international conference on spatial cognition v: reasoning, action, interaction, pp 439–452

Levinson SC (2003) Space in language and cognition: explorations in cognitive diversity, vol 5. Cambridge University Press,

Mani I (2009) SpatialML: annotation scheme for marking spatial expression in natural language. Technical Report Version 3.0, The MITRE Corporation

Mark DM, Frank AU (1996) Experiential and formal models of geographic space. Environ Plan B Plan Des 23:3–24

Porzel R (2010) Contextual computing: models and applications. Springer Science & Business Media,

Pustejovsky J, Moszkowicz JL, Verhagen M (2011) Using ISO-Space for annotating spatial information. In: Proceedings of the International Conference on Spatial Information Theory

Rodriguez MA, Egenhofer MJ (2004) Comparing geospatial entity classes: an asymmetric and context-dependent similarity measure. Int J Geogr Inf Sci 18:229–256

Sen S (2008) Use of affordances in geospatial ontologies. In: Rome E, Hertzberg J, Dorffner G (eds) Towards affordance-based robot control. Lecture notes in computer science 4760. Springer, Berlin

Shariff ARB, Egenhofer MJ, Mark DM (1998) Natural-language spatial relations between linear and areal objects: the topology and metric of English-language terms. Int J Geog Inf Sci 12:215–245

Souza D, Salgado A, Tedesco P (2006) Towards a context ontology for geospatial data integration. On the move to meaningful internet systems 2006: OTM 2006 workshops. Springer, Berlin, pp 1576–1585

Talmy L (2000) Toward a Cognitive semantics: volume 1 concept structuring systems. Cambridge.
    MIT Press, Massachusetts
Zwarts J (2005) Prepositional aspect and the algebra of paths. Linguist Philos 28:739–779

Spatial Prepositions in Natural-Language Descriptions of Indoor Scenes

Stacy Doore, Kate Beard and Nicholas Giudice

Abstract For individuals with significant vision impairment, due to natural aging processes or early vision loss, descriptions of indoor scenes require a high level of precision in spatial information to convey accurate object relations and allow for the formation of effective mental models. This paper briefly describes a single experiment conducted within a larger study investigating the preference of spatial prepositions to describe spatial relations between objects in a vista-scale setting. The goal of the larger study is to identify a small set of spatial concepts and terms that can provide a framework for automated natural-language descriptions of indoor vista-scale environments.

Keywords Spatial prepositions · Indoor scene depictions · Spatial assistance systems

1 Introduction

Over the past 30 years, there is an increase in efforts to represent and communicate spatial information about entities within indoor environments (Riehle et al. 2008; Falomir 2012; Li and Lee 2013). As people in industrial societies spend an estimated 90% of their lives indoors (American Physical Society 2008), the efficient representation of and communication about indoor space has become an active area of investigation for geographic information science. Automated annotation systems

S. Doore (✉) · N. Giudice
Virtual Environment Multimodal Interaction Lab, School of Computing
and Information Science, University of Maine, Orono, ME, US
e-mail: stacy.doore@maine.edu

N. Giudice
e-mail: nicholas.giudice@maine.edu

K. Beard
School of Computing and Information Science, University of Maine, Orono, ME, US
e-mail: kate.beard@maine.edu

© Springer International Publishing AG 2018
P. Fogliaroni et al. (eds.), *Proceedings of Workshops and Posters at the 13th International Conference on Spatial Information Theory (COSIT 2017)*, Lecture Notes in Geoinformation and Cartography, https://doi.org/10.1007/978-3-319-63946-8_41

for indoor environments are being developed for natural-language (NL) processing tasks, such as spatially anchoring events, generating scene descriptions, and interpreting thematic places in relationship to confirmed locations (Lin et al. 2015; Fermuller 2015; Bernardi et al. 2016). However, these systems often lack an explicit grounding in spatial information theory regarding the conceptualization and linguistic description of relations between objects situated within indoor space.

This paper describes a small portion of a larger study on spatial preposition use in descriptions of indoor scenes set in vista-scale spaces. A vista-scale space is defined as a space larger than the human body that can be perceived from a single perspective (Montello 1993). As there are physical and perceptual differences between outdoor and indoor spaces, one question of interest is 'What spatial prepositions do people use to describe spatial relations between objects in a room?'.

2 Background

Previous research has noted differences between outdoor and indoor space. Indoor environments limit observers' field of view and line of sight due to the built environment's physical structure such as walls, doors, and ceilings (Richter et al. 2011). Outdoor space is typically represented in symbolic 2D spaces, while, indoor environments are often represented as 3D multi-level models (Winter 2012). Vertical features such as staircases, elevators, and ramps can interfere with cognitive map development and accurate orientation when navigating (Li and Giudice 2012). Behavioral and neuroscience studies suggest there are differences in the visual and semantic information perceived when viewing indoor and outdoor scenes (Oliva and Torralba 2006; Greene et al. 2016; Henderson et al. 2007).

There is a large body of work in spatial information science and spatial linguistics regarding the alignment of NL spatial relations with formal conceptual models in tabletop and geographic space (Shariff et al. 1998; Schwering 2007; Kordjamshidi and Moens 2015). Many of these studies define a set of principles about how characteristics of the reference object (ground) and the located object (figure) impact the order and use of spatial prepositions (Herskovits 1986; Coventry and Garrod 2004). Contextual factors also influence the use of spatial language, including the location of the observer as well as imprecise use of spatial relations (Herskovits 1980). While this related work provides guidance on the structure of the experiments and the interpretation of the preliminary results, the present study differs in that it is situated firmly in a small room, vista-space setting, rather than tabletop or environmental space.

Fig. 1 Example images for small room (10' × 12') and large room (20' × 30')

Table 1 Participant demographics by setting (n = 90)	Group	Gender distribution	Age range
	UMaine lab setting	20 female/20 male	20–24
	AMT setting	26 female/24 male	24–34

3 Methods

The purpose of the experiments was to test subject responses to a set of predetermined spatial relations (*contact, disjoint, partof*) between indoor structure objects (i.e., walls, windows, doors) and moveable objects (i.e., furniture). The specific room representation, spatial relations and selected spatial prepositions were based on frequency of occurrence in a previous analysis of open indoor scene descriptions (Doore 2017). In the study, the virtual images presented a visual perspective where the entirety of the room could be seen from a single location without motion, except the space behind the participant (Fig. 1).

A total of 90 participants were recruited for the study conducted in two different settings: (1) a lab-based setting, and (2) an Amazon Mechanical Turk (AMT) setting (Table 1). All participants identified themselves as native English speakers and self-reported they were raised in the United States.

All participants were presented with six virtual indoor images with text prompts describing a *contact, disjoint,* or *part of* relation between a moveable object and a structure object. For each of the images, participants were asked to rank their preference of set of 15 spatial prepositions to complete a prompt for the given image context (e.g., 'The bookcase is___the wall.'). The prompts were extracted from an analysis of open scene descriptions of similar real world indoor spaces and objects (Doore 2017). The results were analyzed using similarity matrices, Chi square, Friedman, and Wilcoxon signed ranks tests to test for significant differences in preference for the given terms based on each image prompt.

4 Results

For *contact* relations, there was a statistically significant preference for using *against* instead of *along, next to,* or *touching* ($p < 0.01$). However, there was no statistically significant difference in preference for using *against* versus *on* to describe a *contact* relation. Furthermore, *on* ($p = 0.012$) was not reliably preferred over the other highly rated terms *along* ($p = 0.106$), *next to* ($p = 0.236$), and *touching* ($p = 0.314$). Therefore, in ranking the preference of spatial terms for the *contact* relation between objects, *against* was the most highly preferred term. For *partof* relations, there was a statistically significant preference for the spatial preposition term *on* ($p < 0.01$). The term *on* was the only preferred term to describe the relationship of the window and the wall. Terms such as *in the middle of, connected to,* and *supported by,* were ranked as not preferred ($p < 0.01$). For images with *disjoint* relationships between objects and structures, *near* and *next to* were the only statistically significant terms for preference ($p < 0.01$) and all other terms were ranked as not preferred ($p < 0.01$).

5 Discussion

The results from this study provide support for the refinement of the set of preferred spatial prepositions used in object to object spatial reference expressions (in English) for *contact, disjoint,* and some *partof* relations in indoor scenes. The results of this experiment, in combination with other experiments within the larger study, suggest that the *contact* relation set of spatial prepositions should favor the use of terms *against* and *on* as primary terms for indoor scenes and *along,* and *touching* as secondary terms. For *partof* relations between specific types of room structures such as windows and walls, the results indicate the use of *on* as the preferred term to describe spatial relationships. The preferred terms for *disjoint* relations between objects and structures were *near* and *next to,* which were used interchangeably over other possible terms. While the terms *by* and *to* (*right/left*) can be considered similar alternatives for *disjoint* relations between object and structure pairs, the use of the more vague proximity term *by* or direction term *to* was not strongly supported by the results of the analyses.

The preliminary results confirm previous research regarding the choice of spatial prepositions and how these choices convey key pieces of spatial information such as a *contact/disjoint relation* between the figure and ground and the primacy of objects on the *vertical axis* as spatial references. The use of *on* for *partof* relations between room structure objects, specifically, windows and walls, supports the observation that choice of spatial prepositions conveys information about the *inclusion* of the figure by the ground, as well as the *nature of the support,* if any, provided to the figure by the ground (Feist 2000).

There is evidence to support a gradient of spatial prepositions based on *contact* and *support* sense for the prepositions *on* (Levinson 2003). While the multiple semantics of *on* can be distinguished by the *support* sense and/or the *contact* sense, neither represents the use of *on* in a *partof* relation as was observed in the window and wall relation. The strong preference for using *on* to describe the relationship between a window and a wall suggests there is more to explore in these types of indoor relations.

6 Conclusion

This paper investigates factors that impact preference of spatial preposition use for natural-language descriptions in vista-scale indoor environments. The results identify a small sub-set of spatial prepositions for frequent spatial relations in this type of setting. This sub-set of terms can provide guidance for automated scene description systems that is consistent with spatial information and linguistics theory, as well as point to additional areas for investigation into conceptualization of spatial relations unique to indoor space. Future work includes applications for improved scene descriptions to assist aging populations and other users with significant vision loss, to more effectively orient themselves and navigate within unfamiliar indoor environments.

References

Aditya S, Yang Y, Baral C, Fermuller C, Aloimonos Y (2015) From images to sentences through scene description graphs using commonsense reasoning and knowledge. arXiv preprint arXiv:1511.03292

American Physical Society (2008) Energy future: think efficiency. American Physical Society, Washington, DC

Bernardi R, Cakici R, Elliott D, Erdem A, Erdem E, Ikizler-Cinbis N, Keller F, Muscat A, Plank B (2016) Automatic description generation from images: a survey of models, datasets, and evaluation measures. J AI Res 55:409–442

Coventry KR, Garrod SC (2004) Saying, seeing and acting: the psychological semantics of spatial prepositions. Psychology Press

Doore S (2017) Natural-language semantics of spatial relations for describing indoor scenes. Doctoral thesis. University of Maine, Orono

Falomir Z (2012) Qualitative distances and qualitative description of images for indoor scene description and recognition in robotics. AI Comm 25(4):387–389

Feist M (2000) On in and on: an investigation into the linguistic encoding of spatial scenes. Doctoral dissertation, Northwestern University

Greene MR, Baldassano C, Esteva A, Beck DM, Fei-Fei L (2016) Visual scenes are categorized by function. J of Exp Psych Gen 145(1):82

Herskovits A (1980) On the spatial uses of prepositions. In: Proceedings of the 18th Annual Meeting on Association for Computational Linguistics, pp 1–5. Association for Computational Linguistics

Herskovits A (1986) Language and spatial cognition: an interdisciplinary study of the prepositions in English, studies in natural language processing. Cambridge University Press, London

Henderson JM, Larson CL, Zhu DC (2007) Cortical activation to indoor versus outdoor scenes: an fMRI study. Exp Brain Res 179(1):75–84

Kordjamshidi P, Moens MF (2015) Global machine learning for spatial ontology population. Web Semant Sci Serv Agents World Wide Web 30:3–21

Levinson SC (2003) Space in language and cognition: explorations in cognitive diversity. Cambridge University Press, Cambridge

Li H, Giudice NA (2012) Using mobile 3D visualization techniques to facilitate multi-level cognitive map development of complex indoor spaces. Spat Knowl Acquis Ltd Inf Disp 21:31–36

Li KJ, Lee JY (2013) Basic concepts of indoor spatial information candidate standard Indoor GML and its applications. J Korea Spat Inf Soc 21(3):1–10

Lin D, Fidler S, Kong C, Urtasun R (2015) Generating multi-sentence natural-language descriptions of indoor scenes. In: Xie X, Jones M, Tam G (eds), In: Proceedings of the British Machine Vision Conference (BMVC), pp 93.1–93.13. BMVA Press

Montello DR (1993) Scale and multiple psychologies of space. In: European Conference on Spatial Information Theory, pp 312–321. Springer, Berlin Heidelberg

Oliva A, Torralba A (2006) Building the gist of a scene: the role of global image features in recognition. Prog Brain Res 155:23–36

Richter KF, Winter S, Santosa S (2011) Hierarchical representations of indoor spaces. Environ Plan B Plan Des 38(6):1052–1070

Riehle TH, Lichter P, Giudice NA (2008) An indoor navigation system to support the visually impaired. In: Engineering in Medicine and Biology Society, 2008. EMBS 2008. 30th Annual International Conference of the IEEE. IEEE, pp 4435–4438

Schwering A (2007) Evaluation of a semantic similarity measure for natural language spatial relations. In: Winter S, Kuipers B, Duckham M, Kulik L (eds) Spatial Information Theory: 9th International Conference, COSIT 2007. Springer, Melbourne, Australia, Berlin

Shariff ARB, Egenhofer MJ, Mark DM (1998) Natural-language spatial relations between linear and areal objects: the topology and metric of English-language terms. Inter J Geol Info Sci 12 (3):215–245

Winter S (2012) Indoor spatial information. Intel J 3-D Info Mod 1(1):25–42

Part IV
Spatial Humanities Meets Spatial Information Theory: Space, Place, and Time in Humanities Research

Spatial Humanities Meets Spatial Information Theory: Space, Place, and Time in Humanities Research—Introduction

Benjamin Adams, Olga Chesnokova and Karl Grossner

Humanities disciplines such as history, classical studies, literary studies, and philology have in recent years experienced a "spatial turn" similar to that begun in prior decades within the social sciences and archaeology. Many researchers in these fields are now explicitly recording the spatial and temporal attributes of their data and mapping them for visual analysis and argumentation. In many cases they are also performing spatial or spatial-temporal computations, including but not limited to viewshed, network, and cluster analyses, and agent-based and other models and simulations are increasingly common. The software used for this work is the same as that used for the environmental and social sciences: desktop GIS and specialized spatial and natural language processing libraries for the Python and R languages. These new spatial researchers are experiencing the same representational and analytic challenges in studying geographical dynamics that are well known to other disciplines, but they also face distinctive issues related to the nature of historical humanities data. Furthermore, epistemologies associated with new quantitative approaches must be reconciled with their traditional methodological practices.

Spatial information theorists and geographic information scientists have not normally drawn from humanities research cases for their development of theoretical models or the specific software and systems built upon such models. The SPHINx workshop was envisioned to foster fruitful dialog between these groups. The four papers accepted for presentation represent the work of a diverse group of scholars coming from the digital humanities, history, linguistics, and computer science and

B. Adams (✉)
Department of Geography, University of Canterbury, Christchurch, New Zealand
e-mail: benjamin.adams@canterbury.ac.nz

O. Chesnokova
Department of Geography, University of Zürich, Zürich, Switzerland
e-mail: olga.chesnokova@geo.uzh.ch

K. Grossner
World History Center, University of Pittsburgh, Pittsburgh, PA, USA
e-mail: karlg@worldheritageweb.org

© Springer International Publishing AG 2018
P. Fogliaroni et al. (eds.), *Proceedings of Workshops and Posters at the 13th International Conference on Spatial Information Theory (COSIT 2017)*, Lecture Notes in Geoinformation and Cartography, https://doi.org/10.1007/978-3-319-63946-8_42

covering topics on historical deep mapping, modeling spatio-temporal dialect data, identity in historical geographic data, and narratology of space. In addition to these paper presentations, the workshop included interactive discussions to formulate a future research agenda for the spatial humanities and spatial information theory.

Exploring Deep Mapping Concepts: Crosthwaite's Map and West's Picturesque Stations

Alexander Reinhold, Christopher Donaldson, Ian Gregory and Paul Rayson

Abstract What are the requirements for building a historic deep map using literary data? This is the question we sought to address as part of an exploratory prototype in Lancaster University *Geospatial Innovation in the Digital Humanities: A Deep Map of the English Lake District* project. We created a prototype deep map based on Thomas Wests *A guide to the Lakes*, and a historic map of Derwent Water Lake created by Peter Crosthwaite. Our prototype maps the locations of Wests picturesque viewing stations and creates connections between the literary work and visual representations of the places described. This article describes our approach to building this prototype and discusses what we learned and the issues we revealed about creating a historic deep map.

Keywords Lake district · Deep mapping · Leverhulme · Crosthwaite · West

1 Introduction

This article looks to discuss our investigation into the requirements for and construction of historic deep mapping applications. As part of Lancaster University's *Geospatial Innovation in the Digital Humanities: A Deep Map of the English Lake District* project, we wanted to explore the concept of a Deep Map in a more limited

A. Reinhold (✉) · P. Rayson
School of Computing and Communications, Lancaster University,
Lancaster LA1 4WA, UK
e-mail: a.reinhold1@lancaster.ac.uk

P. Rayson
e-mail: p.rayson@lancaster.ac.uk

C. Donaldson · I. Gregory
Department of History, Lancaster University, Lancaster LA1 4YT, UK
e-mail: c.e.donaldson@lancaster.ac.uk

I. Gregory
e-mail: i.gregory@lancaster.ac.uk

© Springer International Publishing AG 2018
P. Fogliaroni et al. (eds.), *Proceedings of Workshops and Posters at the 13th International Conference on Spatial Information Theory (COSIT 2017)*, Lecture Notes in Geoinformation and Cartography, https://doi.org/10.1007/978-3-319-63946-8_43

scope, to allow for a discussion of the included elements and features, and evaluate their effectiveness. We chose to focus on a single historic map of Derwent Water Lake in the Lake District, created by Peter Crosthwaite in 1783. Crosthwaite's map depicts his representation of Thomas Wests Viewing Stations around Derwent Water. The West stations are picturesque locations in the environment that West visited during his travels.

The prototype our team developed focuses on two concepts that have been included in various definitions of deep mapping. The first is a common concept that appears frequently in the discussion of spatial humanities, space versus place, or a chorography (Rohl 2012). A traditional map is generally a representation of space, while a deep map, under this definition, should be a representation of place (Oxx et al. 2013). The second deep mapping concept we chose to focus on is the 'open-ended exploration of a particular place' (Ridge et al. 2013). Each place represented in our prototype is intended to provide users with the ability to explore the available data pertaining to that place. Our prototype uses these concepts to explore the places identified in Crosthwaite and West's works. This exploration of, and focus on place, makes our prototype an exemplar application of deep mapping.

2 Crosthwaite's Map

The English Lake District is renowned for its literary and artistic history. It is famously a place of poetry and of painting. Crucially, though, it is also a place of maps. True, the region was among the last portions of England to be mapped in the first Ordnance Survey. But the belatedness of the Lakeland's inclusion in the OS is significantly counterbalanced by its rich cartographic history. By the time the first OS maps of Cumberland appeared (in the 1860s), the Lake District was doubtlessly one of the most frequently and widely represented regions in Britain. Admittedly, many of these representations took the form of scenic pictures and verbal descriptions. Many, however, also took the form of maps, plans, and topographical diagrams. Among these maps, plans, and diagrams, few stand out as prominently as Peter Crosthwaite's 'Accurate Maps' of the Lakeland's eponymous lakes.

A native of the parish of Crosthwaite, near Keswick, Peter Crosthwaite (1735–1808) had an exceptionally colourful career. As a master of a gunboat called the Otter, he spent his early adulthood protecting East India Company ships from pirates in the Bay of Bengal. Following his return to England in 1765, Crosthwaite opened a new chapter of his life working as a customs officer on the coast of Northumberland. Then in 1779 he returned to Keswick as a respectable man, married with two children (Fisher Crosthwaite 1778). The year after, he took a house in the Square and quickly set up shop as a purveyor of publications, amusements, and paraphernalia for tourists.

Of all Crosthwaite's business ventures, however, it is his pursuits as a cartographer that are of principal significance. In part, this is because of the peculiar mixed-media character of his maps (which we discuss below); in part, though, this is because his maps were the first designed specifically for Lakeland tourists. Whereas some of the early accounts of the Lakes region included cartographical plans, none of the early guide or tour books did. Even Thomas Wests widely cited and celebrated Guide to the Lakes (1778) did not include a map until the publication of its third edition in 1784 the year after Crosthwaite began publishing his own 'Accurate Maps' of several of the region's key lakes. In this way, Crosthwaite's maps can be seen to complement other aspects of his entrepreneurial enterprises: they rushed to fill a gap in the region's developing tourist trade.

Crosthwaite's 'Accurate Maps' were produced between 1783 and 1794, and thereafter revised and reproduced until 1819. Crosthwaite completed his maps of Derwent Water, Windermere, Ullswater, and Pocklington's Island (in Derwent Water) in 1783. He enlarged this series in 1785 and 1788 with his maps of Bassenthwaite and Coniston Water. Finally, in 1794, he completed his series with a map of the western Lakes, which includes Buttermere, Crummock Water, and Loweswater. Meticulously surveyed and packed with detail, these maps are exquisite productions, and nearly all of them (save those of Windermere and Pocklington's Island) are drawn on the scale of 3 in. to the mile. All of them, moreover, are designed to be handheld; on average, they measure 18 in. by 8.5 in.

These maps, after all, offered much more than information about the locations of places, roads, and routes. They were miniature compendia of topographical and picturesque detail. In addition to marking the principle landmarks, houses, and estates (as well as the names of their owners), Crosthwaite's maps featured vignettes of those landmarks, houses, and estates, as well as passages of descriptive poetry and prose. Beyond this, they also featured the locations of the viewing stations designated by Thomas Wests Guide as those places where the Lakes scenery could be seen to its best advantage. Crosthwaite's creations could therefore be considered as prototype deep maps of their day, and so we considered them to be an excellent first case study in our own exploration of this so far ill-defined concept.

Julia Carlson has recently remarked on this curious 'convergence of cartographic and poetic innovation' (Carlson 2016, p. 44). Crosthwaite's maps, explains Carlson, 'were a new form of print text intended to facilitate a new form of experience [...]. They encouraged physical exploration and demanded the interpretation of intersecting modes of measur[ing]' geometrical, pictorial, and literary (Carlson 2016, p. 44). These composite creations, in short, combined different illustrative media not only to guide the tourist to key locations, but also to inform the tourists feelings about those locations. These maps were tools for instructing not only the eye, but also the mind and the heart.

3 Deep Mapping in the Humanities

Many of the prototype deep map applications discussed in academic papers, including this one, have been primarily map based, focusing on the spatial distribution of data and providing some form of navigation for users to interact with the map and explore the data. A prototype, using Google maps to spatially place data and a content bar to allow users to view details about visible sites, is a good example of the basic implementation of a deep map, because it can provide information to a user that can transform a space into a place. The creators of this implementation of a deep map put the most 'emphasis on one particular user interaction: exploration' (Ridge et al. 2013).

A study using data from Greek Orthodox churches in Indianapolis created a deep map that focused on the 'spatial navigation' of data, rather than the spatial distribution of data. Instead of placing all the data they had available into a map, this team create a Prezi project and placed their evidence into the canvas of the presentation. This structure allows the data to be spatially navigable be viewers of the presentation, but the data does not need to be associated with any geographic coordinate systems. The presentation was also given a storyline so users could be lead through the data in a more curated manor, but users still have the freedom to explore and deviate from the storyline if they want (Oxx et al. 2013). The Prezi project shows a more abstract view of what a deep map could be.

4 Methodology

We wanted to bring the Crosthwaite Derwent Water map,[1] the concept of Wests stations (West 1778), and a selection of texts, by other authors, about the stations from the Lancaster Lake District Literary Corpus[2] into a single environment for users to be able to explore. In addition, we wanted users to have access to a representation of the environment that the text discusses at each station. The initial prototype is intended to provide a framework around which we can carry out interviews with academic colleagues interested in the deep mapping concept and with potential end users in order to gather feedback and requirements for future development. We are loosely exploiting the principles of participatory or co-design in this investigation.

The main page of this application is built around a Google map, centred on Derwent Water Lake, with markers denoting the location of the eight West stations as depicted in Crosthwaite's Map (Fig. 1). A navigation panel on the left side of the map has a toggle for the historic Crosthwaite map, and list two sections, a list of the West stations, and a list of the texts included in the application (Fig. 2). Clicking on a station from the list, or its marker on the map, will take the user to the page for that station, while clicking on a text from the text list will take the user to a text view.

Each station page is a Google Street View of the stations location (or the closest available to the corresponding point on the Crosthwaite Map). The Street View Panorama allows users to explore the site of the station with 360° imagery (Fig. 3). Overlaid on the Panorama is another left navigation (Fig. 4) which contains tabs, one for each text mentioning that station and another tab for a FoamTree.[3] Each tab of text shows the title of the text, a link to the full text, and the section of that text that specifically talks about the station. The FoamTree tab is a Voronoi treemap

---

[1]Crosthwaite Maps http://www.geog.port.ac.uk/webmap/thelakes/html/crosth/ct1fram.htm.

[2]Spatial Humanities: Text, GIS, Places project http://www.lancs.ac.uk/spatialhum.

[3]Carrot Search: Foam Tree https://carrotsearch.com/foamtree/.

Fig. 1 Main page, with markers representing the West station locations on the map. Navigation: contains toggle to historic map, and lists of stations and texts

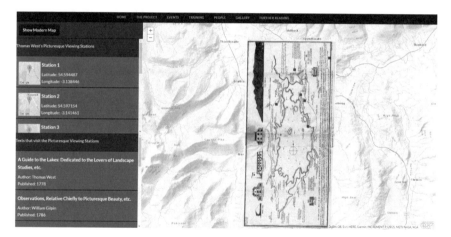

Fig. 2 Historic Crosthwaite map overlaid on topographic map

visualization of the top 50 most used words (excluding standard stop words) in the texts talking about the selected station.

The text view allows users to read the text alongside additional interactive information pulled from the text (Fig. 5). The text view page has a simple navigation in the top right allowing users to flip through the pages of the text. Each page shows an image of the original historic page on the right side and a list of the locations mentioned on that page on the left. If the page of text being viewed mentions a station, a link to that station is shown on the far right. The list of locations for each page were found using the Edinburgh Geoparser, with some manual corrections of

Fig. 3 Station view, panoramic imagery from Google street view of the selected station

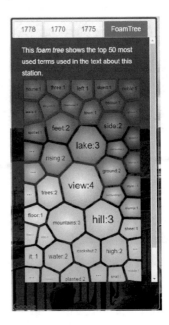

Fig. 4 Station view navigation, tabs containing linked texts and a FoamTree summary of word frequency

the data (Grover et al. 2010; Donaldson et al. 2017). Clicking on location from the list brings users to the main map view and focuses on the selected location.

An underlying database to the application contains the metadata for the stations and text. The database also defines the sections of text, by sentence range or page range, which are associated with a station. Our Crosthwaite Deep Map prototype was

Fig. 5 Text view, showing image of the page, a list of locations mentioned on the page, and link to any mentioned stations

constructed using a Java server to perform data management and database access. The data from the Java server is called via Ajax, by the HTML page and stored by JavaScript.

5 Discussion

The Crosthwaite Deep Map application that we created provides users with an interactive method of engaging with the Derwent Water map and a selection of the literary pieces related to it. The most significant question that must be answered is: is this a Deep Map? A deep map has been described as needing 'quantitative and qualitative data, spatial data, images [...], and virtual representations of places[...]' (Oxx et al. 2013). Under this definition, our map would seem to qualify as a deep map with its inclusion of the texts, stations, and panoramic imagery. Our application provides the user with an array of data related to the Crosthwaite map in an undirected interface, which allows for 'open-ended exploration' of Derwent Water (Ridge et al. 2013).

In the original iteration of this map, the text view showed a plain text rather than the image of the historic page. The decision to add the original page imagery was based on initial user feedback and was an attempt to make the feel of the site more closely linked with the historic source data. Without this imagery, the pages would have little to no visual tie to the original representation of the texts. As these texts are over 200 years old, we decided it was important to make that attribute more explicit, so that users might have a better understanding of the context in which the texts were written, given that the locations West describes in his work may have changed considerably over time.

This project has revealed some issues with the deep mapping of historic data. The Geoparsed locations in the text are dependent on the accuracy of the Natural Language Processing software and the robustness of the gazetteers used, and West's descriptions of the stations were before the advent of GPS, so the true position of each station is subject to interpretation of the text. In this project two of the eight stations did not have fixed locations on the Crosthwaite Map, so we used location data found by the Lake District National Park Authority in a review on West's work (Hardie and Newman 2009). Uncertainty in the data can be difficult to represent on the map, whether it is the accuracy of West's stations or the parsed locations in the text. By placing markers on the map and linking the parsed locations to specific coordinates, the implication to the user is that the locations is absolute.

The use of Panoramas from Google Street View provides a useful visualization of different locations, but they come with some technical problems and concerns about historical representation. The panorama linked to each station is the closest available image from Google and in some cases can be a fair distance away from the position indicated in the Crosthwait map. The panoramas are modern day images of the Station locations, and could have drastically different environments from the historic environment West described. Due to deforesting, reforesting and the introduction of foreign species, the floral environment today is likely to have changed from that of 200 years ago. The quality of the images is subject to the device, and or person that took the images, and on the effectiveness of the registration, or stitching, during processing of the images to create a panorama. Many of the Google Street View Panoramas have significant error.

6 Conclusion

This project has allowed us to explore deep mapping in a historical context. As our overarching project is heavily centred on the literary texts of the Lake District, our Crosthwaite map has demonstrated how the text and spatial elements can be brought together. Understandably, the creation of a historic deep map is complex and will require thought as to which features will bring the most value to the application. The representation of time, and the historic nature of the content, is another key consideration that emerged from our study and will need further thought. As highlighted here, this impacts both on user interface issues (e.g. visual representation of the original pages of the books) as well as side-by-side presentation of materials (such as original drawings, paintings and modern web-based panoramas). This application has revealed that our future deep mapping applications will need to consider how to maintain the historic context of the data using different visualization technologies, and how to represent ambiguity in the spatial features found in the data. Our prototype provides a model for exploration, and focus on linking historic content to the visual representation of place in a deep mapping application.

Acknowledgements This research received funding from the Leverhulme Trust (RPG-2015-230), as part of the 'Geospatial Innovation in the Digital Humanities' project.

References

Carlson JS (2016) Romantic marks and measures: wordsworths poetry in fields of print. University of Pennsylvania Press, Philadelphia

Donaldson C, Gregory IN, Taylor JE (2017) Locating the beautiful, picturesque, sublime and majestic: spatially analysing the application of aesthetic terminology in descriptions of the English Lake District. J Hist Geogr 56:43–60

Fisher Crosthwaite J (1778) Peter Crosthwaite: the founder of Crosthwaite's Museum, Keswick. Trans Cumberland Assoc Adv Lit Sci 3

Grover C, Tobin R, Byrne K, Woollard M, Reid J, Dunn S, Ball J (2010) Use of the Edinburgh geoparser for georeferencing digitized historical collections. Philos Trans R Soc A: Math Phys Eng Sci 368(1925):3875–3889

Hardie C, Newman C (2009) A Review of West's 18th century Picturesque Viewing Stations in the Lake District National Park. Prepared for the Lake District National Park authority (0065/1-09)

Oxx K, Brimicombe A, Rush J (2013) Envisioning deep maps: exploring the spatial navigation metaphor in deep mapping. Int J Hum Arts Comput 7(1–2):201–227

Ridge M, Lafreniere D, Nesbit S (2013) Creating deep maps and spatial narratives through design. Int J Hum Arts Comput 7(1–2):176–189

Rohl DJ (2012) TRAC 2011: Proceedings of the twenty first theoretical roman archaeology conference. Oxbow Books, Oxford

West T (1778) A guide to the Lakes: dedicated to the lovers of landscape studies, and to all who have visited, or intend to visit, the lakes in Cumberland, Westmorland, and Lancashire. Printed for Richardson and Urquhart, under the Royal Exchange, and W. Pennington, Kendal

A Spatio-Temporal Linked Data Representation for Modeling Spatio-Temporal Dialect Data

Johannes Scholz, Emanual Hrastnig and Eveline Wandl-Vogt

Abstract Collections of linguistic and dialect data often lack a semantic description and the ability to establish relations to external datasets, from e.g. demography, socio-economics, or geography. Based on existing projects—the Database of Bavarian Dialects in Austria and exploreAT!—this paper elaborates on a spatio-temporal Linked Data model for representing linguistic/dialect data. Here we focus on utilizing existing data and publishing them using a virtual RDF graph. Additionally, we exploit external data sources like DBPedia and geonames.org, to specify the meaning of dialect records and make use of stable geographical placenames. In the paper we highlight a spatio-temporal modeling and representation of linguistic records relying on the notion of a discrete lifespan of an object. Based on a real-world example—using the lemma "Karotte" (engl. carrot) we show how the usage of a specific dialect word ("Karottn") changes from 1916 until 2016—by exploiting the expressive power of GeoSPARQL.

Keywords Linked Data · Dialect · Linguistics · Spatio-temporal modeling

J. Scholz (✉) · E. Hrastnig
Institute of Geodesy, Graz University of Technology,
Steyrergasse 30, 8010 Graz, Austria
e-mail: johannes.scholz@tugraz.at

E. Hrastnig
e-mail: hrastnig@student.tugraz.at

E. Wandl-Vogt
Austrian Centre for Digital Humanities, Austrian Academy of Sciences,
Wohllebengasse 12-14/2, 1040 Wien, Austria
e-mail: eveline.wandl-vogt@oeaw.ac.at

© Springer International Publishing AG 2018
P. Fogliaroni et al. (eds.), *Proceedings of Workshops and Posters at the 13th
International Conference on Spatial Information Theory (COSIT 2017)*, Lecture Notes
in Geoinformation and Cartography, https://doi.org/10.1007/978-3-319-63946-8_44

1 Introduction and Motivation

Language Geography and Geolinguistics are concerned with the geographic distribution of languages or their constituent elements. It is a field that strives to enhance the usability of digital language databases, and works towards a visual exploration of linguistic data. Languages and dialects are present in space and are mostly represented as language areas (Chambers and Trudgill 1998).

In language geography, and especially dialectology, the basis for creating language maps are field surveys. Examples for this approach include the Wenker Atlas (Schmidt and Herrgen 2001) and the Dictionary of Bavarian Dialects in Austria (Österreichische Akademie der Wissenschaften and Bauer 1985). Field surveys are questionnaires that were answered by teachers or other trained persons from 1887–1888 (Wenker Atlas) and from 1913 onwards (Dictionary of Bavarian Dialects in Austria). Thus, each questionnaire is connected to a specific place—i.e. where the person lives/d and collected evidence. Subsequently, each identified dialect record—at least the word, the pronunciation and its meaning—is connected to a location. Recently, these data have been digitized and stored electronically, using contemporary object-relational database technology. These data can be analyzed by linguists who create maps with isoglosses, dialect continua and finally language dictionaries. Hence, the approach in this paper is concerned with basic data, which are necessary to create more advanced "products," like language maps.

Nevertheless, as these basic data are not opened up for the public, it remains hard to combine other language data sets and/or to compare them with historic socio-economic or demographic datasets. Especially as most linguistic datasets lack a semantic description and the use of shared vocabularies such as e.g. place names. Yet, the ongoing project exploreAT! (exploring Austria's culture through the language glass; Austrian Academy of Sciences; 2015–2019) is going to open up the data sets, interlink existing concepts, make use of semantic technologies and make citizens part of the scientific process. Bird et al. (2009) formulated three fundamental questions concerning the design and distribution of language resources. Of these, the third is of importance for the current paper: "What is a good way to document the existence of a resource we have created so that others can easily find it?" (Bird et al. 2009, p. 407).

We propose a spatio-temporal Linked Data approach to model and publish data on linguistics and dialects. As there are a number of local linguistic data sets in Austria (e.g. dialect database of Upper Austria,[1] dialect database of Salzburg [2]) a Linked Data approach helps integrate different datasets in an ad-hoc manner and facilitates an integrated analysis of different datasets. Based on the Dictionary and Database of Bavarian Dialects in Austria (DBÖ) (Österreichische Akademie der Wissenschaften and Bauer 1985), and the results of the research project "Dictionary Bavarian Dialects in Austria electronically mapped" (e.g. Scholz et al. 2008) our objective is to develop a spatio-temporal Linked Data representation for dialect data.

---

[1] http://www.stifter-haus.at/sprachforschung.

[2] https://www.sprachatlas.at/salzburg.

In addition, we present preliminary results of the Linked Data approach that enable spatio-temporal query capabilities in conjunction with external Linked Data sets, utilizing data originating from the DBÖ and the dialect database of Salzburg.

We elaborate on relevant work in Sect. 2. Section 3 deals with the approach to develop a Linked Data representation for the linguistic and dialect data with a focus on the DBÖ. Subsequently, we elaborate on preliminary results in Sect. 4 and critically discuss them in Sect. 5.

2 Related Work and Background

An overview of mapping techniques in the field of linguistics and dialectology is given in Lameli et al. (2010). Contemporary atlases on dialectology and/or languages present their data using point symbols or thematic maps (e.g. Schmidt and Herrgen 2001). Additional elements often used in language maps are isoglosses and isographs—critically discussed by Pi (2006). Some papers suggest the usage of honeycomb maps around observation points (similar to Voronoi diagrams or Delaunay triangulation) (Goebl 2010; Nerbonne 2010). Rumpf et al. (2010) proposed an analysis of language data using kernel density estimation and elaborated on geographical similarity evaluations on area-class maps.

In the field of linguistics several publications utilize GIScience methods to analyze linguistic data. However, only a handful of papers in GIScience deal with linguistics. Among these are publications by Hoch and Hayes (2010), Sibler et al. (2012), and Scholz et al. (2016). Jeszenszky and Weibel (2015) postulate four research questions to analyze and describe the nature of language boundaries.

Buccio et al. (2014) describe an approach to publish linguistic data of the Syntactic Atlas of Italy, but do not mention any spatial and temporal modeling and/or analysis capabilities. A number of geolinguistic and linguistic projects are of interest for this paper. The first ontology designed to support the publishing and description of linguistic data in the semantic web is mentioned in Farrar and Langendoen (2003). An ontology-based mapping between different linguistic datasets is presented in Chiarcos et al. (2008). Xie et al. (2009) present an outcome of the research project LL-Map, highlighting the integration of language-related data with data from the physical and social sciences with the help of a GIS. In addition the Open Linguistics Working Group is working towards a Linguistic Open Data Cloud, making use of semantic web methodologies (Chiarcos et al. 2011). Lee and Hsieh (2015) present an example of linguistic Linked Data by publishing the Chinese Wordnet as part of the Linguistic Linked Data Cloud. Frontini et al. (2016) report on the transformation of GeoNames' ontology concepts into a GeoDomain WordNet-like resource in English, and its translation into Italian.

3   Linked Data 4 Dialects: Concept and Development Approach

The approach followed here is based on an existing relational database model developed for the DBÖ. Since this database serves as the main storage of linguistic data, we use it as much as possible, avoiding redundant storage of data sets wherever possible. Based on the relational data model we developed an ontology, for modeling and representing geolinguistic resources. The OWL ontology is divided in three parts: derivation, tagging and geographic. The derivation part deals with people speaking a language, whereas the geographic part deals with the locations where a certain language is spoken (and by whom). The tagging part is concerned with language specific classes and properties, like documents, questions, lemma or meaning. The basic structure is given in Fig. 1. The most important classes of the ontology are source, record, lemma, meaning, location, time and geodata. The class *lemma* contains the canonical form, dictionary form, or citation form of a set of words. Each individual of the class *record* is related to a lemma, and has a certain meaning as well. This is necessary, as each usage of a lemma is embodied in a context that influences the meaning. An example is the lemma "mouse", which can be used in the context of computers or biology. The class *source* contains the source (evidence) of each record. The DBÖ relies on a database of 5 million paper sources (vouchers with dialect words), which were digitized from the early 1990s.

The ontology inherits vocabulary from other domains and uses its own domain *dboe*. The inherited vocabularies are *geonames* and *DBpedia* for now. *DBpedia* is used to define the meaning of a dialect record. In the future we plan to include *BabelNet* or *Wikidata* and connect with historical gazetteers, e.g. via the project PELAGIOS.[3] *Geonames* is used to reference place and region names contained in the linguistic datasets.

The spatio-temporal context is related to each source—i.e. to each voucher. Each source has a certain location, as each dialect word is spoken at a specific physical place. In addition, a location has three subclasses, not depicted in Fig. 1: town, community, region. Towns are populated places represented as points, whereas communities and regions are polygons. Towns and communities are inherited from geonames.org. Regions are defined from a linguistic perspective and are not identical with administrative regions. Hence, spatial data on linguistic regions cannot be inherited from an external source. Since language has a dynamic nature (see e.g. Wandl-Vogt 1997; Birlinger 1890; Nerbonne 2010), language phenomena may move, emerge, end, expand or shrink—similar to other real world objects (e.g. Nixon and Stewart Hornsby 2010). Thus we added a valid time span to each source, representing the timeframe a specific word is found at a location. This approach is intended to represent the temporal changes in linguistic phenomena—using a discrete representation.

---

[3] http://commons.pelagios.org.

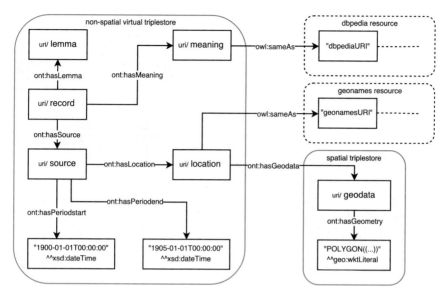

Fig. 1 Excerpt of the developed Ontology—showing only the important classes, their relationships, and the inherited vocabulary from external sources. The spatio-temporal aspect is modeled as the time period of each source, and the spatial data available for each location associated to each source document

Thus, it is not possible to model a gradual change with this approach. Currently, this fulfills the requirements of linguistic phenomena, as surveys are not done in a continuous, high frequent manner.

4 Preliminary Results

The preliminary results are a proof-of-concept implementation and spatio-temporal SPARQL queries (i.e. GeoSPARQL), based on the data present in the DBÖ. The implementation publishes the existing dialect data as a SPARQL endpoint with the help of a virtual RDF graph (Bizer and Seaborne 2004). Figure 2 depicts the architecture of the proof-of-concept implementation. We utilize the existing relational database with the help of a virtual RDF graph using D2RQ.[4] Spatial data on the linguistic regions of the DBÖ are published in a spatial triple store—here Strabon.[5]

Preliminary results are based on existing datasets of the DBÖ and the Salzburger Sprachatlas (dialect database of Salzburg).[6] Here we are focusing on a dataset describing the dialect representation of "carrot" in the province of Salzburg. In the

---

[4]http://d2rq.org.

[5]http://www.strabon.di.uoa.gr.

[6]https://www.sprachatlas.at/salzburg.

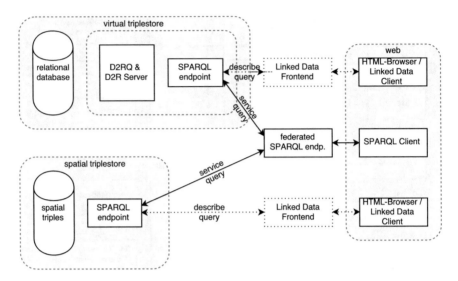

Fig. 2 System architecture of the proof-of-concept implementation

Fig. 3 Spatio-temporal analysis for the lemma "Karotte" (carrot) for 1916–2016. Communities where the dialect word "Karottn" is present in 1916 are colored in *dark red*. In 1966 the communities marked with *orange* were added to the region where "Karottn" is used. From 1966 until 2016 the communities marked in *yellow* were added to the region where "Karottn" is spoken

specific Bavarian dialect a carrot ("Karotte") is represented by the word "Karottn". Nevertheless, the dialect word "Karottn" was not used in Salzburg around 1916. Figure 3 shows the spatial-temporal change of the lemma "Karotte" (engl. carrot) between 1916 and 2016. The communities where the dialect word "Karottn" is present in 1916 are colored in red. The communities that switched to "Karottn" from 1916 until 1966 are colored in orange—hence the region where "Karottn" is used in 1966 contains of red and orange areas. In 2016, the dialect word "Karottn" is used in most communities, where the data on the lemma "Karotte" (carrot) are present. The communities that switched to the usage of "Karottn" from 1966 until 2016 are marked with yellow in Fig. 3. Thus, the region where "Karottn" is used in 2016, are all communities colored in yellow, orange or red.

5 Discussion and Conclusion

The paper presents an approach to modeling and publishing linguistic/dialect data as spatio-temporal Linked Data—based on the Dictionary of Bavarian Dialects in Austria. The approach followed in this paper is based on the development of an ontology for modeling and representing geolinguistic resources—focusing on dialects. This ontology forms the basis for publishing data stored in a relational data model with the help of a virtual RDF graph. The ontology models the dialect records and their associated lemmas, including their meaning. As language is a dynamic phenomenon, we incorporate the spatio-temporal dimension in the ontology. Hence, each source (evidence) has an associated location and temporal validity. This opens up the possibility for making spatio-temporal analyses with dialect data at hand, and to relating those data to other datasets published in the Linked Data cloud. A preliminary example, based on the lemma "Karotte" (carrot), shows the usage of the dialect word "Karottn" for a specific geographic area from 1916 until 2016, utilizing place names inherited from geonames.org. Future research items—especially for the linguistic/dialect application scenario—may include the representation of gradual spatio-temporal change of e.g. linguistic phenomena, with RDF.

Acknowledgements Parts of this work were funded by Austrian Research Fund (Project Nr.: L323-G03) and Austrian Nationalstiftung (Project Nr.: DH2014/22).

References

Bird S, Klein E, Loper E (2009) Natural language processing with Python: analyzing text with the natural language toolkit. O'Reilly Media, Inc

Birlinger A (1890) Rechtsrheinisches Alemannien. Forschungen zur Deutschen Landes- und Volkskunde 4:369–386

Bizer C, Seaborne A (2004) D2rq-treating non-rdf databases as virtual rdf graphs. In: Proceedings of the 3rd international semantic web conference (ISWC2004), Citeseer Hiroshima, vol 2004

Buccio ED, Nunzio GMD, Silvello G (2014) A linked open data approach for geolinguistics applications. Int J Metadata Seman Ontol 9(1):29–41

Chambers JK, Trudgill P (1998) Dialectology. Cambridge University Press, Cambridge

Chiarcos C, Dipper S, Götze M, Leser U, Lüdeling A, Ritz J, Stede M (2008) A flexible framework for integrating annotations from different tools and tagsets. Traitement Automatique des Langues 49(2):271–293

Chiarcos C, Hellmann S, Nordhoff S (2011) Towards a linguistic linked open data cloud: the open linguistics working group. TAL 52(3):245–275

Farrar S, Langendoen DT (2003) A linguistic ontology for the semantic web. GLOT Int 7(3):97–100

Frontini F, Gratta RD, Monachini M (2016) GeoDomainWordNet: Linking the geonames ontology to WordNet. In: Human language technology. Challenges for computer science and linguistics. Springer International Publishing, pp 229–242. doi:10.1007/978-3-319-43808-5_18

Goebl H (2010) Dialectometry and quantitative mapping, vol 2, pp 433–457

Hoch S, Hayes JJ (2010) Geolinguistics: the incorporation of geographic information systems and science. Geogr Bull 51(1):23

Jeszenszky P, Weibel R (2015) Measuring boundaries in the dialect continuum. In: Proceedings of the AGILE conference on geographic information science 2015. Springer International Publishing

Lameli A, Kehrein R, Rabanus S (eds) (2010) Language and space: an international handbook of linguistic variation: language mapping. De Gruyter Mouton

Lee CY, Hsieh SK (2015) Linguistic linked data in Chinese: The case of Chinese wordnet. In: Proceedings of the 4th workshop on linked data in linguistics (LDL-2015). Association for computational linguistics and Asian federation of natural language processing, pp 70–74

Nerbonne J (2010) Mapping aggregate variation, vol 2. Mouton De Gruyter, pp 476–495

Nixon V, Stewart Hornsby K (2010) Using geolifespans to model dynamic geographic domains. Int J Geogr Inf Sci 24(9):1289–1308

Österreichische Akademie der Wissenschaften, Bauer W (1985) Wörterbuch der bairischen Mundarten in Österreich (WBÖ). Verlag der Österreichischen Akademie der Wissenschaften

Pi CYT (2006) Beyond the isogloss: isographs in dialect topography. Can J Linguist/Revue canadienne de linguistique 51(2–3):177–184

Rumpf J, Pickl S, Elspaß S, König W, Schmidt V (2010) Quantification and statistical analysis of structural similarities in dialectological area-class maps. Dialectologia et Geolinguistica 18(1):73–100

Schmidt JE, Herrgen J (2001) Digitaler Wenker-Atlas (DiWA). Bearbeitet von Alfred Lameli, Tanja Giessler, Roland Kehrein, Alexandra Lenz, Karl-Heinz Müller, Jost Nickel, Christoph Purschke und Stefan Rabanus Erste vollständige Ausgabe von Georg Wenkers Sprachatlas des Deutschen Reichs

Scholz J, Bartelme N, Fliedl G, Hassler M, Mayr H, Nickel J, Vöhringer J, Wandl-Vogt E (2008) Mapping languages–erfahrungen aus dem projekt dbo@ ema. Angewandte Geoinformatik 822–827

Scholz J, Lampoltshammer TJ, Bartelme N, Wandl-Vogt E (2016) Spatial-temporal modeling of linguistic regions and processes with combined indeterminate and crisp boundaries. In: Progress in cartography. Springer, pp 133–151

Sibler P, Weibel R, Glaser E, Bart G (2012) Cartographic visualization in support of dialectology. In: Proceeding AutoCarto 2012

Wandl-Vogt E (1997) Alemannisch-Bairische Interferenzen im Dialekt des Tiroler Paznauns. Entwicklung, Verlauf, Beurteilung, Eine Annäherung an Mundartgrenzen

Xie Y, Aristar-Dry H, Aristar A, Lockwood H, Thompson J, Parker D, Cool B (2009) Language and location: map annotation project-a gis-based infrastructure for linguistics information management. In: International multiconference on computer science and information technology (2009), IMCSIT'09. IEEE, pp 305–311

Considering Identification of Locality in Time: Theoretical and Practical Approach

Bogumił Szady and Agnieszka Ławrynowicz

Abstract In this paper, we tackle the problem of determining an identity of a locality evolving with time. Firstly, we discuss origins of this problem, namely how it arises in the everyday research practice of historical geographers. Secondly, we present two contexts of emergence of identities: identities embedded in sources and identities constructed in history. Finally, we discuss how such identities may be captured in information systems with the help of state-of-art technologies regarding ontological modeling and reasoning.

Keywords Historical GIS · Settlement unit · Identity relation · Ontology

1 Introduction. Locality[1] As a Place

The issue stems from the difficulties that arise in the everyday research practice of historical geographers, who apply database solutions in their work. The initial and basic problem is a clear identification of the settlement unit, which takes into account its development and change over time. The basic body of information about a settlement unit, resulting from a geohistorical query, refers to the naming of a

---

[1]There is no equal term in English Language to Polish term 'miejscowoś.ć'. In the official Polish documents 'miejscowość' is defined as follows: "settlement unit or another populated area (territory) which differs from other 'miejscowości' by proper name, and in case of an equal name, by type." In this article two English words will be used as synonyms of 'miejscowość' - 'locality' and 'settlement'.

B. Szady
Tadeusz Manteuffel Institute of History, Polish Academy of Sciences, Warszawa, Poland
e-mail: szady@kul.lublin.pl

A. Ławrynowicz (✉)
Faculty of Computing, Poznan University of Technology, Poznań, Poland
e-mail: alawrynowicz@cs.put.poznan.pl

© Springer International Publishing AG 2018
P. Fogliaroni et al. (eds.), *Proceedings of Workshops and Posters at the 13th International Conference on Spatial Information Theory (COSIT 2017)*, Lecture Notes in Geoinformation and Cartography, https://doi.org/10.1007/978-3-319-63946-8_45

place, its geographical location and territorial scope, character and type (village, city, town, manor, farm etc.), ownership, size (population, number of houses etc.), and its belonging to administrative units. Each of the above attributes could vary over time in relation to one and the same settlement unit.

All the mentioned elements of the description of a locality constitute attributes in a database of the *Historical Atlas of Poland* project. The database currently stores the information on 24.5 thousand settlement units that were located in a territory of the Polish Crown in the second half of the 16th century.[2]

The practical problem of the identification of a locality arose at the time of the introduction of the second timeline for the second half of the 18th century, and aiming at preserving data integrity and structure. It is worth noting that the location of a locality is expressed in point topology, as it is not possible to determine the extent of all localities for such an early period. The most common problematic situations resulting from the lack of sufficient historical sources concern identification of:

1. two localities located in the same geographical position that have different names in two periods while retaining their category,
2. two localities located in the same location that have different names and fall into two categories in two periods,
3. two localities of the same name and type that are located near each other in two periods,
4. two localities of the same name, located close to each other, belonging to different categories in two periods.

In our opinion, the constitutive values for the identification of an individual locality should refer more generally to a theory of place. According to T. Creswell there are three levels or approaches to 'place' interpretation in geography: a descriptive, a social constructionist and a phenomenological level. Despite the fact that the first approach shows the closest relationship with the distinctiveness and individual particularity of place, the "research at all three levels (and the ones in between) is important and necessary to understand the full complexity of the role of place in human life" (Cresswell 2015). Outside the area of interest we leave the implications of broader philosophical discussion on the relationship of space and place. In accordance with the criteria of a 'place' suggested by Agnew (2011), and then broadly adopted, the commune components of place (as general category) are: location, locale and sense of place (Cresswell 2010). Theoretical problems regarding identity and changes in location were also discussed in the context of constructing historical indexes of local names and Web-HGIS systems. A practical review of such applications was conducted by Zedlitz and Luttenberger (2014).

---

[2]http://atlasfontium.pl/index.php?article=korona\Źlanguage=en.

2 Identities: Embedded in Sources and Constructed in History

All the above mentioned elements of settlements should be taken into account in determining the identity of particular villages and towns in time. Identifying and considering the continuity of two settlement units in time does not mean, of course, their complete identity, but rather (and only) the specified set of similarities. What is important, the above enumerated characteristics are objectively accompanying places and localities, independently of a human being exploring or describing them. But in reality, the identity of the villages or other settlements is born in a historical process and is created, and then discovered by individuals or social groups.

Man, as a factor that creates the identity of settlement units, appears in two basic roles. Firstly, as an agent of history creating and constructing the material and non-material components of localities. By natural intuition or sensitivity, individuals reflect the intangible dimension of settlements, the subjective and emotional relationship of man and locality, that corresponds with the idea of place as a concept of 'being in the world' (Withers 2009). Secondly, as a constructor or just as M. Oakeshott wrote assembler, mostly historian, creating the identity of place from dispersed snippets of facts and events related to particular localities (Oakeshott 1999).

2.1 Agent of History and Identity of Locality

A situation when a person determines and expresses directly (in historical sources) the identity (continuity and continuance) of settlement units does not demand a special comment. Without much trouble, this scenario can be applied to recent history–a single person or an entire group of people confirm, often orally or in written documents, the identity of inhabited places. Older residents, remembering the period before World War II, despite the destruction of their village or displacement of the inhabitants, most often confirm its identity. The same concerns the villages moved during the 20th and 21st centuries as the result of large infrastructure investments, for example regarding hydrological structures. As concrete examples we can use the sentences: "After the war **my village** was **moved** to the other side of the river and **changed its name** to Jasionka" or "To construct the dam on the river **my settlement** was **moved** 5 km to the west." Regarding medieval or early modern period, the documents which confirm the identity of a locality directly are infrequent. One of these is, for example, the permission for translocation of Wawolnica town near Lublin dated May 2, 1567 by Sigismundus II Augustus, King of Poland and Grand Duke of Lithuania: "- ut possit et valeat **oppidum praefatum nostrum Wawelnicza dictum** ex eo loco ubi nunc situm ignisque inclementia et voragine funditus deletum est, in alium locum qui sibi commodior, aptior et convenientior videbitur, **transferre** et locare" (AGAD 1566–1569, k. 109–110).[3] The situation, when there are direct documents or other evidence of the continuity and the continuation of the settlement can be defined as the *identity embedded in historical sources*.

2.2   Identity Constructed in History

More complex is the question of identity of a given town or village in time, when the source material does not inform about the continuity of the locality or its components. In this case, *the identity of a settlement is a construct of a historian or a geographer*, which is based on the various premises, derived from different sources. Continuity and change could refer to all together and each individual element of a locality. An additional complication is, especially in the case of medieval history and modern times, the lack of sources and information about basic components of a locality from one historical period. The location of villages in the 14th and 15th century is mainly established with the retrogressive method based on maps from the 18th and 19th centuries. Information about localities presented on the historical maps (not old maps) is the result and combination of information from sources not overlapping chronologically. The exact date of status change (name, location, territorial range, type, content) is often not known. Taking into consideration that most geohistorical projects require teamwork, the process of historical reasoning about the identity of a locality should consider the same criteria and rules described by a domain ontology. It will allow for the development and creation of a uniform and clear interpretative scenario for the information collected in the historical query.

As mentioned previously, all of the basic features of the locality may have changed without disturbing its essential identity. It is quite obvious that it is not possible to identify in time two localities on the basis of information about geographical position, components of infrastructure or name. Popular names like 'Zarzecze', 'Wola', do not indicate unique and concrete villages. Similarly, we cannot automatically treat alike, two localities situated in the same localization, but with different names documented for two historical periods. We can imagine the disappearance of one town and creation of another one in the same place after a period of time. Without the additional research and historical sources, which could confirm the continuity of the other elements of locality, these two centres cannot be identified and recorded in a database with the same unique key. As an initial and cautious proposal to identify two settlements in time, in the absence of historical sources, with "embedded identity," we suggest that one takes into consideration a set of components concerning at least two of the three above listed categories. As a simple and very introductory example we can use the basic elements for all localities: geographical position, settlement

---

[3]"- so he could above mentioned **our town called Wawelnicza** from the present location, where it has been completely destroyed in a fire, **transfer** to another better place.".

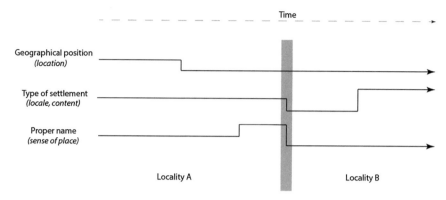

Fig. 1 Geographical position, type of settlement and proper name, and the identification of locality in time. The vertical axis symbolizes changes

type and name. The next steps should develop the problem of continuity and change of these three features of locality, taking into account a variability and different level of credibility of historical information (Fig. 1).

3 Towards Determining Identity of Localities with Ontological Modeling and Reasoning

An information system which aims to (semi-)automatically support the problem of identifying localities in time must cope with at least two problems: (i) determining a set of properties which are sufficient to identify a locality, (ii) determining a consensus between sometimes different claims of various agents regarding the identity of a locality. We will discuss these two issues in the subsequent Sects. 3.1 and 3.2.

3.1 Indiscernibility Properties

The notion of *identity* $\sim$ in ontological representations usually concerns a relation between a and b stating that they are the same entity, or identical to each other ($a = b$). Such notion is grounded in philosophy and follows principles such as 'indiscernibility of identicals' (Principle 1, Leibniz) and 'identity of indiscernibles' (Principle 2), provided below where P denotes a set of all properties.

Principle 1 (Indiscernibility of identicals) $a = b \rightarrow \forall p \in P(p(a) = p(b))$

Principle 2 (Identity of indiscernibles) $\forall p \in P(p(a) = p(b)) \rightarrow a = b$

Those principles essentially state that entities sharing the same properties are the same and there cannot be separate entities which share all their properties. Guarino and Welty, as part of their methodology OntoClean (Guarino and Welty 2000), also considered the aspect of *time*. They distinguished two types of *identity criteria* (i.e. criteria, which are conditions used to determine equality) over time, which they called *diachronic identity criteria* and *synchronic identity criteria*. The former ones are related to how we recognize localities we know as the same locality even though they may have changed over time. The latter ones are identity criteria at a single point in time. Taken the above into account the formula from Principle 2 may be slightly modified to account for time (where E denotes a special purpose predicate used to express that a has actual existence in time t):

$$\forall p \in P(p(a,t) = p(b,t') \wedge E(a,t) \wedge E(b,t')) \rightarrow a = b$$

Then we have a synchronic criterion if $t = t'$, and a diachronic criterion otherwise.

A standard for modeling ontologies the *Web Ontology Language (OWL)* (van Harmelen and McGuinness 2004) follows the above mentioned approach. The identity relation in OWL is modeled with use of a vocabulary term `owl:sameAs` and its semantics is defined as follows, where $\mathscr{I}$ stands for the interpretation function mapping terms to resources, and *EXT* stands for the extension function mapping properties to pairs of resources.

Definition 1 (*Semantics of* `owl:sameAs`)

$$\langle \mathscr{I}(a), \mathscr{I}(b) \rangle \in EXT(\mathscr{I}(\text{owl} : \text{sameAs})) \Leftrightarrow \mathscr{I}(a) = \mathscr{I}(b)$$

Such definition of `owl:sameAs` conforms to the discussed identity relation '='.

However, such definition makes asserting identity very strong and it suffers from several problems. One such problem is that when two entities are asserted as the same, it closes the possibility of asserting new properties to one of them. This renders such solution not sufficient for our purposes since all our constitutive properties (geographical position, name and type of settlement) as well as others would need to be the same for any given two mentions of a locality captured in different time to assert they describe the same entity, which is not the case. Consider, for instance, the following example discussed by McCusker and McGuinness (2010), which due to changes in identity over time violates Leibniz's law regarding the identity of indiscernibles: *"I never made it to Breslau, but I visited Wrocław last week."*

Moreover, if the time argument is omitted for the identity relation (i.e., it is assumed as time invariant) then if two entities are asserted as identical, they are identical forever.

Identity is the strongest *equivalence* relation (where equivalence, denoted $\equiv$ is a relation which is reflexive, symmetric and transitive). Beek et al. (2016) proposed to reinterpret the identity relation $\sim$ as if it were indiscernibility relation $\approx_P$ with a set of its properties implicit in the data.

Then one can make explicit sets of properties to which the identity relation is indiscernible instead of, as in Principles 1 and 2, consider indiscernibility with respect to all of the properties. In this way, besides 'strict' identity, there are other cases of indiscernibility where each of them corresponds to each set of the properties. Consider for instance the following example.

Example 1 (Indiscernibility properties)

$$\text{ex : locality1} \approx_{\{owl:sameAs,ex:type\_of\_settlement,ex:geographical\_position\}} \text{ex : locality2}$$

$$\text{ex : locality1} \not\approx_{\{ex:name\}} \text{ex : locality2}$$

In this example, some locality (described in some historical source with an identifier `locality1` and in another historical source with an identifier `locality2`) could still have the same geographical position and type of settlement, but its name may have changed (e.g., after the war). Thus, indiscernibility properties of `ex:locality1` and `ex:locality2` include `ex:geographical_position`, `ex:type_of_settlement`, and `owl:sameAs`. ∎

We can say that these two entities are *semi-discernible* since their indiscernibility properties are the same. Semi-discernibility is an equivalence relation on pairs of entities which we can use to deduce whether any given two entities are equivalent.

One could also require two of the three constitutive properties to not necessarily match entirely, but to be similar (i.e., match partially, especially in the context of complex properties). In this case, one could apply some notion of *similarity*. Similarity-based identity for historical places was studied by Janowicz (2006, 2009) who argued that place names appearing in historical documents probably refer to the identical place in the real world if they are related via the same or similar predicates to entities which themselves again refer to identical or similar entities (places, events, agents etc.).

3.2 Identity as Claims

Halpin et al. (2010) have identified several varieties of identity regarding ways how identity relation is used including *identity as claims*. According to Halpin et al. (2010), *identity as claims* corresponds to the situation where all statements of identity are treated as *claims*, where any statement of identity is not necessarily true, but it is only stated by a particular agent instead. This may lead to situations where different agents may have different inferences due to the different sets of claims they accept and different identity statements.

In our case, this corresponds to both of the scenarios, but with agents having two different roles. In one case ('entity embedded'), an agent is directly involved to a locality (being a part of the sense of place), influences and shapes the locality and is a witness of the locality. This is also expressed by a historical source.

In another case ('identity constructed'), an agent is a historian who constructs identity from various snapshots of information and various premises. Such agents most often did not see the given locality, and there is no direct interaction between the locality and the agent (e.g., it is the locality from the past not existing anymore).

In the former case, we have a triad:

active agent (participant, direct witness) + locality = source relation

where an identity of the locality is embedded in the relation. In the latter case, we have another triad:

source relation + passive agent (historian, geographer, critical watcher, scientific observer) = locality (its constructed identity)

Another possible approach to how to deal with inconsistent and contradictory knowledge regarding historical facts was pointed by Janowicz (2009) who proposed to use microtheories, i.e., sub-theories that are locally consistent but allow for inconsistencies in the global knowledge base.

How to model co-existence in the source material of different interpretations of the same real-world event was also studied by Kowalczuk and Ławrynowicz who proposed ontology design patterns for reporting events (Kowalczuk and Ławrynowicz 2016). This pattern could be generalized to model not specifically events but also generic things.

4   Conclusions and Future Work

In this paper, we have outlined the problems of determining and identity of a locality, which arise in the daily work of historical geographers. We identified two contexts for identity of localities in time: identity embedded in sources and identity constructed. In both of these contexts, there exists a further issue of possibly conflicting claims of agents concerning the identity of localities, be it direct agents (witnesses) or indirect agents (e.g., historians). We have proposed to assert that two localities are identical if at least two of their three properties (geographical position, name and type of settlement) are the same. Then we discussed the relevant work on ontological modeling and reasoning dealing with determining identity as a special cases of an equivalence relation with use of indiscernibility properties. In future work, we should adapt available ontology design patterns to model our problem and devise procedures for the semi-automatic deduction of an identity relation over time.

Acknowledgements This work has been partially supported from the grant "Ontological foundations for developing historical geographic information systems" (2b H15 0216 83) funded by the National Humanities Development Program.

References

AGAD (1566–1569) The Central Archives of Historical Records in Warsaw, vol 101. Metrica Regni
Agnew JA (2011) Space and place. In: The SAGE handbook of geographical knowledge. SAGE
   Publications Ltd, 1 Oliver's Yard, 55 City Road, London, EC1Y 1SP, United Kingdom, pp
   316–330

Beek W, Schlobach S, van Harmelen F (2016) A contextualised semantics for owl: sameAs. In: The semantic web. Latest advances and new domains—13th international conference, ESWC 2016, Heraklion, Crete, Greece, 29 May–2 June 2016, Proceedings, pp 405–419

Cresswell T (2014) Place. In: Lee R, Castree N, Kitchin R, Lawson V, Paasi A, Philo C, Radcliffe S, Roberts SM, Withers CW (eds) The SAGE handbook of human geography: two volume set. SAGE Publications Ltd, 1 Oliver's Yard, 55 City Road London EC1Y 1SP, pp 3–21

Cresswell T (2015) Place: an introduction, 2nd edn. Wiley-Blackwell, Chichester

Guarino N, Welty C (2000) Identity, unity, and individuality: towards a formal toolkit for ontological analysis. In: Proceedings of the 14th European conference on artificial intelligence. IOS Press, pp 219–223

Halpin H, Hayes PJ, McCusker JP, McGuinness DL, Thompson HS (2010) When owl: sameAs Isn't the same: an analysis of identity in linked data. Springer, Berlin, Heidelberg

van Harmelen F, McGuinness D (2004) OWL web ontology language overview. W3C recommendation, W3C

Janowicz K (2006) Towards a similarity-based identity assumption service for historical places. Springer, Berlin, Heidelberg

Janowicz K (2009) The role of place for the spatial referencing of heritage data. In: The cultural heritage of historic European cities and public participatory GIS workshop, vol 9, p 57

Kowalczuk E, Ławrynowicz A (2016) The reporting event ontology design pattern. In: Workshop on ontology and semantic web patterns, 7th edn—WOP2016

McCusker JP, McGuinness DL (2010) Towards identity in linked data. In: Proceedings of the 7th international workshop on OWL: experiences and directions (OWLED 2010), San Francisco, California, USA, 21–22 June 2010

Oakeshott MJ (1999) On history and other essays. Liberty Fund, Indianapolis

Withers CWJ (2009) Place and the "spatial turn" in geography and in history. J Hist Ideas 70(4):637–658

Zedlitz J, Luttenberger N (2014) Modelling (historical) administrative information on the semantic web. In: Proceedings of the the second international conference on building and exploring web based environments, WEB 2014

Cadmus and the Cow: A Digital Narratology of Space in Ovid's Metamorphoses

Gabriel Viehhauser, Robert Kirstein, Florian Barth
and Andreas Pairamidis

Abstract We apply an interdisciplinary methodology to establish a digital-driven narratology of space on Ovid's Metamorphoses. Two approaches are employed: 1. We highlight the importance of nature in the text by exploring the frequencies of architectural and natural terms. 2. We delve into a single episode of the text (Cadmus) by using a collocation-network-approach that reveals the interrelations between characters and settings. We show that the results can feed an analysis in the light of Lotman's model of space semantics.

Keywords Spatial humanities · Narratology · Network analysis · Classical philology · Ovid's metamorphoses

1 Introduction

As in other disciplines of the humanities, the emerge of the so-called 'Spatial Turn' (Soja 1990) lead to a renewed interest in the category of space in literary studies as well. However, a large amount of work in this field focuses on space as a cultural phenomenon, very often drawing on a metaphorical understanding of the term. In contrast, the modeling of the means by which space is created in narrative texts has received less attention. Efforts to establish a narratology of space are thus still in their

G. Viehhauser (✉) · F. Barth · A. Pairamidis
Department for Digital Humanities, University of Stuttgart, Stuttgart, Germany
e-mail: viehhauser@ilw.uni-stuttgart.de

F. Barth
e-mail: florianbarth@ilw.uni-stuttgart.de

A. Pairamidis
e-mail: andreas.pairamidis@ilw.uni-stuttgart.de

R. Kirstein
Classics Department, University of Tübingen, Tübingen, Germany
e-mail: robert.kirstein@uni-tuebingen.de

© Springer International Publishing AG 2018
P. Fogliaroni et al. (eds.), *Proceedings of Workshops and Posters at the 13th International Conference on Spatial Information Theory (COSIT 2017)*, Lecture Notes in Geoinformation and Cartography, https://doi.org/10.1007/978-3-319-63946-8_46

beginnings (cf., e.g., Ronen 1986; de Jong 2012; Bodenhamer et al. 2015; Barker et al. 2016).

This appears to be due to the fact that narrative space is a complex phenomenon. Rather than constructing a given physical space beforehand, stories tend to evolve their setting in relation to its characters that constitute space through their actions. Therefore, space is not depicted as a continuous and fully determined phenomenon and often evoked implicitly, which makes it hard to map "People read for the plot and not for the map", as (Ryan 2003, p. 238) puts it.

New stimuli for a narratology of space can be expected from the emerging field of digital text analysis, an interdisciplinary methodology that combines literary studies with natural language processing and information retrieval. In our paper we will present first approaches towards such a digital-driven narratology of space by drawing on a central text from Classical Philology, Ovid's *Metamorphoses*.

It has often been pointed out that nature and landscape play a crucial role in the *Metamorphoses*, not only as a setting for the various episodes that Ovid narrates in the 15 books of his work, but also because of the interconnection between the landscape and their characters (cf. e.g. Segal 1969; Hinds 2002). The constant state of change, the *leitmotif* of the *Metamorphoses*, affects settings as well as figures, blurring the boundaries between these two constituents of the narrative. Therefore, the text appears as a promising use case for a digital analysis of space that can also be linked back to an ongoing discussion in traditional literary studies.

We will approach the text in two steps of different scale: In the following section, we will perform a *distant reading* (Moretti 2000) on the macro-level of the *Metamorphoses* by detecting the frequency of space-markers to empirically test assumptions that have been made regarding the spatial structure of the text in traditional literary studies. In a second step, we will focus on a particular episode of the text (*Cadmus*) and explore a network-based approach towards space and its meaning for the basic structure of the episode. Here, we will make use of Juri Lotman's theory of spatial semantics (Lotman 1977), which is one of the few 'classic' approaches towards space in narratology.

Since the application of methods from NLP on ancient languages might pose problems of domain adaptation we used an established modern German prose translation (Albrecht 2012) of the Latin text for modeling purposes; the German translation is relatively close to the Latin original and, above all, consistent in regard to the spatial terminology being applied.[1]

---

[1] By doing so and by presenting the results in English we will operate with no less than three different languages (Latin/German/English). We are aware of the shortcomings connected to the multiplicity of languages. Since the project as presented is designed to blend in a larger project in which comparative linguistics plays a key role we hope to expand our analysis fully to both Latin and German to achieve by comparison a deeper insight into how space is created in Latin and German (literary) texts.

2 Macrolevel: Frequency of Landscape Terms

In the first step of our analysis, we determine the frequency of space markers in the text to verify the observation made in literary studies that nature and landscape are of major importance in the *Metamorphoses*. It has been argued that the landscape is often depicted in a stereotypic way, which has a unifying effect: "In addition to providing a general tonal unity for the poem as a whole, the landscape also helps hold together the narrative material of one or more books" (Segal 1969, p.39).

Spatial markers in a text come in different shapes. To establish a proxy for their frequency, we concentrate on place nouns, which we detect with the help of named entity recognition (NER, performed with the tool[2] and wordlists that have been assembled by manual annotation and by using the semantic word-net GermaNet (Hamp and Feldweg 1997; Henrich and Hinrichs 2010). We distinguish between place names (toponyms, as e.g., "Sparta" or "Styx") and place nouns that are rather unspecified (e.g., "forest" or "cottage"). Both categories are divided into two subcategories in regard to their creation: natural spaces exist without the contribution of men whereas architectural objects are made by them.

The setting in the *Metamorphoses* is often established in remote and unspoiled regions that mostly cover natural space nouns (Kirstein 2015, p. 212). Compared to a corpus of 538 narrative texts (retrieved from the *TextGrid*-Repository, textgridrep. org), we clearly observe a deviation from the norm (cf. Fig. 1, top).[3]

While the occurrence of architectural vocabulary in the *Metamorphoses* (red) closely corresponds to the corpus (purple and blue), much higher values can be observed for terms from the word field of nature (green).[4] The difference is also statistically significant.[5]

Furthermore, a higher variance within the segments of nature can be observed compared to the category of architecture.[6] This corresponds with Kirstein's observation that the variation of landscape entities between episodes is higher than it has been proposed in previous research (Kirstein 2015, p. 213). For toponyms, the proportion of natural terms is still greater, but in this case, the architectural frequencies are higher due to the frequent occurrence of city names (cf. Fig. 1, bottom).

Both distributions are likely to reflect significant structures of the work as a whole: The high density of natural terms especially in the first book indicate that the unifying, 'stereotypic' landscapes are introduced at the beginning of the text and only

---

[2]WebLicht: Web-Based Linguistic Chaining Tool https://weblicht.sfs.uni-tuebingen.de/.

[3]Up to now, the corpus for comparisons mainly consists of texts from the 17th to the 19th century. We plan to include more contemporary texts in future work.

[4]All distributions are based on the relative frequencies of the natural and architectural terms in the *Metamorphoses* and the average of the relative frequencies in the corpus. The 15 segments consist of the 15 books of the *Metamorphoses* and of equally divided chunks from the texts in the corpus.

[5]We tested this with the Wilcoxon rank-sum test (p=1.289e-08) as well as Mood's median test (p=1.289345e-08).

[6]Nature: 0.0860 (variance), 0.2932 (standard deviation); architecture: 0.0085 (variance), 0.0919 (standard deviation).

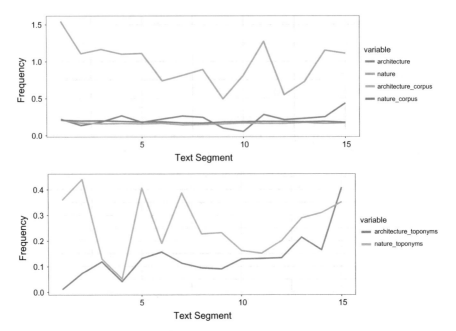

Fig. 1 (*Top*) Relative frequencies of natural and architectural terms in the 15 books of the *Metamorphoses* compared to the average distribution in the corpus. (*Bottom*) Relative frequencies of toponyms in Ovid's *Metamorphoses*

mentioned briefly at a later point. As a consequence, this would mean that the intention of the *Metamorphoses* was a continuous, 'syntagmatic' reception rather than a selective reading of single episodes.

The incline of the architectural vocabulary in book 15 corresponds with the wider temporal horizon of the text that starts with the creation of the world from chaos and ends in Ovid's contemporary Augustan times, reflecting a development towards 'civilization' as well as a more concrete and detailed setting.[7]

As can be seen, the distant reading of place markers in the *Metamorphoses* opens up a variety of research questions. In the next section, we will combine these analyses on the macro-level with further quantitative, but also qualitative research on a more fine-grained micro-level.

---

[7]This view receives support from the Index of passages discussed in (Boyle 2003, p. 299) which shows a clear domination of book 15.

3 Microlevel: A Network of Places and Characters in Cadmus

As a first example for a detailed analysis, we explore the *Cadmus* episode, which opens the third book of the *Metamorphoses* (book 3, v. 1–137). The episode can be divided into three sections: (1) As predicted by the Delphic oracle, Cadmus, who has been banished from his fatherland while searching for his sister Europa, encounters an undomesticated cow on his journeys. In accordance with the verdict of the oracle, he follows her to the place where it settles down. Here he will eventually found his new city, Thebes in the region of Boiotia (cf. v. 129–131). First he speaks a prayer of thanks and kisses the earth, then he starts to prepare a ritual offering in honor of Jupiter (v. 1–27). (2) To gather water for the offering Cadmus sends out his companions. They get ambushed and killed by a monstrous serpent or dragon that belongs to the god Mars and resides in a cave in the nearby woods (v. 32 *Martius anguis*, 'Dragon of Mars', v. 38 *serpens*). Looking for his companions, Cadmus finds the beast and kills it in a fierce fight (v. 28–98). (3) Athena commands Cadmus to sow out the teeth of the dragon. Warriors grow out of the earth, who immediately start to fight each other. While Cadmus watches, only five of the newborn fighters survive and become his companions when they found the new city Thebes (v. 99–137).

For our analysis of the spatial structure of the episode, we choose a collocation-network-approach, which is visualized in Fig. 2.

The network is bimodal, as it includes characters and space-markers as nodes. The nodes have been extracted from the text with the help of NER (performed on the German translation with Weblicht (2012), the nodes have been translated into English for the purpose of this paper) and the aforementioned wordlists. Due to the Latin place and character names in the text the results of the NER had to be refined manually.

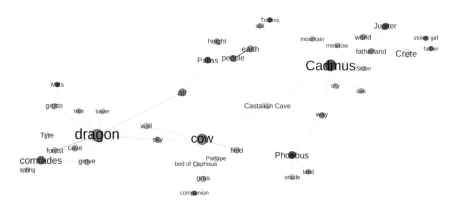

Fig. 2 Network of the *Cadmus*-episode (visualized with the tool GEPHI (Bastian et al. 2009))

In a straightforward approach, relations have been established whenever a character appears in the same sentence as a spatial marker. In the visualization, character nodes are further subdivided into the categories named persons (dark red), anonymous characters or groups of characters (violet) and animals, who often play a major role in Ovid's stories because of their intermediate position between nature and living beings.[8] Spatial markers are divided into toponyms (yellow), nature (green), architecture (light blue) and regions (dark blue).

The visualization does not only give an overview of the setting of the episode and the characters that move in it, but it can also be related to narratological models of space as Lotman's influential theory of spatial semantics (Lotman 1977). At the core of this theory lies the interconnection of space, characters and events: According to Lotman, narrative stories form a so-called *sujet* (i.e., an eventful plot), whenever a character transgresses the borders between two semantically distinct spaces. The idea is that the narration establishes a semantic field, which is divided into two spaces (at least) to which specific semantic values are attributed, and thus constitute the norm/order. The semantic values of the two spaces form a pair of opposition (e.g., "good-bad", "life-death", "friend-enemy", etc.). Usually, characters are not capable of crossing the border (i.e., violating the norm/order), rendering the latter impermeable and classifying the former as immobile characters. Some characters (e.g., the hero), however, do cross the border and are classified as mobile characters consequently. By doing so, they evoke an event.

The semantic value of a space is defined by the description of the inventory (e.g., buildings, plants, furniture, etc.) it hosts. Characters are part of this inventory as well, as they have features that comply to the space they originate (Nitsch 2015).

We will argue that common network measures can be used as a proxy, which can be related to the framework established by Lotman. In Fig. 2, the size of the nodes corresponds with their network-degrees (the number of their relations to other nodes). Cadmus and the dragon, which is turned into a central character by the story (*see below*), are the characters with the highest degree, meaning that they are connected to a wide variety of places, which marks them as the most mobile characters of the episode as well. This indicates that although Cadmus and his opponent are connected to a very distinct spatial setting (Cadmus to his homeland Crete and the dragon to a cave nearby Thebes), both characters also appear to be dynamic and able to transgress their ancestral spaces. Whereas this can be expected from a mobile hero like Cadmus, who has to find a new home on his journey, it is rather surprising in the case of the dragon, who, as a part of nature itself, could be expected to be strongly tied to the place which is about to get cultivated with the augured foundation of Thebes. But far from just being removed and substituted by civilization, the dragon gets active himself, thus changing its status from a passive element of the landscape to a mobile character (this corresponds to similar findings that resulted from an analysis of focalization aspects of the passage in Kirstein 2015, p. 233).

---

[8]Periphrases of proper names (e.g. 'Agenor's son' for 'Cadmus') have been dissolved and subsumed under the same node. In contrast, co-references indicated by pronouns have not been considered yet.

In the case of the Cadmus-episode, the network is further complicated through the aforementioned Sects. 1 and 3 of the plot that edge the main conflict between the hero and the beast. Whereas the story of the dragon seed heroes that appears to be rather static is only reflected in a small section in the upper-middle part of the network (nodes 'people', 'Pallas', 'earth' and 'height'), Cadmus' pursuit of the cow occupies a much larger space. The degree of the node of the cow is almost as high as that of Cadmus and the dragon, which consequently means that the cow has to be regarded an almost as mobile 'character' as the two main opponents.

We will argue that this finding does not reflect a weakness of our model or Lotman's conception but rather reveals a surprising structural similarity that positions the cow as sort of a linking character in between Cadmus and the beast. On the one hand, the cow accompanies Cadmus on his transgressing journey from his homeland via the oracle of Delphi into the foreign wilderness that he will transform into a new home by substituting nature through civilization. On the other hand, the cow is a part of nature itself and resembles the dragon. As the network reveals, this resemblance is also reflected in the spatial relations of the character. Both animals are linked to the spatial markers 'air' and 'sky'. Whereas the dragon's occupation of vertical space is a salient feature of the fight between him and Cadmus that also brings along a change in focalization (Kirstein 2015, pp. 226–228), the cow's linkage to vertical space is uttered no less clearly, but at first glance, in a less noticeable way (Ovid, Met. 3. 20–23):

bos stetit et *tollens speciosam cornibus altis*
ad caelum frontem mugitibus inpulit auras
atque ita respiciens comites sua terga sequentes
procubuit teneraque latus submisit in herba.

In the decisive moment, where Cadmus finds the place for his new home, the *vertical gestures* of the cow lifting first her head with its 'high' horns up to the sky and then sinking *down* on the ground and lowering her body onto the grass have a significant impact for the overall spatial order of the passage. The lifting of the head *up* towards the sky preludes the behavior of the dragon that resides in the wilderness and lifts its body to the sky (cf. v. 43 *leves erectus in auras*, 'raises into thin air') when awakened by the water Cadmus' companions try to draw from the well. In a similar way the sinking *down* of the cow which indicates the right place for the city foundation preludes the kissing of the ground by Cadmus which indicates the hero's willingness to accept the oracle's guidance (cf. v. 24–25 *peregrinaeque oscula terrae/figit*, 'kissing the foreign soil'). But the network analysis also suggests a *horizontal* dimension given to the 'character' of the cow. The cow, by guiding Cadmus and his group from Delphi to future Thebes, establishes and 'embodies' a horizontal link between two central spatial points of the story; furthermore, this horizontal orientation of the cow expands to Cadmus and his people, since both are strongly linked to each other by forming a group of 'companions' (the cow when approaching the right place to settle down looks back to Cadmus and his people who follow her: v. 22 *respiciens comites sua terga sequentes*). The cow, therefore, plays a central role in having both a vertical dimension by means of lifting and lowering its body and thus preluding the 'verticality' of the dragon *and* a horizontal one by connecting Delphi and Thebes.

4 Conclusions

As it is the case with literary cartography "placing a literary phenomenon in its specific space [...] is not the conclusion of geographical work; it's the beginning" (Moretti 1998, p. 7). This holds for our approach as well. It is also the beginning of scholarly interpretation, and we hope to have shown that an interdisciplinary framework that draws on a mixture of different methods appears promising and is able to open new perspectives on texts. Thus the network analysis of Ovid's Metamorphoses 3 on the microlevel has shown that within the Cadmus-episode the two sections dealing first with the predictive cow and then with the monstrous dragon are, by their respective vertical dimensions, much more closely linked to each other than has been argued before. As a result the cow emerges as an important connecting node in the network, and its movement both vertically and horizontally foreshadowing the movement of the main actors, Cadmus and the dragon. This might open new perspectives in the longstanding debate about the unity of the *Metamorphoses* with its multiplicity of different stories, figures, and places.

Of course there are still aspects of our approach that need refinement. This includes the text-base (working with the original Latin text and a better balanced comparison corpus would be desirable) our detection of place markers (refinement of wordlists and NER), the network-based approach (operationalization of the relations between characters and space) and our modeling of Lotman's theory (sentiment analysis of the spheres could be used to determine different semantic fields of narrative space). We plan to tackle these tasks in our future work and expand our analysis to more episodes of the *Metamorphoses* to substantiate our approach towards a digital-driven narratology of space.

References

Albrecht M (2012) Ovid. Metamorphosen. Lat./Deut, Reclam, Stuttgart

Barker E, Bouzarovski S, Pelling C, Isaksen L (eds) (2016) New worlds out of old texts: revisiting ancient space and place. Oxford University Press, Oxford

Bastian M, Heymann S, Jacomy M (2009) Gephi: an open source software for exploring and manipulating networks. In: International AAAI conference on weblogs and social media

Bodenhamer D, Corrigan J, Harris TM (2015) Deep maps and spatial narratives. Indiana University Press, Indiana

Boyle AJ (2003) Ovid and the monuments. Aureal Publications, Berwick, Victoria, A Poet's Rome

CLARIN-D/SfS-Uni. Tübingen. 2012. WebLicht: Web-Based Linguistic Chaining Tool. Online. Date Accessed: 17 Aug 2017. URL https://weblicht.sfs.uni-tuebingen.de/

de Jong I (ed) (2012) Space in ancient Greek literature. Studies in ancient Greek narrative. Brill, Leiden

Hamp B, Feldweg H (1997) GermaNet, a lexical-semantic net for German. In: Proceedings of the ACL workshop automatic information extraction and building of lexical semantic resources for NLP applications, pp 9–15

Henrich V, Hinrichs E (2010) GernEdiT The GermaNet editing tool. In: Proceedings of the seventh conference on international language resources and evaluation (LREC 2010), pp 2228–2235

Hinds S (2002) Landscape with figures. Aesthetics of place in the metamorphoses and its tradition. In: Hardie P (ed) The Cambridge companion to ovid. Cambridge University Press, Cambridge, pp 122–149

Kirstein R (2015) Der sehende Drache. Raumnarratologische Überlegungen zu Ovids Metamorphosen. In: Kugelmeier C (ed) Translatio humanitatis. Festschrift zum 60. Geburtstag von Peter Riemer. Röhrig, St. Ingbert, pp 209–238

Lotman J (1977) The structure of the artistic text. University of Michigan, Ann Arbor, Translated from the Russian by Ronald Vroon

Moretti F (1998) Atlas of the European Novel 1800–1900. Verso, New York

Moretti F (2000) Conjectures on world literature. New Left Rev 1:54–68

Nitsch W (2015) Topographien: Zur Ausgestaltung literarischer Räume. In: Dünne J, Mahler A (eds) Handbuch Literatur & Raum. DeGruyter, Berlin, Boston, pp 30–40

Ronen R (1986) Space in fiction. Poetics Today 7:421–438

Ryan ML (2003) Cognitive maps and the construction of narrative space. Narrative theory and the cognitive sciences (CSLI Lecture Notes 158), pp 214–242

Segal C (1969) Landscape in ovid's metamorphoses. A study in the transformations of a literary symbol. Steiner, Wiesbaden

Soja E (1990) Postmodern geographies. The reassertion of space in critical social theory. Verso, New York

Part V
Computing Techniques
for Spatio-Temporal Data in Archaeology
and Cultural Heritage

Computing Techniques for Spatio-Temporal Data in Archaeology and Cultural Heritage—Introduction

Alberto Belussi, Roland Billen, Pierre Hallot and Sara Migliorini

Archaeological data, and more in general cultural heritage information, are systematically characterized by both spatial and temporal dimensions that are often related to each other and are of particular interest for supporting the interpretation process, that allows to produce new knowledge about artefacts of the ancient times.

For this reason, several attempts have been performed in recent years in order to develop new techniques or to adapt existing ones, tailored to support:

- spatio-temporal data collection and their effective representation for enhancing interoperability;
- the processing of raw data in order to identify artefacts and define their allocation in space and time;
- the reconstruction of ancient structures (buildings, walls, castle, etc.) or their temporal evolution;
- the integrated access and querying of the collected data in different formats.

The main motivation for this workshop stems from the increasing need for bringing together researches of the fields of knowledge representation and discovery with geographical information scientists to share their research results and find effective solutions for user needs in archaeology and cultural heritage applications.

The workshop proceedings contain five papers that are organized in two sections. The first section concerns the *Knowledge representation* topic and is covered by the

A. Belussi (✉) · S. Migliorini
Computer Science Department, University of Verona, Verona, Italy
e-mail: alberto.belussi@univr.it

S. Migliorini
e-mail: sara.migliorini@univr.it

R. Billen
Geomatics Unit, University of Liège, Liège, Belgium
e-mail: rbillen@ulg.ac.be

P. Hallot
Faculty of Architecture, University of Liège, Liège, Belgium
e-mail: p.hallot@ulg.ac.be

© Springer International Publishing AG 2018
P. Fogliaroni et al. (eds.), *Proceedings of Workshops and Posters at the 13th International Conference on Spatial Information Theory (COSIT 2017)*, Lecture Notes in Geoinformation and Cartography, https://doi.org/10.1007/978-3-319-63946-8_47

first three papers: (i) *Immersive Technologies and Experiences for Archaeological Site Exploration and Analysis*, (ii) *HBIM for the archaeology of standing buildings: study case of the church of San Cipriano in Castelvecchio Calvisio (L'Aquila, Italy)* and (iii) *An Analytical Framework for Classifying Software Tools and Systems Dealing with Cultural Heritage Spatio-Temporal Information*. The second section that deals with the *Knowledge discovery* topic and contains the last two papers: (iv) *About the spatiotemporal complexity of cultural heritage information* and (v) *Towards the Extraction of Semantics from Incomplete Archaeological Records*.

The technical program also consists of an invited talk from well-known experts from academia. In particular, in L'Aquila the talk *Representing and Comparing Spatial and Temporal Data in Virtual Heritage Applications* by Sofia Pescarin from ITABC-CNR (Institute for technologies applied to cultural heritage) has been offered to workshop participants.

We would like to thank the authors of all submitted papers. Their innovation and creativity has resulted in an interesting technical program. We are highly indebted to the program committee members, whose reviewing efforts ensured in selecting a competitive set of papers. Finally, we would like to express our sincere gratitude to the invited speaker.

Immersive Technologies and Experiences for Archaeological Site Exploration and Analysis

Jan Oliver Wallgrün, Jiawei Huang, Jiayan Zhao, Claire Ebert,
Paul Roddy, Jaime Awe, Tim Murtha and Alexander Klippel

Abstract Immersive technologies have the potential to significantly improve and disruptively change the future of education and research. The representational opportunities and characteristics of immersive technologies are so unique that only the recent development in mass access fostered by heavy industry investments will allow for a large-scale assessment of the prospects. To further our understanding, this paper describes a project that aims at creating a comprehensive suite of immersive applications for archeological sites, including 360° immersive tours, skywalks, and self-guided explorations for education, and immersive workbenches for researchers.

Keywords Virtual reality · Augmented reality · Cultural heritage · Linked data · Spatio-temporal modeling

1 Introduction

Immersive technologies are becoming a tool of mass communication and as such offer the potential to disrupt education and research in the spatial sciences and beyond. This paper describes an ongoing project that has the goal of creating immersive VR (iVR) experiences of archeological sites for both educational and research purposes. The project focuses on the ancient Maya site Cahal Pech located in Belize. It combines environmentally sensed data, 360° video and photography, and manually created 3D models. Additionally, interactions are being implemented to virtually navigate the site and link additional media resources to create a comprehen-

J.O. Wallgrün (✉) · J. Huang · J. Zhao · P. Roddy · A. Klippel
ChoroPhronesis, Department of Geography, The Pennsylvania State University,
State College, USA
e-mail: wallgrun@psu.edu

C. Ebert · J. Awe
Anthropology, The Pennsylvania State University, State College, USA

T. Murtha
Landscape Architecture, The Pennsylvania State University, State College, USA

© Springer International Publishing AG 2018
P. Fogliaroni et al. (eds.), *Proceedings of Workshops and Posters at the 13th International Conference on Spatial Information Theory (COSIT 2017)*, Lecture Notes in Geoinformation and Cartography, https://doi.org/10.1007/978-3-319-63946-8_48

sive immersive experience for different VR setups: HTC Vive, mobile VR solutions, and augmented reality (AR). Intended applications for education are virtual field trips which allow students to experience and learn about the site in a more effective and realistic way compared to classic media. For scientists, digital workbenches will allow analyses and investigations with 3D models and associated data sources while being immersed in the site. We discuss the background of this project (Sect. 2), our data capture methods and content creation approaches (Sect. 3) as well as first prototypes for both educational and scientific iVR applications (Sect. 4). We also describe our plans for an underlying linked data based information system that stores all the heterogeneous spatio-temporal data about the site in a way that is suitable to drive the different applications and use cases (Sect. 5).

2 Background

Cahal Pech is located in the Belize Valley of the west-central portion of Belize on top of a natural limestone escarpment. Archeological investigations at Cahal Pech have been ongoing since the late 1980s. Excavations conducted in the site core in Plaza B identified contexts representing the earliest permanent settlement at Cahal Pech dating to 1200–900 BC. By the Classic Period (AD 300–900), the presence of temple pyramids, stone monuments, and the elaborate royal burials identify Cahal Pech as the seat of an important regional kingdom governed by a dynastic lineage (Awe et al. 2016). The monumental center at Cahal Pech is composed of at least 34 buildings (Fig. 1a), the largest of which, Structure B1 (Fig. 1b), is approximately 24 m tall. Plaza B is the largest plaza at the site, measuring approximately 50 × 30 m, and is bounded to the east by a triadic temple group.

Immersive technologies, as an interactive communication medium, are seeing a resurgence in popularity thanks to massively improved and more cost-effective products. The spectrum of immersive technologies has been described with different terms including mixed reality as a summary concept with AR at one end of the spectrum and VR at the other (Milgram and Kishino 1994). Here we use xR to refer

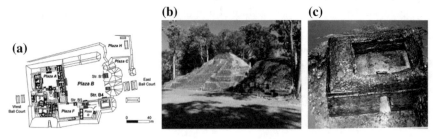

Fig. 1 **a** Map of the Cahal Pech archeological site. **b** Structure B1 and east side of Plaza B. **c** SfM 3D model of Mayan Sauna at Cahal Pech

to the vast spectrum of technologies that are becoming available. On the scientific side, xR research and visual analytics share a common interest in creating intuitive interfaces and digital immersive analytics workbenches (Simpson et al. 2016). xR technologies similarly take advantage of our innate understanding of physical reality within software environments (Bowman and McMahan 2007) and have proven to be very useful in visualizing spatial data in practice (Donalek et al. 2014).

3 Data Capture and Modeling

Developments in xR are accompanied by unprecedented advancements in environmental sensing and modeling technologies. In the following, we describe current approaches at our field site Cahal Pech.

Structure from Motion (SfM): 3D archeology is not new. Using image-based modeling, or Structure-from-Motion (SfM), to build photorealistic 3D models is, however, a more recent development (e.g., López et al. 2016). SfM can be applied to constructing models of large areas such as entire archeological sites, but also to create 3D models of smaller objects, such as individual artifacts. In our project, we have been using Agisoft PhotoScan Pro. Figure 1c shows a 3D model of a Mayan Sauna at Cahal Pech.

360° Photography and Videography: Photos taken by 360° cameras allow users to immerse themselves in a scene. We are using high resolution still imagery (Panono camera with 108 MP), 4K video (Nikon KeyMission), as well as flexible viewpoints using mega-tripods.

3D Modeling: Classic manual 3D modeling plays an important role for archeology in general and in the context of this work. The availability of such manually created models opens up the possibility for educational and scientific iVR applications to incorporate a temporal dimension in the form of interactive timelines or animations. We are using SketchUp for hands on modeling. In the addition, we plan to also make use of LiDAR data of terrain and objects to feed into the 3D modeling process.

4 Immersive VR Experiences and Research Tools

4.1 iVR Field Trip Experiences

Empirical studies have shown the potential of iVR in the teaching-learning process (e.g., Barilli et al. 2011). One of the biggest advantage of iVR for learning is that it affords learners a direct feeling of objects and events even if they are in the past, future, or imaginary. We briefly describe first prototypes of educational iVR experi-

(a)   **(b)**

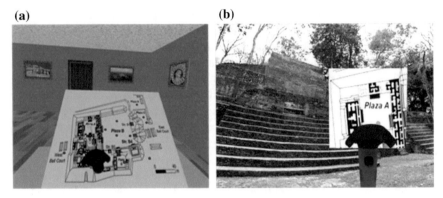

Fig. 2 **a** Overview perspective showing a map in a museum room. Users can point the laser emitted by the controller at a map point to view a 360° photo or select a path for a 360° video. **b** Image perspective with zoom-in map and current position indicated as a *green point*

ences created for the HTC Vive and Android-based mobile phones, all developed in Unity3D.

4.1.1 HTC Vive Experience

We are using the HTC Vive as it allows for room-scale experiences. Users can physically walk around, perceive the simulated space using a 1:1 body scale, and interact with digital content to extract, for example, geometric data from the visualization of realistic models. Figure 2a shows how users select a point or path on the map to immerse themselves in 360° scenes. A minimized overview map is attached to the controller for navigation purposes. The user's current position and visited places are marked in different colors to foster spatial awareness. A zoom-in map is attached to the backside of the overview map displaying the user's current position as a static or animated (in the case of a 360° video) point (Fig. 2b). Users can also navigate between different images or go back to the overview map to select other points.

4.1.2 Mobile VR Experience

The mobile version of our iVR experience is shown in Fig. 3a. Only the orientation of the display can be tracked based on the phone's accelerometer, screen space is limited, and no controller is available. Gaze control based on a reticle placed in the center of the screen is the main technique used to realize interactivity. The reticle will grow when placed on an element that can be clicked with the only button available by typical devices such as the Google Cardboard. In the image view, most of the interaction is realized via a simple popup menu that is also gaze-controlled.

(a)   **(b)**

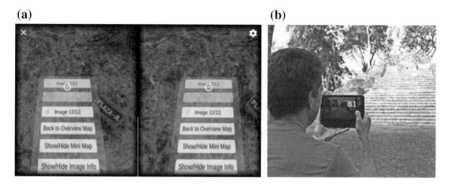

Fig. 3 **a** Image perspective with gaze-controlled navigation menu. **b** A simple AR application displaying additional information about buildings at Cahal Pech

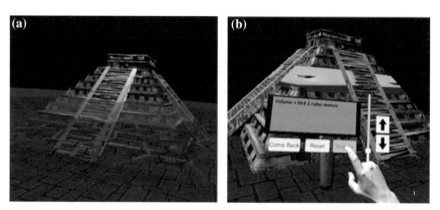

Fig. 4 *Top* and *bottom* faces of the volume detector cut the temple into three parts (**a**). A mesh is constructed along the surface of the middle part of the temple between two cross sections (**b**)

4.2 iVR Digital Workbench for Researchers

We are developing a digital workbench to provide archeologists with an immersive experience with information retrieval functionality and different measuring and interaction tools. One of the available tools displays general information of the site and allows for model manipulations such as vertical positioning, rotating, and scaling. Three other tools are available for measuring distances, areas, and volumes. Figure 4 illustrates interface and application to measure the volume of some part of a temple. More tools are currently under development and will be added to the workbench. We expect that some of them will find their way into the educational iVR applications as well to support exploration and experimentation.

4.3 AR Applications

In addition to creating iVR applications, the same approach and input data can be used to realize AR applications to improve on-site experience by providing complementary information about environmental features, buildings, or artifacts, or by even changing buildings. Manually created models for past times can be superimposed to create vivid animations of the historic developments of an area over time. We are just starting to tap the potential of AR for archeology and cultural heritage. A first prototype of an AR application is shown in Fig. 3b: Information about names of buildings are displayed when the building is looked at by the tablet's camera.

5 Sketch of an Underlying Site Information System

Creating and updating xR experiences described above currently requires substantial amounts of work. For the case of a xR visual analytics workbench for researchers, flexibility is required to allow for exploring all individual data, configure how objects and data are displayed, and modify the data if needed. We are therefore aiming for a more flexible approach in which archeological xR experiences and workbench for researchers are based on a central information system containing all data, media, and additional information regarding the particular site (see Luczfalvy et al. 2016 for the general idea of an archeological information system). This information system needs to support queries that can be spatial, temporal, and semantic at the same time. Table 1 lists a few use cases for querying the information system using natural language examples.

Table 1 Exemplary queries to the information system

Query	Modality
Provide all **360° Images** taken on *Plaza B*	Semantic + spatial
Provide all **Aerial Imagery** taken after *Jan 1, 2015*	Semantic + temporal
Provide all **3D Models** of **Historic Artifacts** found within 100 m of *Cahal Pech* that are from the *Postclassic Period*	Semantic + spatial + temporal
Provide a linked data graph view of *Artifact ID23415* with all **Media Sources** directly linked to it	Semantic, RDF

In the following, we briefly discuss the main components of the central information system we are envisioning, consisting of a central linked data storage with associated ontology, semantic-spatial-temporal query interface, VR experience generation software, and VR interface to the stored linked data. We also list challenges for which solutions need to be found when realizing the different components.

Central linked data storage: To support the automatic creation of VR experiences for education and research, the information in the information system needs to be heterogeneous (including sensor data, imagery, vector GIS data, archeological objects, general historic background information, external literature and media resources), multi-dimensional, and related to particular times and locations in space. In addition, as the query examples from Table 1 show, being able to access the data via complex mixed semantic-spatial-temporal queries is a key requirement. Hence, the goal is to store all information as a linked data database organized based on an ontology particularly designed for our purposes. Linked data approaches have recently become very popular to represent archeological information (e.g., Tudhope et al. 2011).

Ontology: The ontology for organizing and querying the linked data storage needs to cover general archeological and historic concepts, concepts related to Mayan history, and more application-specific concepts and relationships related to observation data, models, and media resources. Ontological modeling for archeology and cultural heritage is an active field and has led to approaches such as the CIDOC Conceptual Reference Model (CRM) ISO standard (Doerr et al. 2003) which could form a basis for our ontological modeling work. Challenges here include adequately dealing with metaphysical problems related to existence and temporal change, and with temporal and spatial information available at different scales.

Query interface supporting qualitative and quantitative relations: SPARQL is the de-facto standard for querying linked data. Queries can use quantitative (typically proximity based) relations as well as qualitative relations (spatial "on" and temporal "after", "from"). Spatial and temporal extensions of SPARQL (such as stSPARQL and GeoSPARQL) supporting qualitative relations have been proposed but expressivity and dealing with uncertainty are still major challenges (Belussi and Migliorini 2014).

Automatic VR experience generation: As indicated above, one of our goals is a high-level approach in which content and functionality is specified declaratively and the application is then created automatically from that specification drawing content from the central information system. Such a specification could, for instance, be based on a combination of state machine-like view graphs, rules for view transitions and interactions, and associated queries to the information system.

Flexible VR interface to linked data: To provide archaeologists with the means to conduct analyses and explore the available information in an immersive VR application, things need to be kept very flexible. The idea therefore is to provide researchers with configuration tools and a querying/exploration interface that allows them to choose what is shown and how, and directly interact with the linked data. While visualization and browsing tools for semantic networks exist in 2D, being inside a VR environment opens up new possibilities to interact with linked data.

6 Conclusions and Outlook

We presented work and ideas on creating immersive experiences and immersive digital workbenches for researchers for archeological sites using the example of the ancient Mayan site Cahal Pech. We discussed data capturing and 3D modeling methods as well as first xR applications and interaction methods we designed. These should be seen as first steps that will be improved and extended as part of future work. Finally, we described our plans for basing these applications on an underlying information system based on linked data representation and query technology. While challenges remain regarding the application of these semantic web methods to the archeological domain, we believe the ontology-driven linked data approach is the most suitable one to realize the envisioned xR applications.

Acknowledgements Support for this research by the National ScienceFoundation under Grant Number NSF #1526520 is gratefully acknowledged.

References

Awe Jaime J, Zender M, Chan K'inich K, Bahlam K (2016) Royal titles and symbols of rulership at Cahal Pech. Belize. Mexicon 38(6):157–165

Barilli ECVC, Ebecken NFF, Cunha GG (2011) The technology of virtual reality resource for information in public health in the distance: an application for the learning of anthropometric procedures. In: Ciênc. saúde coletiva 16:1247–1256

Belussi A, Migliorini S (2014) A framework for managing temporal dimensions in archaeological data. In: Proceedings of 21st international symposium on temporal representation and reasoning (TIME 2014), pp 81–90

Bowman DA, McMahan RP (2007) Virtual reality: how much immersion is enough? Computer 40(7):36–43

Doerr M, Hunter J, Lagoze C (2003) Toward a core ontology for information integration. J Digital Inf 4(1)

Donalek C, Djorgovski SG, Cioc A, et al (2014) Immersive and collaborative data visualization using virtual reality platforms. In: IEEE international conference on big data, pp 609–614

Luczfalvy Jancsó A, Billen R, Hoffsummer P, Jonlet B, Hallot P, Poux F (2016) CASTLE4D: an archaeological information system based on 3D point clouds. In: Lerma JL, Cabrelles M (eds) Proceedings of the Arqueologica 2.0—8th international congress on archaeology, computer graphics, cultural heritage and innovation, pp 247–252

López JAB, Jiménez GA, Romero MS, Esquivel J (2016) 3D modelling in archeology: the application of Structure from Motion methods to the study of the megalithic necropolis of Panoria (Granada, Spain). J Archeol Sci Rep 10:495–506

Milgram P, Kishino F (1994) A taxonomy of mixed reality visual displays. IEICE Trans Inf Syst E77-D(12):1321–1329

Simpson M, Wallgrün JO, Klippel A, Yang L, Garner G, Keller K, Bansal S (2016) Immersive analytics for multi-objective dynamic integrated climate-economy (DICE) models. In: Hancock M, Marquardt N, Schöning J, Tory M (eds) Proceedings of the 2016 ACM companion on interactive surfaces and spaces, pp 99–105

Tudhope D, Binding C, Jeffrey S, May K, Vlachidis A (2011) A STELLAR role for knowledge organization systems in digital archeology. Bull Am Soc Info Sci Tech 37:15–18

HBIM for the Archaeology of Standing Buildings: Case Study of the Church of San Cipriano in Castelvecchio Calvisio (L'Aquila, Italy)

Romolo Continenza, Fabio Redi, Francesca Savini, Alessandra Tata and Ilaria Trizio

Abstract This paper describes the application of a BIM (Building Information Modelling) process to a medieval building analysed through the most classic tool of architectural archaeology, namely the stratigraphic analysis of structures. The case study, conducted by a multi-disciplinary team, attempts to shed light on the procedure's strong points and shortcomings, and to widen the range of digital tools employed during the preparatory stages of restoration projects. The ultimate aim is to acquire in-depth knowledge of the asset and the history of its construction through the integration of documentary sources and on-site material findings, with the firm belief that proper data analysis and processing may lead to correct decisions concerning the asset's conservation and enjoyment.

Keywords HBIM · Architectural archaeology · Stratigraphic analysis · Survey · parametric 3D modeling

R. Continenza · A. Tata
Department of Civil Engineering, Construction—Architecture, Environmental,
University of L'Aquila, L'Aquila, Italy
e-mail: romolo.continenza@univaq.it

A. Tata
e-mail: alessandra.tata@virgilio.it

F. Redi · F. Savini
Department of Human Sciences, University of L'Aquila, L'Aquila, Italy
e-mail: fabio.redi@cc.univaq.it

F. Savini
e-mail: savini_francesca@libero.it

I. Trizio (✉)
Institute for Construction Technology, Italian National Research Council, L'Aquila, Italy
e-mail: ilaria.trizio@itc.cnr.it

© Springer International Publishing AG 2018
P. Fogliaroni et al. (eds.), *Proceedings of Workshops and Posters at the 13th International Conference on Spatial Information Theory (COSIT 2017)*, Lecture Notes in Geoinformation and Cartography, https://doi.org/10.1007/978-3-319-63946-8_49

1 Foreword

The process of knowledge acquisition and documentation aimed at protecting and preserving the cultural heritage—in its broadest sense—is still a very topical issue, and it encompasses a wide range of concerns that are confronted by multiple disciplines, often with different approaches.

The collection, processing, transmission and use of data pertaining to the historical, architectural and archaeological heritage are facilitated nowadays by digital technologies (hardware and software) that have enabled increasingly powerful and accurate investigation techniques; the research team of archaeologists and engineers from the University of L'Aquila and researchers from ITC-CNR are exploring precisely this avenue, experimenting with innovative technologies for the collection, documentation and analysis of cultural heritage data.

This collaboration has already proved successful, resulting in the possibility of fine-tuning the three-dimensional stratigraphic analysis through the processing and post-processing of digital photogrammetry models (Giannangeli et al. 2015, 2017), of constructing parametric models on which to map and document the state of conservation of historical artefacts (Continenza et al. 2016), and of importing digital photogrammetry models into a 3D GIS in order to carry out three-dimensional stratigraphic analyses and propose diachronic evolution hypotheses for the sites (Giannangeli et al. 2017).

In particular, this paper illustrates the potential of integrating the lexicon and data peculiar to architectural archaeology within the framework of a parametric digital model; in other words, of employing a BIM process to conduct the stratigraphic analysis of an architectural asset. The scientific literature concerning the use of BIM to document a historical building provides several examples (for a summary of the state of the art, v. Mingucci et al. 2016), as does the debate on the methodology applied to cultural assets (Acierno et al. 2017). Experiments are already under way in the archaeological sphere for the purpose of documenting, managing and disseminating excavation data (Garagnani et al. 2016; Scianna et al. 2015), or in the context of analysing stone and brickwork masonry (Spallone et al. 2016). Finally, it is worth noting that the experimentation with BIM processes applied to historical buildings and archaeological artefacts runs in parallel with 3D GIS applications, as shown by several examples, including recent ones (Dell'Unto et al. 2016), and by detailed evaluations aimed at comparing both systems (Dore and Murphy 2012; Saygi and Remondino 2013).

2 The Church of San Cipriano: Historical Context

The building that constitutes the object of experimentation is located at the gates of the village of Castelvecchio Calvisio, L'Aquila, on the Aquilan face of the Gran Sasso mountain (Fig. 1). It is one of the oldest buildings of the historic Barony of

Fig. 1 Location of the church of San Cipriano to Castelvecchio Calvisio (L'Aquila, Italy)

Carapelle, mentioned as early as 779 in the *Chronicon Vulturnense* and in some papal bulls and privileges (Celidonio 1911; Faraglia 1888). The church is built upon a previous classical structure, probably belonging to a Roman villa (Staffa 2000) or to an Italo-Roman shrine (Mattiocco 1988), a hypothesis supported by the presence on its walls of numerous reused elements from the classical period.

Analysis of the structure has revealed a total of seven construction stages, from the 9th to the 20th century. The earliest stage is suggested by a small remnant of masonry in the apse, but, pending archaeological excavations, the paucity of data precludes further interpretation. The second stage may be dated to the 10th or 11th century and comprises the construction of a single-apse structure with a rectangular hall divided by pillars or columns into three naves, with very narrow lateral naves, and a tower-portico on the façade in the Carolingian tradition that finds a few scattered counterparts in the territory. The next stage took place in the Romanic period proper, and involved widening the hall, eliminating the lateral naves and replacing the tower-portico with the still visible, tall bell gable. In the course of the 14th and 15th centuries, the church was complemented with the addition of a chapel next to the entrance, a ciborium on pillars covered by a groin vault that replaced the original apse conch, and a second frescoed chapel where the date 1429 can still be read. In modern times, in order to counter the state of disrepair mentioned in the sources (Archivio di Stato dell'Aquila 1876), and perhaps after one of the several historical earthquakes, the church underwent conservation work which was also meant to solve some problems of statics by buttressing the right side wall, slightly shoe-shaped, and the bell gable, the asymmetry of which alters the layout of the main front. The last stage corresponds to work carried out in the 19th and 20th century to restore the roof and the interior.

3 Data Acquisition and Modelling

The building was digitized through a direct scan, integrated with photogrammetric processing using Agisoft's Photoscan Professional (release 1.2.6). The digital photogrammetric model, georeferenced and suitably scaled, was stored in *.las format and imported into Autodesk Revit 2016 (release 16.0.428.0); thus, the software-indexed point cloud constituted the basis for the subsequent parametric modelling.

Before undertaking the parametric 3D-modelling of the structure, the semantic architecture of the model had to be designed, adding new parameters to the usual ones—classified into categories, as is standard practice—in order to meet the requirements of a multidisciplinary work group. The parametric 3D model thus obtained (Fig. 2) allows for photo-realistic renderings of the structure that employ photographic planes from the photogrammetric model and their integration with archaeological and documentary data (state of conservation, brickwork techniques, stratigraphic relationships, dating, and construction stages) topologically correlated to each of the model's elements.

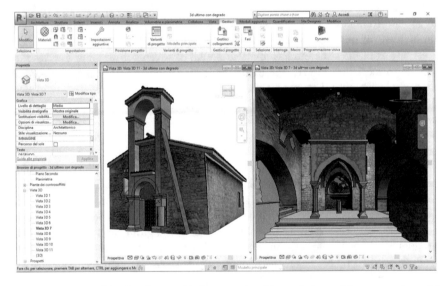

Fig. 2 Screenshot of Revit software with the external and internal views of the parametric model of the church of San Cipriano

4 Three-Dimensional Stratigraphic Analysis in a BIM Environment

The building itself and its various construction stages account for a manifold dia-chronic complexity eminently suited to experimentation with Autodesk Revit for the representation and management of data concerning the Wall Stratigraphic Units.

From a methodological point of view, parametric modelling offers a choice between two alternative modes: the first calls for the volumetric modelling of each of the Stratigraphic Unit that constitute the walls through the creation of specific families that meet the requirements of stratigraphic analysis. Each Stratigraphic Unit was individually modelled through profile modification, and assigned metric and descriptive parameters and a photo-realistic texture directly extrapolated from the respective photographic frontal plane (Fig. 3). The second mode consists in modelling extremely thin surfaces that, following the Wall Stratigraphic Units contours, overlap with the wall surfaces analysed and are representative of the stratigraphically analysed units. The choice of mode was dictated by the widespread diffusion—in the difficult context of the post-earthquake reconstruction of L'Aquila and its territory—of parametric models applied to historical buildings for the pur-poses of deliberate reconstruction, in most cases independently of stratigraphic analyses; it was therefore considered appropriate to integrate into those models the information obtained from such analyses.

Both modes enable to be assigned Stratigraphic Units their characteristic information, such as their number, masonry type, construction stage, dating, pres-ence of reused elements, state of conservation, restoration work, etc. (Fig. 4).

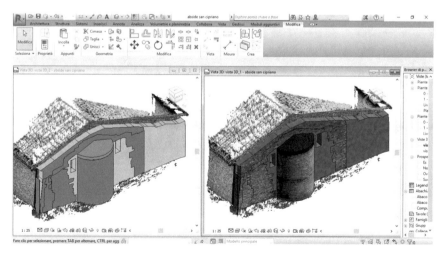

Fig. 3 Screenshot of Revit software with point clouds imported for the parametric modeling of individual Wall Stratigraphic Units

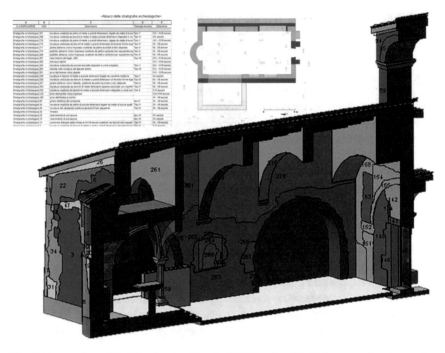

Fig. 4 Parametric model of the church which the Wall Stratigraphic Units were colored according to the identified phases and the detailed graphs of the archaeological data

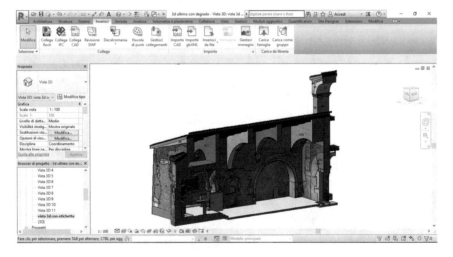

Fig. 5 Screenshot of Revit software with thematic view of the model with photorealistic texture

An emerging problem that deserves further discussion is the connection between the stratigraphic relationships and the corresponding Harris matrix, a difficulty that has been circumvented by incorporating the parameters that define the relationships within the detail charts, and through the inclusion of theme views (Fig. 5).

5 Data Export and Diffusion

The potential of parametric modelling software is well established, and the preceding discussion underlines the effectiveness of this technology for applications concerning historical and archaeological structures, where a multitude of heterogeneous data may be processed within a single BIM modelling environment, thus avoiding the interoperability problems that would arise if multiple tools were employed for reaching the same goals. There are several alternatives for exporting the model and related data so that they can be employed by users who do not have the same software installed. Among these, the more subjective ones involve compiling summary tables, cutaways, axonometric projections and charts from the recorded data and producing video sequences showing the stratigraphic mappings. Other, more objective methods that are not so dependent on the operator's choices and present complete information, are made possible by Revit applications such as the A360 platform for online model visualization and the free BIM Vision software for the remote visualization of *.IFC files (Fig. 6).

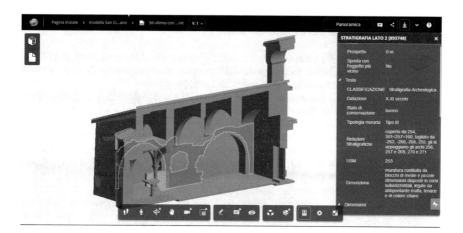

Fig. 6 BIM model of the church of St. Cipriano visible on the A360 platform

6 Conclusions

In addition to the positive features described above, experience has brought out some weaknesses related to the characteristic unevenness of archaeological sites and historical buildings that renders them hardly amenable to automated software processing. The software employed is mainly oriented to exploit the geometric regularity that characterizes contemporary architecture, and its capacity to deal effortlessly with repetition, seriality and equality of proportions is ill-suited to historical contexts; hence, it is often necessary to resort to manual procedures (such as the implementation of ad hoc families) which slow down the process.

We hope that the ongoing research may lead to methodological improvements and to further development of the semantic model's fourth dimension, namely asset management in time; this temporal dimension is already incorporated in planning an asset's preservation, and may be fruitfully applied to the enhancement, interpretation and communication—construed as historical reconstruction—of a cultural asset.

References

Acierno M, Cursi S, Fiorani D, Simeone D (2017) Architectural heritage knowledge modelling: an ontology-based framework for conservation process. J Cult Heritage 24(2017):124–133

Archivio di Stato dell'Aquila (1876) Intendenza, Serie I, Categoria IV

Celidonio G (1911) La diocesi di Valva e Sulmona, Volume III (dal 1100 al 1200)

Continenza R, Trizio I, Tata A, Giannangeli A (2016) HBIM per il progetto di restauro: l'esempio della chiesa di San Cipriano a Castelvecchio Calvisio (L'Aquila). DisegnareCon 9(2016):15.1–15.9

Dell'Unto N, Landeschi G, Leander Touati A-M, Dellepiane M, Callieri M, Ferdani D (2016) Experiencing ancient buildings from a 3D GIS perspective: a case drawn from the Swedish Pompeii project. J Archaeol Method Theory 23(2016):73–94

Dore C, Murphy M (2012) Integration of HBIM and 3D GIS for digital heritage modelling. digital documentation 22–23 October, 2012, Edinburgh, Scotland

Faraglia NF (1888) Codice diplomatico Sulmonense 43–45

Garagnani S, Gaucci A, Gruška B (2016) From the archaeological record to archaeobim: the case study of the etruscan temple Of Uni In Marzabotto. Virtual Archaeol Rev 7(15):77–86

Giannangeli A, Marchetti A, Redi F, Savini F, Trizio I (2017) La chiesa di San Cipriano a Castelvecchio Calvisio (AQ) nella Baronia di Carapelle: documentazione speditiva e analisi stratigrafica 3D del manufatto. Archeologia dell'Architettura, XXII (in press)

Giannangeli A, Marchetti A, Redi F, Savini F, Trizio I (2017) Le emergenze medievali nel complesso archeologico delle terme di Vespasiano a Cittaducale (RI). L'acquisizione fotogrammetrica e la gestione in ambiente GIS 3D delle mesh. Archeologia dell'Architettura, XXII (in press)

Giannangeli A, Marchetti A, Trizio I (2015) Dalla fotogrammetria digitale ai Pdf 3D: considerazioni sulle potenzialità offerte dalle metodologie di rilievo speditivo basate sullo SfM. Atti del XVI convegno ANIDIS - L'ingegneria sismica in Italia. L'Aquila, 13–17 settembre 2015

Mattiocco E (1988) Le antiche chiese della terra di Carapelle. In: AAVV (eds) Homines de Carapellas. Deputazione di Storia Patria, L'Aquila, pp 101–122

Mingucci R, Brusaporci S, Cinti Luciani S (eds) (2016) Le dimensioni del B.I.M. DisegnareCon 9

Saygi G, Remondino F (2013) Management of architectural heritage information in BIM and GIS. State of the art and future perspectives. Int J Heritage Digital Era 2(4):695–713

Scianna A, Serlorenzi M, Gristina S, Filippi M, Paliaga S (2015) Sperimentazione di tecniche BIM sull'archeologia romana: il caso delle strutture rinvenute all'interno della cripta della chiesa dei SS. Sergio e Bacco in Roma. Archeologia e Calcolatori, supplemento 7:199–212

Spallone R, Piano A, Paino S (2016) B.I.M. e beni architettonici: analisi e rappresentazione mul-tiscalare e multidimensionale di un insediamento storico. Il caso studio di Montemagno, Borgo Nuovo piemontese. DisegnareCon 9:13.1–13.13

Staffa AR (2000) Le campagne abruzzesi fra tarda antichità ed altomedioevo (secc. IV–XII). Archeologia Medievale XXVII:17–99

An Analytical Framework for Classifying Software Tools and Systems Dealing with Cultural Heritage Spatio-Temporal Information

Andrea Luczfalvy Jancsó, Benoît Jonlet, Patrick Hoffsummer, Emmanuel Delye and Roland Billen

Abstract This paper presents an analytical framework for classifying software tools and systems dealing with spatio-temporal information developed for applications in the cultural heritage field. These can be numerous and quite different from one another, depending for the purpose they were developed for. In order to assess if one of the already existing software tools and systems can be used or modified to fit our research goals, a list of needs was established. Starting from those, requirements were defined and the software tools and systems dealing with spatio-temporal information are then compared and evaluated based on those characteristics. This analytical framework is an important step in our research since using a tool that is not appropriate for the study object will not be able to provide valuable information to answer the main research questions.

Keywords Analytical framework · Spatio-temporal information · Cultural heritage · Medieval castles · 3D point clouds · Semantics

A. Luczfalvy Jancsó (✉)
Geomatics Unit, Department of Geography, European Archaeometry Centre,
University of Liège, Allée Du 6 Août 19 (B5a), 4000 Liège, Belgium
e-mail: aljancso@ulg.ac.be

B. Jonlet · R. Billen
Geomatics Unit, Department of Geography,
University of Liège, Allée Du 6 Août 19 (B5a), 4000 Liège, Belgium
e-mail: bjonlet@ulg.ac.be

R. Billen
e-mail: rbillen@ulg.ac.be

P. Hoffsummer · E. Delye
European Archaeometry Centre, University of Liège,
Allée Du 6 Août 19 (B5a), 4000 Liège, Belgium
e-mail: phoffsummer@ulg.ac.be

E. Delye
e-mail: Emmanuel.Delye@ulg.ac.be

© Springer International Publishing AG 2018
P. Fogliaroni et al. (eds.), *Proceedings of Workshops and Posters at the 13th International Conference on Spatial Information Theory (COSIT 2017)*, Lecture Notes in Geoinformation and Cartography, https://doi.org/10.1007/978-3-319-63946-8_50

1   Introduction

For the study of a series of medieval castles located around the city of Liège in the east of Belgium, an archaeomatic approach using technology such as 3D laser scanner and information systems was chosen. Through this application, we strive to deepen our understanding of their history, the reciprocal influence between the fortress and their respective landscape as well as the stakes that were involved in the choice of their location and the defence of those cultural heritage sites.

As for the available data on those castles, it comes in various forms, such as textual testimonies and visual documents, e.g. prints, wash drawings, photographs from different periods and ancient postcards. Additionally, during the past century, studies and excavations have been conducted on site according to the methodologies of those times. Therefore, the state of the documentation linked to those interventions can be inconsistent, insufficient or even missing. In addition to this, every castle in this research is in a different conservation state and the data available for each site differs in quality and quantity.

As stated in Luczfalvy Jancsó et al. (2016), a 3D point cloud will serve as a base for the current research. Acquired by lasergrammetry and photogrammetry, these 3D models are objective digitisations that will be completed with the already existing data in order to obtain new information to further the knowledge about medieval castles from Belgium and northeastern France as well as the relationship between them and their physical environment.

2   Assessing Our Needs and Requirements

The starting point of our research is a 3D digitisation of every one of the study objects. This leads to the necessity to use a software tool or a system that supports not only the integration of a 3D point cloud, but that also allows to interact with it, be it through segmentation or adding data to those parts (Luczfalvy Jancsó et al. 2016). Therefore, in order to deal with the three dimensional data, it is required that the software tool or system provides a **3D input**.

As these cultural heritage sites date back to the Middle Ages, the architecture mostly does not follow any standards or symmetry. The understanding of their architectural history is complicated through the frequent modifications and repair works. The identification of the different construction and/or destruction phases is the key to tell the story of those cultural heritage sites since for most of them, there are rarely testimonies with concrete information on the various interventions that the castle underwent. These **temporal** and **spatial** characteristics have to be taken into consideration and managed by the software tool or the system selected for this research. Therefore, as already mentioned in Luczfalvy Jancsó et al. (2016), the software tool or the system has to be able to cope with all the particularities of an archaeological research, such as the multivocality (Cripps 2013;

Luczfalvy Jancsó et al. 2016) that concerns both the spatial and the temporal components and such as the uncertainty, imprecision, ambiguity and incompleteness of archaeological data (De Runz 2008; Luczfalvy Jancsó et al. 2016). Temporality also has to cope with the fuzziness inherent to this discipline (Belussi and Migliorini 2014, 2017; De Roo et al. 2013a, Luczfalvy Jancsó et al. 2016).

These features also intervene when dealing with past excavations. Some of those sites have been subjected on and off to archaeological digs executed by various directors over the course of the last 100 years. These have been documented to the best of their periods, techniques and methodologies. However, archaeological techniques have evolved quite a bit during the second half of the last century, which is mostly due to new technical possibilities, to the joint work between different disciplines of the science and the humanities sectors and to the development of new research interests. Additionally to the missing information that was not deemed necessary at the time or that were not uncovered the same way that nowadays excavations and recording techniques would, the data resulting from those past digs is not always available as it could have been lost, never been published or not even brought to paper if the operations director did not do so. Added to other data sources, such as texts or images, all of these different document types, also taking into account the particularities of archaeological data, need a **semantic** framework allowing linking them with the 3D model.

Moreover, those past excavations are not always part of the 3D digitisation as the trenches have sometimes been backfilled. Only plans, cross-sections and photographs can be used to analyse those sections. Therefore, the software tool or the system needs to include some **3D modelling/reconstruction** possibilities. This could be helpful for the understanding of conclusions and results from those ancient studies. This would also be beneficial to the spatial understanding of this underground, inaccessible and lost evidence due to the destructive nature of archaeological research. The spatial relocation of the findings could also be helpful to verify the chronology that has been established based on the stratigraphic evidence.

As the research currently being carried out deals with the castles and their surrounding territories, it is not always necessary that every detail available is also accessed. Indeed, be it minor architectural decorations and some internal building elements that have no impact on the relationship between the landscape and the cultural heritage site or on a construction volumes analysis do not have to be visible. Also, the documentation that is specifically linked to them does also not have to accessible, as they would not provide relevant information. To limit the visualization of the 3D model and the data access, the integration of **Levels of Details (LoD)** would be an interesting feature. This way, each LoD could be defined in order to provide the needed visual support and documentation for specific studies.

The software tool or system dealing with spatio-temporal information for our research should be able to handle large 3D point clouds as well as spatial, temporal and semantical aspects. As archaeological data consists of different supports, in digital or physical form, the **import** function should allow for multiple file formats to be inserted into the software tool or the system. The **storage** must also be thought of as the 3D point cloud on its own already weighs quite a lot and the rest of the data can vary from only a few to a great number of documents. The **export** format should also be chosen in order to permit for interoperability with other softwares and tools.

3 Requirements Definition

In order to evaluate the usefulness of existing software tools and systems dealing with spatio-temporal information, we have determined a series of requirements. They are linked to the needs we have previously established.

General Approach This first item is less a requirement than basic information about the main components of the software tool or the system. This gives a first indication of the features that will be included. In the Table 1, it seems that the general approach is often based on one of these three possibilities: database, GIS and CAD/BIM.

3D Input The 3D input is vital to our research as a 3D digitisation is our primary working tool for the study of each of the castles. If such an input is possible, it is also interesting to list the actions that can be undertaken on the 3D model such as the opportunity to take measurements, to model and/or reconstruct missing parts.

Spatiality This is one of the key features while working with cultural heritage objects and their environment. They are characterised by their location, their dimensions and their volumes. This entry has been divided into three subparts: vector, raster, 3D point cloud. This separation has been added since all three kinds of spatial representation are not always taken into consideration by the various software tools and systems. For the vector entry, a difference is marked by adding the information if the data is structured or not. This means to differentiate the spatial data that only is provided with a geometric model, in contrast to the spatial data that is also structured with topological information. This difference does not apply to raster data as it consists of georeferenced imagery. As research on 3D point clouds is not currently far enough to apply topological structures, we are limited to indicate if the software tools and systems support the integration of 3D point clouds. The possibility to add vector, raster or 3D point cloud data will affect storage, querying and analysis, further down during the investigations.

Spatial Standard The use of standards or not can infer on the interoperability with other softwares and systems and as well as provide a common ground for every researcher involved with the study.

Temporality Temporality can be approached in different ways, such as an attribute in the database or through phasing the visual support with different versions of a same 3D model with or without attributes. It is however not excluded that another possibility exists to integrate a temporal feature. Indeed, as the temporal feature is one of the most complicated aspects to handle in such a software or system since it is not always an absolutely defined date and it might be subject to fuzziness and changes (Belussi and Migliorini 2014; Belussi and Migliorini 2017;

Table 1 Non-exhaustive overview of software tools and systems dealing with cultural heritage spatio-temporal information

Name	Description	Available	General approach	3D input	Spatiality		Spatial Standard	Temporality		
Arches Project	Web-based inventory for the recording and management of CH	Yes	GIS	No	Vector	Yes, unstructured	?	Object attributes		Attribute in database: time filter for querying database
					Raster	Yes				No
					3D point cloud	No, but could be added		Versioning		
ARK (Archaeological Recording Kit)	Web-based archaeological data management system/framework	Yes	Spatially enabled Database GIS (in progress)	Modelling/Reconstruction	Vector	Yes, unstructured	No data standard but can be added if wanted	Object attributes		Yes, attributes in database?
					Raster	Yes				No
					3D point cloud	?		Versioning		
ArkeoGIS	Web-based software for the inventory of archaeological and geographical data	Yes, only online and upon free subscription	GIS (only a map viewer; editing and publishing has to be done with CAD or GIS software)	?	Vector	Yes, unstructured	?	Object attributes		Attribute in database: time filter for querying database
					Raster	Yes				No
					3D point cloud	?		Versioning		
CASTLE3D	Computer aided system for labelling archaeological excavations in 3D	No?	Database	Measurement Modelling/Reconstruction	Vector	Yes, unstructured	?	Object attributes		?
					Raster	Yes				?
					3D point cloud	Yes		Versioning		

(continued)

Table 1 (continued)

Name	Description	Available	General approach	3D input	Spatiality			Spatial Standard	Temporality	
					Vector	*Raster*	*3D point cloud*		*Object attributes*	*Versioning*
CityGML + Cultural Heritage ADE	Open standardised data model and exchange format to store digital 3D models of cities and landscapes	Yes	Database	Modelling/Reconstruction	Yes, unstructured	Yes	?	GML < OGC ISO 19136	Yes	No?
Heritage BIM	BIM for CH parametric	Yes	CAD/BIM	Measurement Modelling/Reconstruction (mesh is main working component)	Yes, unstructured?	Yes	Yes	GDL	Attributes in database?	Yes: phasing/4D modelling
QueryArch3D tool/MayaArch3D	Tool chain for web-based visualisation and interaction with a reality-based multi-resolution 3D model – 3D WebGIS	Yes?	GIS	Modelling/Reconstruction	Yes, unstructured?	Yes	Yes? Maybe only mesh integration	?	Yes	?
REVEAL	Platform for the documentation of archaeological excavations (documentation and data capture, archive, and analysis)	Yes	Database	Measurement Modelling/Reconstruction	Yes, unstructured	Yes	Yes	?	?	?

(continued)

Table 1 (continued)

Name	Temporal standard	Semantics		Semantic standard	LoD	Import formats	Storage	Export formats	Cost	Open source
Name	Temporal standard	Semantics		Semantic standard	LoD	Import formats	Storage	Export formats	Cost	Open source
Arches Project	?	*Object attributes*	Yes	CIDOC CRM CIDOC CDS ISO 21127:2014 Dublin Core Metadata Element ISO 5836-2009	?	*.arches file	PostgreSQL PostGIS SQL GeoJSON KML Shapefile	Can be integrated with GeoServer so that may be consumed with GIS applications	Free	Yes
		Specific objects	?							
ARK (*Archaeological Recording Kit*)	No data standard but can be added if wanted	*Object attributes*	?	CIDOC CRM (available soon) ISO 21127:2014	?	Various	Apache MySQL PHP	XML/CSV/Web services (RSS, ATOM, WMS, WFS)	Free	Yes XHTML CSS
		Specific objects	?							
ArkeoGIS	?	*Object attributes*	?	?	?	CSV	PostgreSQL PostGIS No database: software works as database feed aggregator	CSV	Free	Yes
		Specific objects	?							
CASTLE3D	?	*Object attributes*	Yes?	?	?	XML Different formats of point clouds	Database as back end Octree	XML Segmented point cloud	?	?
		Specific objects	Yes?							
CityGML + Cultural Heritage ADE	GML comprises a time primitive	*Object attributes*	Yes	CityGML OGC	Yes, up to LoD4	XML?	GML (stores CityGML in a DB) XML	XML	Free	"Yes"
		Specific objects	Yes							

(continued)

Table 1 (continued)

Name	Temporal standard	Semantics		Semantic standard	LoD	Import formats	Storage	Export formats	Cost	Open source
Heritage BIM	?	*Object attributes*	Yes	?	Yes, up to LoD6 (AEC (UK) BIM Technology Protocol (v2.1.1) aligned with PAS 1192-2:2013) or up to LoD4 for historical applications (Metric Survey Specifications for CH)	Every type of digital document can be added, mainly: ASCII TIFF/JPF DXF/DWG DXF/OBJ	?	Various proprietary formats IFC (ISO 16739:2013) COBie (subset of IFC)	Yes/Free	No/Yes
		Specific objects	Yes							
QueryArch3D tool/MayaArch3D	?	*Object attributes*	Yes?	In process of being mapped to CIDOC CRM ISO 21127:2014	Yes, up to LoD4	?	PostgreSQL/PostGIS → unique data management system PHP (communication with external database)	?	?	Yes?
		Specific objects	Yes?							
REVEAL	?	*Object attributes*	?	?	?	Various?	MySQL PHP	Yes, but format?	Free	Yes
		Specific objects	?							

De Roo et al. 2013b; Katsianis et al. 2008). Also, as the authors in (Katsianis et al. 2008) state, up to six time categories can be considered. Some standards have been developed but not every software or system relies on them.

Temporal Standard The use of a standard for the temporal component would be of help to assure interoperability and comparison between the different study objects. However, it will have to be verified if the used standard can be correctly applied to each case.

Semantics As the 3D model is meant to be the interface through which the available data will be accessed and queried, it is necessary to integrate a semantic framework that will be used as basis to combine archaeological data and sources with the 3D point clouds. A difference can be made between attributes that link to an object, such as a polygon linked to a function, or specific objects that are linked to each other in order to create a semantic model.

Semantic Standard The semantic standard that is the most used is the CIDOC CRM (ICOM's International Committee for Documentation Conceptual Reference Model) which is based on the ISO 21127:2014 standard. The CIDOC CRM standard provides a common ground for the description and integration of the various types of cultural heritage data into information systems and conceptual models (http://www.cidoc-crm.org/). However, based on the type of semantic that is incorporated into the software tool or system, other standards can also be considered.

Level of Detail (LoD) Depending on the current research goal, it is not always necessary to render all of the details of the 3D model. Therefore, working with LoDs can be taken into consideration. However, each used LoD repartition differs from one system to another: the specificities of each level as well as the number of levels can vary.

Import/Export Formats The import and export file types cover a wide range of possibilities. Some of the export formats follow international standards.

Storage The storage is mostly handled through databases. The spatial component is sometimes managed separately from the rest of the data. This feature will impact the way the data is accessed and the actions and queries the data could be subjected to.

Cost/Open Source Additionally, even if this point does not directly answer the previously mentioned needs, it can provide useful information on the adaptability and the interoperability of the software tool or the system as well as the financial impact as the cost can greatly impact the integration of these softwares and systems as a research tool. The open source working allows for the addition of extensions and plugins that managed needed aspects, which are not taken care of by starting pack of the software or the system.

4   Some Existing Software Tools and Systems Dealing with Cultural Heritage Spatio-Temporal Information

Once our needs have been ascertained and the requirements established, a first non-exhaustive list of software tools and systems dealing with cultural heritage spatio-temporal information was established.

The Arches Project This system (Carlisle et al. 2014; http://archesproject.org/; Myers et al. 2016) was developed by the Getty Conservation Institute and the World Monuments Fund as an inventory and management tool for cultural heritage.

ARK (Archaeological Recording Kit) This collection of tools (http://ark. lparchaeology.com/) is destined for the "collection, storage and dissemination of archaeological data".

ArkeoGIS This software compiles archaeological and environmental data (http://arkeogis.org/fr/arkeogis-de-lhypothese-a-la-carte/; Bernard et al. 2015).

CASTLE3D CASTLE3D stands for Computer Aided System for Labelling Archaeological Excavations in **3D** (Houshiar et al. 2015; Nuechter et al. 2015). This system focuses on the recording of archaeological evidence.

CityGML with the Cultural Heritage ADE CityGML (https://www.citygml. org/ade/; https://www.citygml.org/; http://www.3dcitydb.org/3dcitydb/ 3dcitydbhomepage/) is a data model as well as an exchange format. It deals with 3D models of cities and landscapes. An Application Domain Extension (ADE) (Costamagna and Spanò 2013; De Roo et al. 2014; Finat et al. 2010) has been developed specifically for cultural heritage means. Combined with basic CityGML, it is better suited for the specificities of an archaeological research.

Heritage BIM This is a BIM version that is better adapted for the work on a cultural heritage site (Antonopoulou 2017; Chiabrando et al. 2016).

QueryArch3D Tool/MayaArch3D This tool (Agugiaro et al. 2011; http://www. mayaarch3d.org/language/en/sample-page/) allows to visualize and to query multiresolution 3D models.

REVEAL The acronym stands for **R**econstruction and **E**xploratory **V**isualization: **E**ngineering meets **A**rchaeo**L**ogy (Sanders 2011; http://ark.lparchaeology. com/). It is a software that helps with the documentation recording process of an archaeological excavation as well as with the analysis of this data.

Although this list might be non exhaustive and the analysed software tools and systems were primarily chosen based on their ability to integrate 3D data as well as on their availability for testing, this first try allows us to verify if some potentially interesting working tools already exist. Additionally, we had the opportunity to develop an analytical framework based on our previously established requirements for an information system that fits our research.

In order to evaluate the usefulness of these existing software tools and systems, the previously determined requirements based on the needs discussed in the first part of this paper were applied to each of the selected software tools and systems.

Indeed, each of them was entered into the Table 1 to get a better overview of their characteristics and to facilitate comparisons between them. Currently, the entries into the table are provisional as the framework has been set, but the analysis is still ongoing.

As we have tested those existing software tools and systems against our requirements, it is obvious that some cells are left blank. Currently, we preferred to add a question mark instead of leaving an empty unit. This decision was made because the absence of information about a specific entry does not necessarily mean that this characteristic is not supported by the software or the system. However, it may not be mentioned in the software or system description as it does not have an impact on its use and its smooth running. Also, although the content added in the storage column does not always directly relate to it, we decided to maintain it there as these entries provide information on how the data will be accessed. If deemed necessary, a new section regarding the main technologies and programming tools will be added. However, as we did not define a requirement for a specific technology or programming tool, it should be considered as a descriptive column.

5 Conclusion

With our requirements in mind for the study of a series of medieval castles, this analytical framework provides a sound basis to compare the different available software tools and systems dealing with cultural heritage spatio-temporal information.

In order to complete some of the missing information in the table cells, some trials are probably the best way to provide precise information about some of those requirements.

These preliminary results show that some softwares or systems (e.g. CityGML with the cultural heritage ADE, the QueryArch3D tool and the Heritage BIM) seem to be the most appropriate since they fulfil most of our requirements. However, even if at first glance they seem the most suited, a test drive will also be necessary to assess their actual usefulness. Additionally, these tests will also be the opportunity to check if they are able to cope with the particularities of a medieval building structure as well as with those sections for with there is no 3D digitisation available, e.g. the ancient excavations that have been backfilled.

Further research will also be carried out to complete the table with other approaches, this time providing an exhaustive list and a state-of-the-art of the existing software tools and systems dealing with spatio-temporal information for cultural heritage.

References

3D Geoinformation group at TU Delft: Application Domain Extensions (ADE). https://www.citygml.org/ade/

3D Geoinformation group at TU Delft: CityGML homepage. https://www.citygml.org/

Agugiaro G et al (2011) QueryArch3D: querying and visualising 3D models of a maya archaeological site in a web-based interface. Geoinform FCE CTU 6:10–17

Antonopoulou S (2017) BIM for heritage: developing a historic building information model

L-P Archaeology - ARK. An open source solution to project recording. http://ark.lparchaeology.com/

ArkéoGIS. http://arkeogis.org/fr/arkeogis-de-lhypothese-a-la-carte/

Belussi A, Migliorini S (2014) A framework for managing temporal dimensions in archaeological data. In: Cesta A et al (eds) Proceedings of the 2014 21st international symposium on temporal representation and reasoning. IEEE, pp 81–90

Belussi A, Migliorini S (2017) A spatio-temporal framework for managing archeological data. Ann Math Artif Intell 80:175–218

Bernard L et al (2015) ArkeoGIS, merging geographical and archaeological datas online. In: Giligny F et al (eds) CAA2014, 21st century archaeology: concepts, methods and tools: proceedings of the 42nd annual conference on computer applications and quantitative methods in archaeology. Archaeopress, Oxford, pp 401–406

Carlisle PK et al (2014) The arches heritage inventory and management system: a standards-based approach to the management of cultural heritage information. In: CIDOC (International Committee for Documentation of the International Council of Museums) conference: access and understanding–networking in the digital era, Dresden, Germany, pp 6–11

Chair of Geoinformatics, Technische Universität München: The CityGML Database. 3D City DB. http://www.3dcitydb.org/3dcitydb/3dcitydbhomepage/

Chiabrando F et al (2016) Historical buildings models and their handling via 3D survey: from points clouds to user-oriented HBIM. ISPRS—Int Arch Photogramm Remote Sens Spat Inf Sci XLI-B5:633–640

Costamagna E, Spanò A (2013) CityGML for architectural heritage. In: Abdul Rahman A et al (eds) Developments in multidimensional spatial data models. Springer, Berlin, pp 219–237

Cripps P (2013) Places, people, events and stuff; building blocks for archaeological information systems. In: Archaeology in the digital era, vol II. e-Papers from the 40th annual conference of computer applications and quantitative methods in archaeology (CAA), Southampton, 26–29 March 2012. Amsterdam University Press, Amsterdam, pp 487–497

De Roo B et al (2013a) On the way to a 4D archaeological GIS: state of the art, future directions and need for standardization. In: Proceedings of the 2013 digital heritage international congress, pp 617–620

De Roo B et al (2013b) The temporal dimension in a 4D archaeological data model: applicability of the geoinformation standard. ISPRS Ann Photogramm Remote Sens Spat Inf Sci II-2/W1, W1:111–121

De Roo B et al (2014) Bridging archaeology and GIS: Influencing factors for a 4D archaeological GIS. Lect Notes Comput Sci Subser Lect Notes Artif Intell Lect Notes Bioinforma 8740:186–195

De Runz C (2008) Imperfection, temps et espace : modélisation, analyse et visualisation dans un SIG archéologique. University of Reims Champagne-Ardenne

Finat J et al (2010) GIRAPIM. A 3D information system for surveying cultural heritage environments. ISPRS—Int Arch Photogramm Remote Sens Spat Inf Sci XXXVIII-4, W15:107–113

German Archaeological Institute, GIScience Research Group: MayaArch3D—3D Visulisation and Analysis of Maya Archaeology. http://www.mayaarch3d.org/language/en/sample-page/

Houshiar H. et al (2015) CASTLE3D—a computer aided system for labelling archaeological excavations in 3D. ISPRS Ann Photogramm Remote Sens Spat Inf Sci II-5/W3:111–118

International council of museums, International committee for documentation: CIDOC CRM. http://www.cidoc-crm.org/

J. Paul Getty Trust. World monuments fund: arches project cultural heritage inventory and management software. http://archesproject.org/

Katsianis M et al (2008) A 3D digital workflow for archaeological intra-site research using GIS. J Archaeol Sci 35(3):655–667

Luczfalvy Jancsó A et al (2016) CASTLE4D: an archaeological information system based on 3D point clouds. In: Lerma JL, Cabrelles M (eds) Proceedings of the ARQUEOLÓGICA 2.0 8th international congress on archaeology, computer graphics, cultural heritage and innovation. Universitat Politècnica València, Valencia, pp 247–252

Myers D et al (2016) The Arches heritage inventory and management system: a platform for the heritage field. J Cult Heritage Manage Sustain Dev 6(2):213–224

Nuechter A et al (2015) Das Castle3D Framework zur fortlaufenden semantischen 3D-Kartierung von archäologischen Ausgrabungsstätten. Allg Vermess-Nachrichten 06(07):233–247

Sanders DH (2011) Enabling archaeological hypothesis testing in real time using the REVEAL documentation and display system. Virtual Archaeol Rev 2(4):89

The Institute for the Visualization of History: REVEAL—reconstruction and exploratory visualization: engineering meets archaeology. http://www.vizin.org/projects/reveal/setting.html

Considering Rich Spatiotemporal Relationships in Cultural Heritage Information Management

Pierre Hallot

Abstract This paper describes the use of rich spatiotemporal relationships in cultural heritage information modelling in order to increase the information organization and extraction. The idea is to tailor the spatiotemporal state of identity model, i.e. a model based on spatiotemporal identity that takes into account relationships between non-existing and non-presents objects, to the management cultural heritage information. In doing so, we propose an enhancement of the knowledge representation for cultural heritage.

Keywords Spatiotemporal relationships · Cultural heritage information · Life and motion configuration · Identity · Existence · Presence

1 Introduction

Cultural Heritage information management is impacted by the evolution of the acquisition techniques, mainly by the arising of the laser scanner technology (Yastikli 2007). The digital transformation conduces to an increase of the acquisition performance, the completeness of acquired information and the evolution of the represented information acquired from such technologies. Managing digital heritage data requires systems that are able to integrate the complexity of cultural heritage information, i.e. managing the several levels of temporality involved in the life-cycle of a represented element (Meyer et al. 2007). Geographical information science is, among others, attached to the definition of theoretical concepts that are implemented in information management systems. Following the same approach, the theoretical definition on relationships leads to a refinement of concepts closely tied to space and time that the first step for semantic enrichment of cultural heritage information.

For years now, the development of spatiotemporal relationships models was predominantly dedicated to the management of geospatial information (Stock

P. Hallot (✉)
Faculty of Architecture, LNA-DIVA, Liège Université, Liège, Belgium
e-mail: p.hallot@ulg.ac.be

© Springer International Publishing AG 2018
P. Fogliaroni et al. (eds.), *Proceedings of Workshops and Posters at the 13th International Conference on Spatial Information Theory (COSIT 2017)*, Lecture Notes in Geoinformation and Cartography, https://doi.org/10.1007/978-3-319-63946-8_51

1997). Although, cultural heritage information owns a spatial component, the main difference resides in the temporal dimensions that is a fundamental descriptor for such kind of information. The description of the evolution of an entity is based on the study of the succession of spatial and temporal states that symbolize the life cycle of the object.

Moreover, the accessibility to the information is rather limited or incomplete. Accessing to a cultural heritage information is performed through one of the following ways:

- Identity: the identity of represented cultural heritage element is known. There is no ambiguity on its identification. The information can therefore be structured around object's identity. E.g. the pyramids of Giza in Egypt.
- Physical realization: the cultural heritage element is only known by a physical structure visible in the real world. The spatiality can be represented but is not necessarily linked to a defined or unique identity. The link between the physical observed elements and the object identity is not sure. E.g. an unidentified ruin.
- Spatiality: the spatiality of the object is known without manifestation in the real world. Information related to the geometry of the element are known but there is no physical manifestation of the element in the real world. The spatial position can be known or not. E.g. the plan of La Bastille in France.

Each of the information accessibility way faces to an uncertainty. Indeed, information related to cultural heritage objects results of sources interpretation performed by a operator who has a vision of the world that relying on its experience. Several interpretations or evolution scenario are frequent in cultural heritage information management. Developed information systems have to consider this ambiguity to fully encompass the cultural heritage information.

In a previous research (Hallot et al. 2015; Hallot and Billen 2016), we proposed a spatiotemporal states (STS) model that defines spatial and temporal relationships between (geographical) objects in considering their whole range of possible existence states. The commonly used spatial relationships only deals with two objects that exists at the time of the analysis. Starting from a description of the possible states of an object in regards of its identity, spatiality and presence, we defined a set of possible states of existence and presence for a geographical object. This classification is applied to several object to deduce extended spatiotemporal relationships. For example, the STS model proposes anachronistic relationships between an object that exist with another object that does not exist physically anymore.

In this paper, we throw the bases for an application of the STS model to cultural heritage information management. Our research hypothesis is that considering refined spatiotemporal relationships between heritage objects ensure a better description of the complex cultural heritage information. We postulate that spatiotemporal relationships based on the STS model will carry more semantic for the relationships description.

The remaining of the paper is structured as follows. First, we remind the concept of spatiotemporal states of existence and presence. Then we expose the main issues in cultural heritage information management. The second part shows the adaptation of the STS model to deals with cultural heritage information in considering object identity as a starting point. This section symbolizes the life cycle of cultural heritage object in the regard of the evolution of its identity. Finally, we introduce the concept of relationship observer that refines the proposed relationships by considering the temporal point of view of a relationship between objects. Then we conclude.

2　Spatiotemporal States of Identity

The identity of an object is the property intrinsic to each object which allows it to be differentiated from all others (Khoshafian and Copeland 1986). The complete definition of a geographical object involves the definition of its identity and spatiality, which can vary over time. Research focusing on the identity of geographical objects (Hornsby and Egenhofer 2000) focuses on the concept of a geographical object and its modeling as a unique object. Campos et al. (2003) proposes a set of relationships and their associated semantics describing the representation of geographical objects in a virtual environment. Although this approach can be linked to cultural heritage information management, the model does not take into account the possibility to establish relationships between an object present at the moment of the analysis as against an object that existed in the past or an event existing in the future. Spatiotemporal relationships such as that proposed by Claramunt (Jiang 2001; Theriault 1996) define a set of spatiotemporal relationships that integrate topological reasoning with the temporal logic. These models describe, by means of qualitative operators, the relationships that exist between moving objects which are only existing and visible during the analysis.

　　The spatiotemporal states of identity model (STS-I) has been initially developed to fully encompass the complexity of geographical information (Hallot et al. 2015; Hallot and Billen 2016). Indeed, there exist a lot of situation in the geographical domain where the spatiality of a geographical object cannot be acquired, at least for some period of time. In order to propose a continuous analysis of the geographical object over time, the STS-i model has been developed focusing on the objects identity. Several spatiotemporal states have been outlined in regards of spatial, temporal and identity vision. The Fig. 1 shows the organization and the transition between spatiotemporal states. Since it is not possible to model every objects that exists or that have been existing over time, represented objects are limited to the ones that are pointed by a relationship. The existence of an object begins as soon as a semantic or spatial relationship is established for this object and that it ceases to exist when there are no more relationships with it. Once the object identity is proposed, the object switch from the "non-existent" state to the "existent" state. This state does not define a physical realization or a spatial design of the object. The materialization operation provides an extension in the physical world for the object. At that moment, the object owns an identity, a spatiality and a physical representation. We qualify this state of "present". When the physical realization of the object

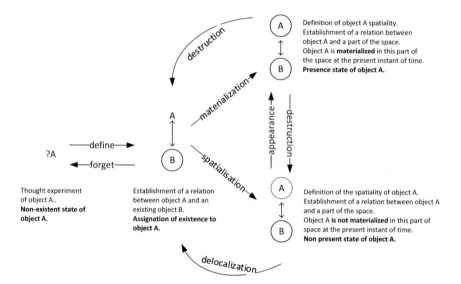

Fig. 1 Conceptual schematization of the different spatiotemporal states of an object A as defined in (Hallot and Billen 2016). The *arrows* represent the transition events between these different states. The destruction corresponds to an omission of the location of the object; the delocalization marks the loss of the realization in the physical space without omission of its position. The destruction symbolize the loss of spatial extension, which is either with loss of spatiality or not

is destroyed or if the spatiality is planned but not yet materialized, the object is in a non-present state. This state gathers either the objects that are destroyed with a souvenir of their spatiality or the objects for which spatiality is planned but not yet visible in the physical world. These different cases do not entirely correspond to a lack of information or vague information because the identity of the object very often remains known. Loss of information about the temporary spatial extension does not destroy the existence of the geographical object.

When combined together in regards of two objects, the succession of spatiotemporal states of identity of two objects defined the Life and Motion Configuration (LMC). The LMC is an iconic language that symbolize the evolution of spatiotemporal relationships between two objects considering their relationship at each temporal succession.

3   Application of STS-I to Cultural Heritage Objects

The application of STS-I model to cultural heritage objects requires some adaptations to fit to the complexity of cultural heritage concepts as described in the introduction. When dealing with geographical information, the succession of spatiotemporal states of identity is largely linear, starting from non-existence,

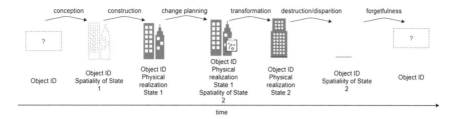

Fig. 2 Conceptual schematization of successive spatiotemporal states of a heritage building. The figure shows, for each state, the knowledge on the identity, the spatiality and the physical realization of the object. The *arrows* represent the transition events between these different states

existence, non-presence and presence. The destruction scheme follows the inverse succession. However, building archeology or conservation/restauration research either leads to the discovery of unidentified physical fragments or faces to sources asserting the identity and/or the spatiality of objects without physical realization. The lifecycle of cultural heritage buildings is complex due to the multiple transformation, affectation change or the ageing due to time. The representation of cultural heritage information with the prism of spatiotemporal states of identify gives a formal classification of every states of the historical elements. Figure 2 shows the evolution of an object for which the historical knowledge is complete, this means that each of the transformation, or planned transformation is documented and can be established by undoubtable sources. The first step in time represents the identity of the object. At that time, there is not yet a definition of any spatial information of the element. The second step is obtained after a stage of architectural conception, which is attached to the pacification, and the definition of the spatial extension of each part of the building. The construction phase leads to the first state of the edifice. Most of spatial and temporal reasoning starts at that step in the object definition. Long-term evolution representation implies to consider multiple changes that occurs to modelled objects. In cultural heritage information modelling, each historical sources provides information on physical or thematic changes that affects modelled object. When planning a transformation, there is a concurrent existence between the physical realization and the spatial definition of the next state. Once the transformation is done, the new physical realization appears and the object switch from state 1 to state 2. At the end of the lifecycle, the physical realization of the edifice disappears; the spatiality continues to exist for a while. The heritage building is qualified of non-present. At the ends, when no information on the former spatiality of the edifice remains, e.g. the plans are lost, only the identity of the elements continues to exist.

The organization of heritage information around the spatiotemporal states allows sketching what information was available at a fixed time in history. Moreover, it refines the traditional spatiotemporal representation in allowing the concurrence of what is visible in the physical world with what is planned for the future. In the same way, each previous states can be stored and represented independently of their temporality, i.e. in proposing a view of all states of the heritage edifice evolution.

When considering incomplete knowledge on the object evolution, the information acquisition is done through sources that have to be interpreted (Ruymbeke et al. 2014). The representation of incomplete knowledge and fussy knowledge remain an open issue (Belussi and Migliorini 2017; de Runz and Desjardin 2010). The spatiotemporal states are then subjective results that can leads to several concurrent interpretation scenarios. By recording every planned transformation, even if they are not all implemented, we allow to store and organize comprehensive knowledge on cultural heritage elements. Planned others spatiality's is of a high importance for researchers to fully understand the lifecycle and the history of the heritage buildings and to allow a classification to every sources related to build heritage. There is actually no degrees of confidence related to the interpretation of the spatio-temporal states. We plan to consider this perspective in a future research.

4 Anachronistic Relationships for Cultural Heritage Objects

The transformation or the restauration of the built heritage has to take into account the history of the building and surrounding environment it works on. More specifically, if the built heritage has been transformed or partially destroyed, the vision of the former space has to be considered during the conception of the architectural project. For example, the restoration of a heritage building which is, among others, composed of an empty space at the present time has to integrate the former role and the former geometry of what that space was. The proposed project will have to consider the past state, usually by proposing a symbolization of that former spatiotemporal state. In doing so, we draw a spatiotemporal relationship between built heritage at different period of time. This is what we define as an anachronistic relationship, i.e. the spatiotemporal relationships between objects states that were not existing in during a concurrent period of time.

Going deeper in the refinement, we assume cases where the anachronistic relationships points on the object spatiality, the object physical extension or only to the object identity. It is important to propose information management systems that allows to store and describe such kind of relationship because of their fundamental aspects to understand architectural planning. Figure 3 represents two kinds of anachronistic spatiotemporal relationships. The blue area shows an architectural project which is still in conception phase.

This means that the identity of the project is known, the spatiality is being designed but the project does not have an extension in the physical world. This object is in relation with a heritage building (represented by a laser scanner acquisition in the figure). The second part of the Fig. 3 shows some elements (the doors) that do not have a physical extension but only a spatiality at the present time. Their representation gives an interpretation of the relationships of their actual spatiotemporal state (non-present) with the building (present). Such kind of

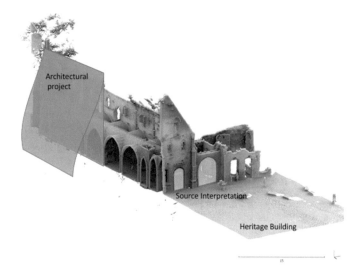

Fig. 3 Anachronistic representation of several states on a heritage building representation. The architectural project is a planned spatiality of the building evolution. The source interpretation is an anachronistic representation of past-existing doors of the building added a contemporary representation of its actual shape

representations helps at understanding the relationship that was existing between the two elements at the time when they were both present (i.e. with a spatial extension).

The previous examples show that a spatiotemporal relationship are used to conceptualize the relations that exists at every time between buildings elements. This helps first at organizing the historical sources or their interpretation, and secondly to understand and plan future architectural projects which has to consider the historicity of the context. These two approach lead to think at the targeted goals of establishing such spatiotemporal relations. The role of the operator is significant in order to define at what time the relationship is observed and consequently what will be the use of the relationships. Let assume that two objects A and B share a spatiotemporal relationship, which remain stable over time (the objects does not change of state). The relation can be observed in three ways:

- as it was in the past: this view of the relation is used each time when an historical study is performed. We try to retrieve what was the relationships in the past.
- as it is now: this view of the relation corresponds to an actual observation of the spatial organization of past elements. For example, the relation is used to explain the actual spatial organization of a former medieval city, which is influenced by the link that exist between defense walls and the surrounding houses.

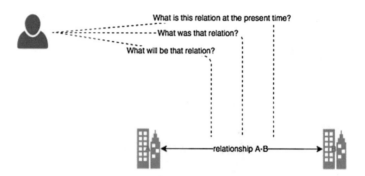

Fig. 4 Temporal interpretation of spatiotemporal relationships depending on the observer. The relationships between objects can be studied as it was in past (at the time objects were both present), as it is actually (what does the relationships implies in the present), as it will be in the future (how the relationship will be considered in the future)

- as it will be for the future: this view of the relation is used in planning when transforming one of the object participating to the relation. It will affect the spatiotemporal relation (Fig. 4).

Anew, the conceptual refinement of a relationship by considering the temporality of the observation helps at organizing the cultural heritage information and leads to a better sources interpretation. This helps at understand the spatial choices that are continuously taken in architectural design.

5 Conclusion

The cultural heritage information management has to consider multiple historical sources. Their interpretation conduces to different scenario that can sometimes be contradictory. To fully understand the lifecycle of built heritage, relationships between actual and former states of elements has to be drawn. In this paper, we proposed to adapt a spatiotemporal relationships model based on object identity to test if it could fit to cultural heritage information modeling. This first conceptual step aims at describe the general concept where the STS-I model can be applied and the uses that can be done in this context. The first results are promising and should be now defined formally before being implemented and tested. This next research step will be considered soon with real cases examples on the "Hotel Rigo" in Liège, Belgium. This case is highly interesting since is should be destroyed in favor of an urban development.

References

Belussi A, Migliorini S (2017) A spatio-temporal framework for managing archeological data. Ann Math Artif Intell 80(3):175–218

Campos J et al (2003) A temporal model of virtual reality objects and their semantics. J Vis Lang Comput 14(5):469–492

Claramunt C, Jiang B (2001) An integrated representation of spatial and temporal relationships between evolving regions. J Geogr Syst 3(4):411–428

Claramunt C, Theriault, M.: (1996) Toward semantics for modelling spatio-temporal processes within GIS. Adv GIs Res I pp 27−43

Hallot P et al (2015) Les états spatiotemporels d'existence et de présence. Int J Geomatics Spat Anal 25(2):173–196

Hallot P, Billen R (2016) Enhancing spatio-temporal identity: states of existence and presence

Hornsby K, Egenhofer MJ (2000) Identity-based change: a foundation for spatio-temporal knowledge representation. Int J Geogr Inf Sci 14(3):207–224

Khoshafian SN, Copeland GP (1986) Object identity. SIGPLAN Not 21(11):406–416

Meyer É et al (2007) A web information system for the management and the dissemination of cultural heritage data. Cult, Herit

de Runz C, Desjardin E (2010) Imperfect spatiotemporal information analysis in a GIS: application to arch{æ}ological information completion hypothesis. In: Jeansoulin R et al (eds) Methods for handling imperfect spatial information. Springer, Heidelberg, pp 341–356

Van Ruymbeke M et al (2015) Towards an archaeological information system: improving the core data model. In: Giligny F et al (eds) CAA 2014 21st century Archaeology : Concepts methods and tools : Proceedings of the 42nd annual conference on computer applications and quantitative methods in archaeology, Archaeopress, Paris, France pp 245–253

Stock O (1997) Spatial and temporal reasoning. Kluwer Academic Publishers, Dordrecht, Boston

Yastikli N (2007) Documentation of cultural heritage using digital photogrammetry and laser scanning. J Cult, Herit

Towards the Extraction of Semantics from Incomplete Archaeological Records

S. Migliorini and P. Grossi

Abstract The Archaeological Geographical Information System of Verona (SITAVR) has been developed since 2011, starting from an existing and well-consolidated system, the Archaeological Geographical Information System of Rome (SITAR). The main objective of both projects is collecting information about the archaeological findings regarding the two Italian urban centres with the aim to support a complete archaeological analysis and allow for easy data reuse and dissemination. Data collected in both projects come not only from current excavation campaigns, but they have also been extracted from very old documents describing excavations performed during the 18th century. In this last case, many details about the location and the time of existence of findings are not known. The SITAVR project has lead to the definition of a domain model for archaeological data, called $\mathscr{S}tar$, and now it is focused on the mapping between this model and other international standards, such as the CIDOC CRM (2016) and its extension CIDOC-CRM$_{archeo}$ (Felicetti et al. 2016). This paper presents a general approach for guiding the mapping of a dataset regarding ancient archaeological data towards CIDOC-CRM$_{archeo}$, especially for what concerns to spatial and temporal properties that are only partially specified.

Keywords Incomplete archaeological data · Spatio-temporal archaeological data · Ontology · CIDOC-CRM · Semantic mapping

S. Migliorini (✉)
Department of Computer Science, University of Verona, Strada le Grazie 15,
37134 Verona, Italy
e-mail: sara.migliorini@univr.it

P. Grossi
Department of Culture and Civilization, University of Verona,
Viale Dell'Università 4, 37129 Verona, Italy
e-mail: piergiovanna.grossi@univr.it

© Springer International Publishing AG 2018
P. Fogliaroni et al. (eds.), *Proceedings of Workshops and Posters at the 13th International Conference on Spatial Information Theory (COSIT 2017)*, Lecture Notes in Geoinformation and Cartography, https://doi.org/10.1007/978-3-319-63946-8_52

1 Introduction

In the city of Verona very well preserved monuments can be found; some of them are still used nowadays for exhibitions and events (e.g., the Roman theatre and the amphitheater). For this reason and for a distinctive continuity of life from pre-Roman age until today, the city has been listed since November 2000 in the UNESCO World Heritage List. Its archaeological heritage has been subjected to many and important cataloguing and studying initiatives in the past years, for example the archaeological map of Lanfranco Franzoni or other works focusing on specific time periods or sectors of the city, for instance the documents of Cipolla (1880). All these works led anyway only to paper-based artifacts that, for their own nature, cannot be easily processed and are characterized by an intrinsic uncertainty regarding space and time location. In this paper the term uncertainty is used in its wide sense to denote a imprecise, incomplete or not totally reliable information.

Given the need to revise, digitalize and publish online these archaeological data, the SITAVR project started its activities in 2011. The first activity of this project was the definition of a common data model for representing archaeological data, called $\mathscr{S}tar$ (Spatio-Temporal ARchaeological model) (Belussi and Migliorini 2014, 2017; Migliorini et al. forthcoming). Based on this conceptual model, some web applications and analysis tools have been developed in order to collect, share, disseminate and encourage the study of archaeological information (Migliorini et al. forthcoming; Basso et al. 2013, 2016; Bruno et al. 2015). Currently, the SITAVR database contains about 330 information sources (referring mainly to excavation of the last 50 years, but also to ancient documents about very old excavation campaigns) and 720 archaeological evidences that describe row archaeological findings. In the last years a pilot activity has been started with the support of the SITAR project (Serlorenzi and Ainis 2011) and in cooperation with the ARIADNE European infrastructure, for mapping the $\mathscr{S}tart$ data model towards CIDOC Conceptual Reference Model (CIDOC-CRM) (2016) and its extension $CRM_{archaeo}$ (Felicetti et al. 2016). Notice that the mapping has been performed considering only the $CRM_{archaeo}$ extension, whose purpose is to support the archaeological excavation process and which takes advantages from the concepts defined in CRM_{sci}, since they are those involved in the ARIADNE project. Conversely, other extensions such as CRM_{geo} and CRM_{inf} were out the scope of the project, while CRM_{ba} has not been considered since it regards only ancient buildings and not the representation of general findings and excavation processes.

During this activity three main issues have emerged: (i) the standard CIDOC-CRM is composed of an ontology for the description of data, it consists of a hierarchy of classes with the root superclass E1 CRM Entity and several different approaches for mapping the source database are admissible; however, not all the applicable mappings preserve the same amount of information or allow an effective reasoning process on the mapped data. (ii) Space and time dimension can be represented in CIDOC-CRM, but, as cited above, in different ways; in particular, as regards to the space aspect when it is represented as a geometry in WKT or GML, it cannot be specifically recognized by the ontology. (iii) The accuracy of a temporal instant or of a location or shape on the Earth surface cannot be explicitly specified in the ontology. The accuracy of spatial and temporal information is particularly relevant when partial and imprecise information is collected from bibliography describing ancient excavation campaigns.

The mapping towards standards like the CIDOC-CRM is an important step in the SITAVR project. The main intent of SITAVR was to leverage on the experience of SITAR for both data and application design, in order to allow interoperability between these twin systems and between them and other repositories of archaeological data. The final overall goals of these projects are: supporting full archaeological analysis of the data, data dissemination, and data reuse in different contexts, in Italy or abroad, for urban planning, safeguard, preservation, etc. The work presented in this paper aims to describe the lesson learned during the activity concerning the mapping of the $\mathscr{S}tar$ conceptual schema described in UML towards CIDOC-CRM, especially for what regards space and time dimension and their accuracy.

2 Related Work

The need for a spatial-temporal model suitable for describing archaeological data is widely recognized in literature. In Wheatley and Gillings (2002) the authors evaluated the applicability of GIS and its related spatial technologies in the archeological context. Even if such technologies can have a powerful role in archaeological analysis and interpretation, there is also the need to incorporate the temporal dimension in data representation. Likewise, in De Roo et al. (2014) the authors proposed the idea to develop a comprehensive 4D GIS tailored to archaeology, where the fourth dimension is the temporal one. Such system was centered on the ability to analyse data for archaeological investigations.

In Belussi and Migliorini (2014) we proposed an extension of the ISO Standard 19108 with fuzzy constructs, in order to incorporate the inherent uncertainty of archaeological time. Moreover, we investigated the applicability of currently available automatic reasoning techniques to derive new temporal knowledge or to reduce the uncertainty of some dates, and in general to guide the dating process. In Belussi and Migliorini (2017) we extended such model to derive new temporal knowledge starting from available spatial and stratigraphic information. Finally, in Migliorini et al. (forthcoming) we explored the problem of developing interoperable systems for archaeological data based on the proposed $\mathscr{S}tart$ model, in order to promote the sharing, dissemination and analysis of archaeological data. However, in order to really promote interoperability between different archaeological agencies, it is necessary to refer to widely recognized standards such as CIDOC-CRM. Therefore, in this paper we perform a step forward, by considering the mapping of the $\mathscr{S}tar$ model towards CIDOC-CRM and its extension CIDOC-CRM$_{archaeo}$.

In Binding et al. (2008) the authors proposed an attempt to improve interoperability in the archaeological domain through a mapping towards CIDOC-CRM. The main aim of the work was to demonstrate the potential benefits in cross searching data expressed as RDF and conforming to a common conceptual data structure schema. The authors defined some extensions to the core CIDOC-CRM classes in order to correctly represent archaeological data. Conversely, in this work we try to use only the core classes and its official extension CIDOC-CRM$_{archaeo}$. Moreover, our attention concentrates on the correct representation of spatial and temporal aspects even when such information is imprecise, incomplete or not totally reliable.

3 Mapping Approaches

Modeling languages and tools. As already mentioned above, since its inception our project has always put an emphasis on interoperability and data sharing. This goal proceeds on two parallel tracks: on one side, the necessary IT technologies, and on the other side the definition of a domain model for archaeology in order to connect, relate and standardize contents coming from different conceptual frameworks. For instance, the term "digging" has different meanings, depending if it indicates an extensive excavation or a small sampling of an area; similarly, "segment of matter" may refer to completely different data attributes.

For the definition of the conceptual schema specification, we have adopted the UML as formal language, and in particular the constructs that allow one to define class diagrams. The using of the UML approach for describing data at conceptual level, namely independently from any specific technology, is common in many international Standards (e.g., ISO Standards, like ISO/TC 211 Geographic Information/Geomatics). More specifically, in this project we have adopted the tools of the GeoUML methodology. GeoUML (Belussi et al. 2006; Pelagatti et al. 2009) is an extension of UML which includes some predefined constructs for dealing with the geometric representation of the spatial properties of data. Specifically, it helps the data designer to incorporate in the class diagram concepts that have been defined using the constructs of the ISO/TC211 Standards, mainly ISO 19107 and ISO 19109. The GeoUML methodology includes a series of tools which support both the definition of conceptual schemas, the mapping towards different technologies, producing the corresponding physical schemas (e.g., SQL for PostgreSQL or Oracle, XML schema, ShapeFile, etc.) and the validation of spatial constraints defined at conceptual level.

This paper describes the experience performed during the mapping of the $\mathscr{S}tar$ model, formalized using the GeoUML Catalogue, towards the Standard CIDOC-CRM. More specifically, we targeted the mapping of the conceptual schema to the general classes of CIDOC-CRM (2016) and to its extension CIDOC-CRM$_{archeo}$ (Felicetti et al. 2016), and we exploited the Art & Architecture Thesaurus of Getty Research Institute (Getty Res. Inst. www), through Simple Knowledge Organization System (SKOS) (SKOS www), to map terminology.

Mapping techniques. In order to describe the different possible mapping techniques, the mapping process can be formalized as follows.

Definition 1 (*Conceptual schema*) Let $\mathscr{S}$ be a conceptual schema specified as a UML class diagram. $\mathscr{S}$ contains a set of classes, $\mathscr{S} = \{C_1, \ldots, C_n\}$, and each class $C_i \in \mathscr{S}$ has a set of properties represented as attributes of the class $C_i.attributes = \{a_1, \ldots, a_m\}$ or as roles toward other classes $C_i.roles = \{r_1, \ldots, r_p\}$. Each attribute has a domain and each role has a target class belonging to the schema.

Similarly, the target CIDOC-CRM ontology can be formally described as follows as a set of reference classes.

Definition 2 (*Target ontology*) Let $\mathscr{O}$ the target ontology. It contains a set of reference classes $\mathscr{O} = \{RC_1, \ldots, RC_n\}$ organized in a hierarchy so that each class can have one or more super-classes and one or more sub-classes. Each reference class has some properties $RC_i.properties = \{p_1, \ldots, p_q\}$, where each property refers to a target class or a basic domain.

A mapping between a conceptual schema and a target ontology can be defined by using one of the following techniques: (1) *Bottom-up* (from attributes (roles) to properties): given the attributes (or roles) of a class in $\mathscr{S}$, the mapping consists in specifying for each attribute (role) the corresponding property, if exists, belonging to a reference class of the ontology. After this initial mapping at the bottom level, we can derive the mapping between classes by applying the following derivation rule: a class $C_i \in \mathscr{S}$ is mapped into a reference class $RC_j \in \mathscr{O}$, if at least on property of C_i has been mapped to a property of RC_j. Notice that in this test the properties inherited by RC_j are considered as local properties. (2) *Top-down* (from classes to reference classes): in this case, it is firstly defined the mapping between the classes in $\mathscr{S}$ and the reference classes in $\mathscr{O}$, then given a pair (C_i, RC_j) the mapping between attributes (roles) of C_i and properties of RC_j is defined.

Following the first technique might seem to be easier than the second one. However, when we deal with the mapping towards an ontology this approach can lead to incorrect mappings. For example, let us consider a class $C_{person} \in \mathscr{S}$ having an attribute a_{name}. Then, if we apply the *Bottom-up* technique, we find in the ontology of CIDOC-CRM the corresponding property p_{p_1} "*is identified by*" of the reference class RC_{E1} "*E1 CRM Entity*". This attribute is inherited by all classes of the ontology, thus choosing the mapping $a_{name} \rightarrow p_{p_1}$, the class C_{person} may be mapped to every class of $\mathscr{O}$, which is not correct. Clearly, by considering attributes that are more specific, the mapping of classes can be more effective. Regarding the second technique, starting with the mapping between classes can avoid the situations described above, but again the choice of the target class should be addressed toward more specialized classes. For instance, considering again the above example, the class $C_{person} \in \mathscr{S}$ might be again mapped to the target class $RC_{E1} \in \mathscr{O}$, but a better mapping should have chosen the reference class RC_{E21} "*E21 Person*".

Finally, independently from the applied technique, in some cases there could be properties in a source class that have no mapping in the target ontology, since there is no target properties that can match with it directly. However, these situations can be overcome, indeed, there could be in the ontology a possible way for representing the same information by means of some intermediate concepts.

These issues appear in particular when the space and time dimensions have to be represented in the data that are subject to the mapping process. We deal with this problem during the SITAVR project and we propose the mapping techniques presented in the following section.

4   Space and Time Dimensions

When dealing with archaeological data, the investigation always need to collect evidences about the space and time dimensions regarding a specific finding. Starting from row data it is necessary to define with the maximal accuracy the interval of time the finding dates back, but also the location of the finding, that could be different from the place it was found in, or the shape it had in the ancient times.

Sometimes this kind of information is hidden in the archaeological records where: (i) the accuracy of time span is represented implicitly by means of intervals; (ii) the spatial properties contained in the records are not clearly referred to a period of time (i.e., the specified location is the place where it is guarded or where it has been found?); (iii) finally, the spatial accuracy is frequently not specified at all.

In the SITAVR project (Migliorini et al. forthcoming; Basso et al. 2016; Bruno et al. 2015) this kind of information and meta-information has been explicitly defined inside the conceptual model (Belussi and Migliorini 2014). The resulting model, called $\mathscr{S}tar$, has been presented in Belussi and Migliorini (2014), Belussi and Migliorini (2017), Migliorini et al. (forthcoming). The $\mathscr{S}tar$ model is based on three main concepts: Information Source (IS), Archaeological Partition (AP) and Archaeological Unit (AU). For the purposes of this paper, we refer to the first two.

An IS represents an activity leading to the knowledge about a specific archaeological context. Therefore, an IS can describe an excavation, a field survey, a bibliographic research, and so on. Through a set of related concepts, details are provided such as the type of research being performed, the place and time it was conducted, the related documentation, some geo-positioning data, etc. Conversely, an AP describes a piece of knowledge limited in time and space. Therefore, an AP represents a segment of matter, a physical material in a relative stability of form (substance) within a specific space-time volume. An AP can then be a structure, in a more or less complete form, a movable element, a stratigraphic or a geological substrate, which is important in an historical perspective. Through a set of related concepts, details are provided such as the place and time of the finding, its dating and measurements, related documents such as pictures, maps, etc.

The following approaches have been defined for representing spatial properties and their accuracy.

Geometric attributes containing vector data: The attributes describing the location of an instance of IS or AP class, have been mapped to CIDOC-CRM as

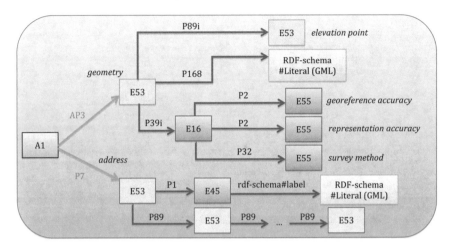

Fig. 1 The mapping of the spatial properties. The identifiers used in the figures are taken from the CIDOC-CRM$_{archaeo}$ ontology

described in Fig. 1. Notice that: (i) a geometric attribute can be mapped into an instance of the *E53 Place* class with a property *P168 place is defined by*. This property has as target value the GML representation of the instance location specified as *rdf-schema#Literal*. (ii) The "geo-reference accuracy" of this geometric value can be mapped to an instance of the *E55 Type* class, since the geo-reference accuracy is a vocabulary in *𝒮tar* model, with values like: exact, approximate, derived. The same approach has been followed for the "representation accuracy", which is described by a *𝒮tar* vocabulary, with values like: detailed, schematic. Both *E55 Type* instances are linked using an intermediate class *E16 Measurement* by means of the property *P2 has type*. *E16 Measurement* is connected to *E53 Place* by means of the property *P39 measured (was measured by)*. Notice that the use of property *P2* is a solution that allows one to include any characteristic of an information instance but weakens its semantics: indeed, the *E55* instances used for representing the accuracy of the measurement cannot be recognized as such by any reasoning tool. (iii) The "survey method" is also mapped to an instance of *E55 Type* class, through the intermediate class *E16 Measurement*; also the survey method is a *𝒮tart* vocabulary, with values like: based on vector cadastral maps, based on aerial photogrammetry. (iv) Finally, only for geometries of AP, property *P89i contains* is specified by linking the elevation points that are spatially contained in the represented geometry. Each elevation point is mapped to other instances of *E53 Place*.

Geometric attributes containing addresses. The attribute describing an address can be represented directly as an instance of the *E53 Place* class and using the property *P168 as above* for storing the complete address as *rdf-schema#Literal*. When also a hierarchy of administrative units is available, the property *P89 falls within* can be used to indicate that the address is contained in a municipality and the municipality is in a province and so on.

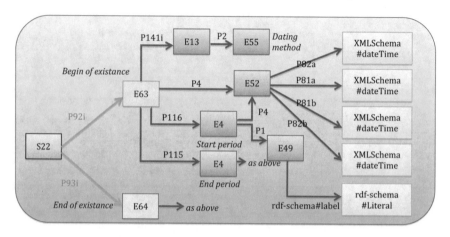

Fig. 2 The mapping of the temporal properties. The identifiers used in the figures are taken from the CIDOC-CRM$_{archaeo}$ ontology

Similarly, the following approaches have been defined for representing temporal properties and their uncertainty. The proposed mapping is shown in Fig. 2 describing the mapping of the begin and end of existence properties of a *S22 Segment of Matter*. In particular, an intermediate instance of *E63 Begin of existence* and of *E64 End of existence* are introduced, which represent the corresponding events, and then starting from *E63* (*E64*) the temporal dimension can be specified as follows.

Temporal attributes as intervals. In this case an instance of *E52 TimeSpan* is used. The temporal interval of the *E52* is specified through the properties *P82a begin of the begin*, *P81a end of the begin* and *P81b begin of the end*, *P82b end of the end* that allow one to define the start and end instant of an interval with some uncertainty, expressed by means of a specific range for the start instant (*P82a*, *P81a*) and for the end instant (*P81b*, *P82b*). Notice that, these properties are an extension proposed inside the ARIADNE project to the standard properties of *E2 Time Span*.

Temporal attributes as periods. If the temporal attribute is specified by means of a start period and an end period, it is mapped using two instances of *E4 Period*. This class has a property *P4 has time-span* towards the class *E52 TimeSpan*, so that the uncertainty for a period described by an *E4* instance can be expressed using the previous mechanism. Moreover, *E4* has a property *P1* towards *E49 Time appellation*, for explicitly assign a period name. Notice that, classes *E63* and *E64* have been connected to an intermediate class *E13 Attribute assignment*, through the property *P140 assigned attribute to (was attributed by)*. This intermediate class allows one to represent the dating method using the property *P2* which links to class *E55*. Again this information cannot be recognized as a dating methods by a reasoning tools.

5 Conclusions

The SITAVR project has collected a considerable amount of archaeological records as result of several investigation campaigns previously collected only in the archives of the local archaeological agency. In order to obtain the maximum benefit from this data collection, it is necessary to enable interoperability processes that involve this data source. This can be achieved only by means of a mapping and transformation process that has as target an international standard like CIDOC CRM. We can also observe that in many cases archaeological records are incomplete, in particular for what regards the specification of space and time dimensions, thus it is necessary to correctly represent the spatial accuracy and/or the temporal uncertainty that characterized the collected data.

This paper presents the result of the mapping from the SITAVR conceptual schema to the CIDOC-CRM. In particular, it concentrates on the different approaches for representing spatial and temporal properties together with their accuracy or uncertainty, using the constructs currently available in the ontology. Future work could be the specification of general rules for the representation of spatio-temporal data in CIDOC-CRM and the application of reasoning tools, like the ones in Podestà et al. (2007), Belussi and Migliorini (2012, 2017), that consider the uncertainty of the data represented as an RDF file produced by means of the defined general rules.

References

Basso P, Belussi A, Grossi P, Migliorini S (2013) Towards the creation of an archaeological urban information system: data modeling of the historical heritage of Verona. In: Proceedings of AGILE 2013, Workshop on integrating 4D, GIS and cultural heritage, pp 1–3

Basso P, Grossi P, Bruno B, Manasse GC, Belussi A, Migliorini S (2016) The SITAVR project (Archaeological territorial information system of Verona). The tale of an example of reuse and virtuous collaboration in scope of public administration. Archeologia e Calcolatori (Suppl 8): 72–79

Belussi A, Migliorini S (2012) A framework for integrating multi-accuracy spatial data in geographical applications. Geoinformatica 16(3):523–561

Belussi A, Migliorini S (2017) A spatio-temporal framework for managing archeological data. Ann Math Artif Intell 80:175–218

Belussi Alberto, Negri Mauro, Pelagatti Giuseppe (2006) Modelling spatial whole-part relationships using an iso-tc211 conformant approach. Inf Softw Technol 48(11):1095–1103

Belussi A, Migliorini S (2014) A framework for managing temporal dimensions in archaeological data. In: 2014 21st international symposium on temporal representation and reasoning (TIME), pp 81–90

Binding C, May K, Tudhope D (2008) Semantic interoperability in archaeological datasets: Data mapping and extraction via the CIDOC CRM. In: Research and advanced technology for digital libraries, 12th European conference, ECDL, pp 280–290

Bruno B, Basso P, Grossi P, Belussi A, Migliorini S (2015) SITAVR project—an archaeological charter for Verona. Archeologia e Calcolatori (Suppl 7):155–167

Cipolla C (1880) Notizie degli scavi di antichità. Atti della Regia Accademia dei Lincei

De Roo B, Ooms K, Bourgeois J, Maeyer P (2014) Bridging archaeology and GIS: influencing factors for a 4D archaeological GIS. In: Proceedings of the 5th digital heritage. Progress in cultural heritage: documentation, preservation, and protection, pp 186–195

Felicetti A, Masur A, Kritsotaki A, Hiebel G, May K, Theodoridou M, Doerrand M, Ronzino P, Hermon S, Schmidle W (2016) Definition of the CRM$_a$rchaeo. An extension of CIDOC CRM to support archaeological excavation process

Getty Res. Inst. Art & Architecture Thesaurus. http://www.getty.edu/research/tools/vocabularies/aat/

ICOM/CIDOC CRM Special Interest Group (2016) Definition of the CIDOC conceptual reference model, version 6.2.2. http://www.cidoc-crm.org/

Migliorini S, Grossi P, Belussi A (forthcoming) An Interoperable spatio-temporal model for archaeological data based on ISO standard 19100. ACM J Comput Cult Heritage (JOCCH)

Pelagatti G, Negri M, Belussi A, Migliorini S (2009) From the conceptual design of spatial constraints to their implementation in real systems. In: Proceedings of the 17th ACM SIGSPATIAL international conference on advances in geographic information systems, GIS '09, New York, NY, USA, 2009. ACM, pp 448–451

Podestà P, Catania B, Belussi A (2007) Using qualitative information in query processing over multiresolution maps, pp 159–186

Serlorenzi M, Ainis D (2011) SITAR: sistema informativo territoriale archeologico di roma. Iuno, Roma

Simple Knowledge Organization System (SKOS). https://www.w3.org/2004/02/skos/

Wheatley D, Gillings M (2002) Spatial technology and archaeology: the archaeological applications of GIS. Taylor & Francis

Printed in the United States
By Bookmasters